Electrochemistry

Grenoble Sciences

The aims of Grenoble Sciences are double:

▸ to produce works corresponding to a clearly defined project, without the constraints of trends or programme,
▸ to ensure the utmost scientific and pedagogic quality of the selected works: each project is selected by Grenoble Sciences with the help of anonymous referees. In order to optimize the work, the authors interact for a year (on average) with the members of a reading committee, whose names figure in the front pages of the work, which is then co-published with the most suitable publishing partner.

(Contact: Tel.: (33)4 76 51 46 95 - e-mail: Grenoble.Sciences@ujf-grenoble.fr
Information: *http://grenoble-sciences.ujf-grenoble.fr*)

Scientific Director of Grenoble Sciences
Jean BORNAREL, Emeritus Professeur at the Joseph Fourier University, France

Grenoble Sciences is a department of the **Joseph Fourier University** supported by the **French National Ministry for Higher Education and Research** and the **Rhône-Alpes Region**.

Electrochemistry - The Basics, with Examples is an improved version of the original book
L'électrochimie - Fondamentaux avec exercices corrigés
by Christine LEFROU, Pierre FABRY and Jean-Claude POIGNET
EDP Sciences, Grenoble Sciences' collection, 2009, ISBN 978 2 7598 0425 2.

The Reading Committee of the French version included the following members:

▸ **Michel CASSIR**, Professor - ENSCP, Paris
▸ **Renaud CORNUT**, PhD - Grenoble INP
▸ **Christophe COUDRET**, Researcher - CNRS, Toulouse
▸ **Guy DENUAULT**, Senior lecturer - Southampton University, United Kingdom
▸ **Didier DEVILLIERS**, Professor - Pierre et Marie Curie University, Paris VI
▸ **Bruno FOSSET**, Professor - Henri IV High School, Paris
▸ **Ricardo NOGUEIRA**, Professor - Phelma, Grenoble INP

Translation from original French version performed by
Lauren AYOTTE, Isabel PITMAN and Jean-Claude POIGNET

Typesetted by **Centre technique Grenoble Sciences**
Cover illustration: **Alice GIRAUD**

Christine Lefrou • Pierre Fabry • Jean-Claude Poignet

Electrochemistry

The Basics, With Examples

 Springer

Christine Lefrou
LEPMI
Saint Martin d'Heres Cedex
France

Pierre Fabry
Meylan
France

Jean-Claude Poignet
Saint Martin D'Heres
France

Originally published in French: L'électrochimie - Fondamentaux avec exercices corrigés by Christine Lefrou, Pierre Fabry and Jean-Claude Poignet, EDP Sciences, Grenoble Sciences' collection, 2009, ISBN 978-2-7598-0425-2

ISBN 978-3-662-50719-3   ISBN 978-3-642-30250-3 (eBook)
DOI 10.1007/978-3-642-30250-3
Springer Heidelberg New York Dordrecht London

Cover design: Grenoble Sciences, Alice Giraud

Printed on acid-free paper

Springer is part of Springer Science+Business Media (www.springer.com)

PREFACE

The emerging constraints related to energy production, which are already shaking our economies, will undoubtedly increase. Our societies will not only have to produce the tens of terawatts of energy they require while resorting less and less to fossil fuels (a fact that implies that electrical energy will dominate), but will also need to find adequate ways to use and store the transient electrons thus produced. These are considerable challenges that our present world is not ready to fulfill with its current technologies. New technologies will have to be envisioned for the efficient management of the considerable fluxes required, and to this end, *Electrochemistry* seems to provide some of the most promising and versatile approaches. *Electrochemistry* will be involved in solar cells, electrolytic cells for the production of hydrogen through water electrolysis or the reductive recycling of carbon dioxide, supercapacitors and batteries for the storage of electricity produced intermittently by solar cells and windmills, as well as in the use of electrons as chemical reagents, and so on. This is a vast program that will require the dedicated and skilled competence of thousands of researchers and engineers, which is in stark contrast with the present status of electrochemistry in many industrial countries, where its main focus is the never-ending fight against corrosion or improvement lead car batteries.

There will be a requirement for much more knowledgeable and versatile electrochemists than are currently trained in our universities and engineering schools, which is tantamount to saying that our teaching of electrochemistry must evolve drastically. Indeed, even if today one can easily foresee the great challenges that electrochemists will face, nobody can know for sure which sustainable and economically viable solutions will emerge, be selected and even how they will evolve. But to occur all of this will necessarily be rooted on a deep understanding of the fundamental principles and laws of electrochemistry. Future electrochemical researchers and engineers will unquestionably adapt, but this can only happen provided that their knowledge is firmly and confidently mastered. We should recall the great Michael FARADAY's answer to the Prime Minister of his time, who asked him about the purpose of understanding electricity and electromagnetism: *Sir, I certainly don't know, but I am sure that within thirty years you will be taxing its applications.* To paraphrase him: Today we do not know how electrochemistry will solve the great challenges ahead, but we do know that nothing will be possible without a deep understanding of this science.

Within this context, it is a great pleasure to see the present increasing number of new electrochemistry textbooks, though sadly many of them continue to be written not to provide students with a deep understanding, but rather with operational conceptual recipes; this is certainly handy and useful knowledge, but it is ultimately rooted on sand. So it is my great pleasure to see that a few colleagues, the authors of this book among them, have undertaken a deeper pedagogical questioning to produce a new type of electrochemistry textbook for students in their freshman years.

This book offers new approaches to the teaching of electrochemical concepts, principles, and applications. It is based on a translation and improvement of a previous version written by the same authors for French-speaking students, so its efficiency has already been tested in excellent French universities and engineering schools. In fact, these new approaches were primarily elaborated and refined by one of the authors during the electrochemical classes she taught to student engineers of Grenoble INP, one of the major French educational centers, where electrochemistry is integrated as one of its major courses.

The rigorous but pedagogical approaches developed in this textbook will unquestionably provide its readers with a strong knowledge base. Yet in this case, «rigor» is not synonymous with «painful» or «nerdy». Indeed, the original presentation and the possibility of different reading levels will make this textbook accessible and pleasant to all, irrespective of their initial level. I have absolutely no doubt that students initiated and trained through clever use of this book will benefit from sound foundations upon which they will be able to build up the more specialized knowledge that they will acquire during either their follow-up studies or scientific careers.

Christian AMATORE, HonFRSC
Membre de l'Académie des Sciences
Délegué à l'Education et à la Formation

FOREWORD

Electrochemistry is a branch of science that focuses essentially on the interfaces between materials. Therefore it is also a science that lies at the interface between other scientific disciplines, namely physics and chemistry. These two disciplines use specific concepts as well as specialised vocabulary which can sometimes be confused. Today, with the fast-growing spread of new technologies, specialists from various sectors are finding themselves increasingly drawn together to collaborate on research and development projects, including synthesizing and elaborating materials as well as in areas such as analysis, the environment and renewable energies. As a consequence, certain notions need to be clarified to ensure that the interested reader is able to understand, whatever his or her core education.

Electrochemistry is taught as part of many scientific courses, from basic lessons in physical chemistry to science for engineers. However, for a long time it was hard to find books focused exclusively on electrochemistry and its specific concepts, especially in France. Over the last few decades several textbooks have been published on electrochemistry, each of these presenting different yet equally valid approaches. Without calling into question the overall quality and originality of these texts, there are nonetheless several points in each case which have remained obscure, or even sunk into oblivion. This could be explained by the ever pressing need to respond to the demands of the fast-growing field of technology. Whatever the case, it has had serious consequences, namely potentially preventing the scientist from gaining a full understanding of the subject, and moreover leading to approximations or even errors.

This book owes a lot to the method developed by Christine LEFROU on the university course that she gives to engineering students at the Grenoble Institute of Technology. It presents several novel developments as well as helping to bring the reader to a more profound understanding of the fundamental concepts involved in the different phenomena that occur in an electrochemical cell. Rather than focusing on an in-depth study of electrode mechanisms (other books give a detailed account of this subject), this book develops in particular the movement of species in complete electrochemical systems. It is divided into four chapters, giving a progressive approach. The few redundancies that might be spotted are therefore not fortuitous and should be viewed as part of a specific pedagogical method aimed at improving the scientific level in gradual steps.

The authors wish to invite the reader on «a fascinating electrochemical journey between two electrodes», with the following little piece of advice, in the form of a maxim: *the traveller should know that if he moves too fast, he will miss out on the chance of appreciating to the full the landscapes he encounters, and he will prevent himself from gaining a proper understanding of the life and customs of the inhabitants in the land he is exploring…*

READER GUIDELINES

Here are a few guidelines to help you make the most out of this voyage…

First of all, there are two main reading itineraries to choose from. If you stick to the main path, then follow the main paragraphs focused on the basic notions. However, if you take the other path, then you will be going into more rough terrain, exploring the back-country the paragraphs are written in smaller characters, and the content goes into more detail, usually giving examples to illustrate the topic. Therefore, these in-depth paragraphs regularly feature issues which are solved in numerical terms, and can be seen as a list of applied exercises, laid out in an original fashion (the question is immediately followed by the solution, including a commentary) so as not to lose the thread. These exercises and descriptive diagrams often give numerical values that should be simply viewed as teaching examples. Although the cases covered are plausible in technical terms, they do not refer to any particular real experimental data.

The appendices give more lengthy and developed calculations, which are not described in detail elsewhere in the main body of the text. They also provide further reading, which is kept apart at the end so as not to disrupt the overall pedagogical approach of this book. A good half of these appendices unveil novel developments and original material that have never been published before. Throughout the book, the reader can also find numerous footnotes, comments, added clarifications and cross-references between sections.

The first chapter focuses on the basic notions that need to be mastered before being able to go on and tackle the following chapters. The reader is reminded of the basic concepts, all defined in precise detail, as well as being introduced to certain experimental aspects. This chapter is therefore meant more or less for beginners in electrochemistry. The common electrochemical systems are described in the second chapter, which introduces the elementary laws so that they can be applied immediately by the reader. This chapter does not therefore provide any in-depth demonstrations. However, it is the last two chapters and the appendices that go into greater depth to tackle the key notions in a thorough and often original way. The third chapter focuses on aspects related to thermodynamic equilibrium, and the fourth chapter deals with electrochemical devices with a current flow, and which are therefore not in equilibrium.

Summary tables can be found at the end of the book recapping the key features of each chapter. Finally, in order to give the reader the opportunity to carry out a self-assessment, each chapter ends with a series of related questions (the answers can be found at the back of the book).

This book does not aim to give a detailed account of electrochemical applications. However, certain electrochemical applications are mentioned in illustrated boards in order to show that the concepts covered are not disconnected from technological reality. These explanations can be read separately from the core of the text. To find them in the table of contents, their titles are shaded in and designated by the symbol ▶▶.

Finally, the bibliography indicates the main titles examined by the authors in the course of writing this book. Therefore, the list is centred on books (both in French and English) that include a presentation of the fundamental laws of electrochemistry.

ACKNOWLEDGEMENTS

We would like to thank all the people who have helped in working out this book.

First of all, we are indebted to the members of the reading committee for all the care that they brought to their task. Their suggestions and questions, always delivered with great tact and modesty, helped to enrich and inspire our work so as to ultimately improve the content and writing.

Our thanks also go to the members of the Grenoble Sciences editorial team, its director Jean BORNAREL, and also Laura CAPOLO, Sylvie BORDAGE, Julie RIDARD, Anne-Laure PASSAVANT and Isabel PITMAN. Their suggestion to include illustrated boards was highly appreciated, since the result is that they make for more enjoyable reading, and we would like to express our gratitude to all those who helped compile the content of those illustrated boards. We also heartily acknowledge the invaluable help of Lauren AYOTTE and Guy DENUAULT, and their contribution towards improving this work.

Finally, we would like to mention all of the students we have had the pleasure of working with over the years while developing this project. Although they are too numerous to be named individually, they equally have all played a role in contributing to this book. Their questions, as much as their misunderstandings of our lectures as teachers, have all helped to refine our own thinking, and even shake up our certainties!

The authors

Contents

1 - BASIC NOTIONS

1.1 - INTRODUCTION

1.1.1 - ETYMOLOGY

The word electrochemistry derives from the terms electricity and chemistry. It applies to a scientific discipline as well as to a sector of industry. Ordinary dictionaries define it as the science which describes the interactions between chemistry and electricity, or the chemical phenomena that are coupled with reciprocal exchanges of electric energy.

More precisely, it is a science that analyses and describes the transformations of matter on the atomic scale by shifts of electronic charge which can be controlled by means of electric devices. Such transformations are called oxidation-reduction reactions[1]. It is therefore a matter of controlling oxidation-reduction reactions with an electric current[2] or with a voltage[3]. Therefore electroforming, which consists of forming an object by making a deposit with an oxidation-reduction reaction, belongs to the field of electrochemistry. On the other hand, electro-erosion or EDM (electrictrical discharge machining), in which matter is removed by electric discharges, is not considered to be a part of electrochemistry.

One of the advantages of electrochemistry over chemistry when taken in its broadest sense lies in the additional, adjustable degree of freedom offered by the voltage or current. Indeed, it is possible to vary the energy of the active species in a continuous and controlled manner, and also therefore, at room temperature for example, to attain a highly selective reactivity with acute control of the reaction and of its extent.

By extension, the term electrochemistry stretches to include systems which have no controlled exchange of electrical energy with the exterior. The overall electric current is zero: the electrochemical system is at open circuit. The term combines two very different situations. The first applies to any system in thermodynamic equilibrium, that is in which no transformation of matter occurs. This is the case with many potentiometric sensors. The second situation covers systems that are likely to react spontaneously, namely with a transformation of matter and the internal exchange of electric energy, such as in corrosion. The concepts of electrochemistry are the suitable tools to describe such systems.

The scope of the scientific field of electrochemistry, and consequently of this document, can be summarized as the search for links between current and voltage at any given

---

[1] These notions are defined in section 1.2.

[2] These notions are defined in section 1.3.

[3] These notions are defined in section 1.5.

time, for a given electrochemical system. Understanding these links allows one to anticipate the behaviour of electrochemical devices and improve their performance.

1.1.2 - THE HISTORICAL DEVELOPMENT OF IDEAS

The origins of electrochemistry in the history of science are rather difficult to determine. They are often attributed to the end of the 18th century with GALVANI's work on animal electricity. In actual fact, even before electrostatic machines were developed, similar observations on the excitation of muscles in contact with metals of a different nature had already been made by SWAMMERDAN in the middle of the 17th century. However, the link with electricity had not been clearly shown at the time. GALVANI published his results in 1791, laying out how he considered living muscles to be sorts of LEYDEN jars storing electricity which would be discharged if metals were set between two points. At that time, the link between electricity and life raised a number of questions and many scientists took an active interest in fish capable of striking down their victims by electric discharges. History records several experiments made by GALVANI which led him to observe the influential impact of the presence of two different metals. The anecdote of the frog legs hanging from the iron railing of a balcony by a copper wire is often cited: the frog legs contracted convulsively when they came into contact with the iron railing, as if exposed to an electric shock.

GALVANI's works attracted the attention of several other scientists, among whom SULTZER, who discovered the acidic taste on his tongue when put in contact with two different short-circuited metals, Pb and Ag. VOLTA also made numerous tests of the same kind, on his tongue, his ears, his eyes, his nose and his skin, either smooth or scratched. He gave a lot of himself to help science progress. Most of all he was the first to pile up two different metals in stacks, each layer separated by a wet sheet (tissue, paper, etc.) impregnated with substances such as salts or acids. From his experiments, VOLTA was able to grade the various metals according to the intensity of the electric pulses that he felt. He also connected elemental cells in series and/or in opposition and realised that the voltages were additive. He unveiled his classification of metals in 1794. We must nevertheless note that if he was indeed right to refute GALVANI's theory of animal electricity, he was himself mistaken in his interpretation. Until the end of his life, he clung to the idea that this phenomenon was only due to the difference in the nature of the metals at the metallic junction. In his view the only role of the electrolyte was to equalize the potentials. It was thanks to these discoveries, empirical as they were, that the true foundation of electrochemistry could be laid. During a conference given by VOLTA in 1801 at the *Institut de France*, he was awarded a gold medal by Napoléon BONAPARTE, thanks to his demonstration of a working battery. This was an important political event, knowing that Italy was waging war against France.

▶ Among the major events and discoveries which followed in the field of electrochemistry, one can cite the following:

1800   NICHOLSON and CARLISLE achieved the first electrolysis of water by means of a battery and observed gas evolution, revealing the production of dihydrogen. It can be noted that VOLTA had also achieved similar findings but he did not come to any conclusion.

1807 GROTTHUS put forward a theory on the electrolytes and the movement of charges (separation of the charges on H and O in water molecule).

1807 DAVY discovered potassium using a battery (with 2 000 elements!) by electrolysing molten potash. He then discovered sodium and calcium. Moreover, he was the first to identify the role played by the reactions at the electrodes and the decomposition of the electrolyte.

1824 DAVY made use of zinc to protect against the corrosion of copper or iron parts in ships.

1826 BECQUEREL observed the polarisation effects of electrodes caused by hydrogen evolution. He then proposed the use of depolarizers in two-compartment batteries.

1833 FARADAY, a student of DAVY, introduced the vocabulary of electrochemistry[4] (electrode, anion-anode and cation-cathode) and observed the link between the mass of compound produced or consumed and the amount of charge passed (laws of electrolysis).

1836 DANIELL made up the two-compartment battery called the DANIELL cell, which is still the main reference example given of an electrochemical battery in a number of educational books. Anyway this is its only use since the electric power that this battery can provide is quite negligible.

1837 JACOBI invented galvanoplasty, which has numerous applications today.

1839 GROVE discovered the reversibility of water electrolysis reactions, and laid the basis of the first fuel cell, which was not to undergo any significant development until the NASA program in the 1960's.

1859 PLANTÉ invented the lead-acid battery, which is still widely used because it can deliver high levels of electric power at a low cost. Of course, its manufacturing process underwent many improvements, but its main principle remained unchanged.

1868 LECLANCHÉ discovered the saline battery based on zinc and manganese dioxide which is also still very successful today. Incidentally, it is interesting to note that LECLANCHÉ, having failed to secure funding in France to develop his project, expatriated himself to Belgium, where he then made his fortune.

1874 KOHLRAUSCH wrote his theory on the conductivity of electrolytes.

1886 HALL in the United States, and HÉROULT in France both developed the aluminium electrolysis process. The simultaneous nature of these discoveries did entail a certain degree of polemics, but what is even more unsettling is the fact that both men were born the same year and also died the same year. Would they now be together in Heaven with beautiful, gleaming aluminium wings?

1887 ARRHENIUS developed his theory on acido-basic reactions and on ionic dissociation.

1889 NERNST worked out the thermodynamics of electrochemistry.

1897 BOTTGER developed the hydrogen electrode (first measurements of pH).

1899 The first electric car (*JAMAIS CONTENTE*) was developed [5]. It reached a record speed of 100 km h^{-1} (over the stretch of only a few kilometres).

1902 COTTRELL wrote the equations which rule the electrode kinetics with mass transport by diffusion.

1905 TAFEL found an empirical law of electrode overpotential as being a function of the current on various metals.

---

[4] *It may be worth knowing the etymology of these terms, which are so familiar in electrochemistry. The term electrolysis means splitting a compound, namely the electrolyte, which can be unbound, and decomposed. The suffix -ode means the path: the anode is thus literally the 'path towards a hill'. It is with this electrode that the current enters the system. Before deciding on anode, FARADAY could have also used, for example, the term 'eisode' (entrance for the current), in which case the cathode should have been called 'exode'... To finish, the word ion comes from the verb 'to go' in Greek: the cations are species which move towards the cathode while the anions move towards the anode.*

[5] *See the illustrated board entitled 'The first electric vehicles'.*

1906   CREMER invented the glass bulb pH electrode, which is still widely used.

1914   EDISON developed the Ni/Fe alkaline secondary battery.

1922   HEYROVSKY worked out the theory for the mercury electrode in polarography, an electrochemical analysis method which, after a few improvements, meant that ultra-traces could be analysed in heavy metals for instance. He was awarded the NOBEL prize for his work in 1959.

1924-1930 BUTLER and VOLMER laid the foundations of the charge transfer theory at an electrode.

Other more recent, important events could also be mentioned here, although it is really during these two centuries that the fundamental basis of electrochemistry was shaped. It is interesting to note that most concepts relating to the existence of ions and the reactions involving the exchange of charge were put forward before the atomic theory of matter was fully accepted. It was in 1803 that DALTON reintroduced the concept of the atom, which had been previously buried for centuries. THOMSON's work on the electron was carried out in1887, and the introduction of the BOHR model dates back to 1913.

Without the VOLTA battery, which delivered a direct current, could it ever have been possible to spot the magnetic effects of an electric current? Would FARADAY have discovered the dynamo all the same? Had GALVANI's works not existed, would VOLTA have shown any interest in these issues? What were those frog legs doing on GALVANI's balcony? It is obvious that all these discoveries are interdependent and chance plays a great part in the history of science.

1.1.3 - SOCIOECONOMIC IMPORTANCE

The industrial applications of electrochemistry can be classified under seven large categories: electrosynthesis, surface treatments, energy storage and conversion, analysis and measurements, the environment, corrosion and bio-electrochemistry.

▶▶ *Electrosynthesis*

Electrosynthesis is a process used in heavy industry because, depending on the material being produced, its energetic yield is higher than that found in thermal synthesis processes. Moreover, the processes used are selective and easy to control by means of the voltage, the current and the amount of charge, which is a very accurate indicator of the advancement rate in production. The raw materials produced in the greatest quantities by electrosynthesis are aluminium, dichlorine and sodium hydroxide.

Today, the annual world production of aluminium by means of electrosynthesis has been seen to reach up to about 38 Mt (data given for 2007) [6].

Dichlorine is a raw material used for many manufactured products such as plastics and detergents, etc. The total industrial production quantity of dichlorine through the electrosynthesis process is nowadays about 50 Mt (data given for 2007). About 56 Mt of sodium hydroxide are produced simultaneously. There exists three main dichlorine and sodium hydroxide electrosynthesis processes from aqueous solutions containing

---

[6]   See the illustrated board entitled 'Industrial production of aluminium in France'.

sodium chloride: the diaphragm process, the membrane[7] process and the mercury cathode process. Figure 1.1 shows the distribution of world production for the different processes.

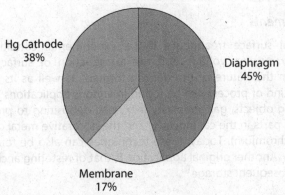

Figure 1.1 - Distribution of the world production of dichlorine
using the different electrolysis processes (data given for 2007)

Difluorine, sodium, lithium and magnesium are also mainly produced through electrosynthesis using molten salts (figure 1.2).

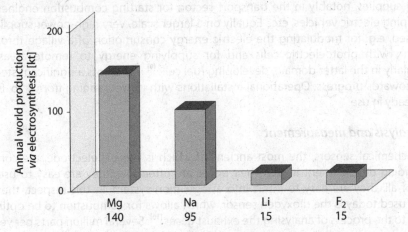

Figure 1.2 - Annual world production
via electrosynthesis of a selection of metals and halogens in 2007

High purity dihydrogen is likewise produced by electrolysing water. It is also worth mentioning here the purification of certain metals, such as copper, zinc and aluminium, by an electrorefining process involving anodic dissolution and cathodic deposition by a selective electrolysis.

Thanks to their high selectivity, electrochemical processes also enable complex molecules to be synthesized (as used in the pharmaceutical industry, biotechnology,

_____

[7]   More precisely an ionic conducting membrane, see section 1.3.2.2.

perfumery, and in artificial flavouring in the food industry, etc.). Also worthy of mention is the selective synthesis of adiponitrile which is a precursor molecule in the synthesis of nylon.

▶▶ *Surface treatments*

In electrochemical surface treatments, if the experimental conditions are suitably managed (namely current, voltage and the introduction of surfactants), then it is possible to govern the nature of the deposit formed, as well as its quality (porosity, sheen, etc.). This kind of process can be found in various applications such as polishing and electroforming objects, galvanoplasty (e.g., zinc depositing to protect against the corrosion of metal parts in the car industry) and the decorative metal coating of objects (silver, gold and chromium). Today, these techniques can also be found in the micro-electronic industry. Another original application is that of restoring ancient artefacts and managing their subsequent storage [8].

▶▶ *Energy storage and conversion*

Batteries play an essential role in modern society. These days the use of low-power applications is increasingly on the up in portable electronic apparatus (telephones, computers, MP3 players, etc.) or in the medical sector (hearing aids, pacemakers, micro-injectors, etc.). On a domestic scale there is a growing demand for more powerful energy supplies, notably in the transport sector, for starting combustion engines, and developing electric vehicles, etc. Equally on a larger scale, very high-power supplies are also used, e.g. for modulating the electric energy consumption of a village through a coupling with photoelectric cells and for supplying energy to remote areas, etc. Particularly in the latter domain, developing fuel cells [9] represents a significant stepping stone towards progress. Operational installations with power ranging from 1 to 10 MW are already in use.

▶▶ *Analysis and measurement*

Electrochemical sensors, the most ancient of which is the pH electrode, are currently undergoing development in that their prices are often low, they are easy to use and, most of all, they are easy to insert into a regulation system. In this respect, the most widely used today is the dioxygen sensor, which allows for combustion to be optimised thanks to the process of analysing the exhaust gases [10]. Several million parts per year are produced, notably for the car industry. In the biomedical field, electrochemical sensors are also used to monitor glucose and pH, and to measure out certain cations. New developments are also being made in the field of pollutant analysis.

Polarography is an electrochemical method using a mercury drop electrode which permits analysis of a very high number of chemical species. It is mainly used for analysing metal cations in aqueous solutions. The usual detection limits are about 10^{-6} mol L^{-1} but they can be lowered to 10^{-12} mol L^{-1}, with inexpensive equipment.

---

[8]   See the illustrated board entitled 'Conservation of archaeological artefacts'.

[9]   See the illustrated board entitled 'Fuel cells'.

[10] See the illustrated board entitled 'Regulating of fuel engines'.

▶▶ *The environment sector*

In the environment sector, electrochemistry is still of little use on a large scale, however predictions point towards significant growth in this field in the future. Electrochemical techniques can be used for:

▸ separation, e.g., brackish water desalination through electrodialysis (the membrane processes is capable of producing up to 2 000 m³ per day), for supplying fresh water to remote areas[11],

▸ recovery, e.g., electrodepositing metallic elements such as copper, nickel, zinc, cobalt, silver and gold, etc.,

▸ concentrating or purifying effluents through electrodialysis, or cathodic deposition processes, etc.,

▸ destroying pollutants, e.g., the oxidation of cyanide ions into carbon dioxide and dinitrogen, disinfection by means of producing oxidizing species in situ, e.g., dichlorine or sodium hypochloride, for example for disinfecting air, or swimming pool water.

▶▶ *Corrosion*

Corrosion is the phenomenon whereby a metallic part is destroyed[12]. It generally occurs spontaneously, e.g., *via* a reaction with dioxygen dissolved in water in the case of wet corrosion. Attempts are generally made to fight against this phenomenon which engenders a considerable economic cost and also poses security and toxicity problems. Such is the case for the corrosion of lead tubing which can cause serious health problems. The economic cost of corrosion is estimated today as being 2% of the GNP[13] in developed countries. Sometimes however, corrosion can be beneficial. For example, in the process of dismantling of nuclear plants, corrosion can be used to reduce the quantity of contaminated matter that has to be stored.

▶▶ *Bio-electrochemistry*

Biology is an additional field in which electrochemistry plays an important role in the development processes. A significant number of phenomena in the living world involve oxidation-reduction reactions or controlled ionic movements through membranes. In addition to its list of increasing widespread uses in the field of biosensors[14], bio-electrochemistry is likely to grow in other sectors, such as in the development of new processes.

1.2 - OXIDATION-REDUCTION

Oxidation-reduction is a notion that has been developed over the course of centuries, as outlined in the brief historic account given above. In the first notion introduced by

---

[11] *See the illustrated board entitled 'Electrodialysis'.*

[12] *See the illustrated board entitled 'Corrosion of reinforced concrete'.*

[13] *GNP: Gross National Product.*

[14] *See the illustrated board entitled 'Electrochemistry and neurobiology'.*

LAVOISIER, oxidation was considered to be the reaction between chemical species and dioxygen, e.g.:

$$2\,Hg + O_2 \longrightarrow 2\,HgO$$

The modern notion which we will define and use in the following section was rooted at the start of the 20th century, with the discovery of the electron.

1.2.1 – THE MODERN NOTION OF OXIDATION-REDUCTION

An oxidation-reduction reaction (redox reaction) involves transforming matter *via* electron shifts at the atomic level. When a species, or more exactly a chemical element of this species, loses one or more electrons, this species is said to undergo oxidation. When it gains electrons, it is said to undergo reduction.

Such a transformation is called an oxidation-reduction half-reaction or a redox half-reaction[15]. It concerns two species for which a given element exists under two different forms. These two species are called the oxidant (or oxidizing agent) and the reductant (or reducing agent) denoted by Ox and Red respectively. They make up a redox couple usually denoted by Ox/Red.

The overall equation of the redox half-reaction for the Ox/Red couple is the following:

$$Red \underset{reduction}{\overset{oxidation}{\rightleftharpoons}} Ox + n\,e^-$$

where Ox is the oxidant, i.e., the form that is capable of gaining electrons,
     Red is the reductant, i.e., the form that is capable of giving electrons.

Although the terms oxidant and reductant (or oxidizing and reducing agents) are the most commonly used, one can also use the terms oxidized form and reduced form of the couple. It is then necessary to keep in mind the fact that Ox is both an oxidizing agent and the oxidized form of the couple Ox/Red, whereas Red is the reducing agent and the reduced form of the couple.

When the direction of the transformation actually occurring is not the main interest[16], which is the case here, then the direction in which the reaction is written should be considered immaterial. This reaction can be written by choosing the direct orientation for oxidation as well as for reduction[17].

---

[15] *According to the authors, the terms half-reaction or oxidation-reduction reaction can be used for this electron exchange reaction. The term half-reaction, which is often used in this document, stresses the fact that if a redox couple reacts, e.g., in the direction of oxidation, at least one other couple must react in the direction of reduction. This term is derived from redox-chemistry in solution-based reactions, however it is still of educational interest for electrochemists given that at least two half-reactions are always occurring simultaneously, one at the anode and the other at the cathode.*

[16] *Predicting the direction of the redox half-reactions actually occurring in an electrochemical system is an important question which will be dealt with in section 2.4.*

[17] *Certain authors are in favour of the reaction being written in the direction of reduction for a redox couple, because of the highly observed convention which places Ox in first position for writing the couple Ox/Red. We will not systematically adhere to this convention here but rather we will show how an entire algebraic form of relationships stemming from thermodynamics can allow one to*

As in other areas of chemistry, key numbers appear in these balanced equations. In thermodynamics or in kinetics, stoichiometric numbers[18] are therefore defined for the reaction, which are algebraic and denoted by v_i. They are set up as being positive for the products (on the right side of the balanced reaction) and negative for the reactants (on the left side of the balanced reaction), as shown in the following generalised expressions, as well as in the ensuing examples, with A_i representing the chemical constituents in the reaction:

$$0 \;\Longleftrightarrow\; \sum_i v_i A_i$$

or most commonly:
$$\sum_{i\,=\,\text{reactants}} |v_i| A_i \;\Longleftrightarrow\; \sum_{j\,=\,\text{products}} |v_j| A_j$$

The redox half-reaction of the Cu^{2+}/Cu couple can be written for example in the direction of reduction:

$$Cu^{2+} + 2\,e^- \;\Longleftrightarrow\; Cu$$

The stoichiometric numbers of the species are then:

$$Cu \qquad v_{Cu} = +1$$
$$Cu^{2+} \qquad v_{Cu^{2+}} = -1$$
$$e^- \qquad v_e = -2$$

This example shows that the usual form of writing the balanced reaction uses absolute values for the stoichiometric numbers. In some books, one finds a more mathematical form of these equations, in which algebraic stoichiometric coefficients are used directly:

$$0 \;\Longleftrightarrow\; \sum_i v_i A_i$$

In the case of the Cu^{2+}/Cu couple, this writing form therefore corresponds to:

$$0 \;\Longleftrightarrow\; Cu - Cu^{2+} - 2\,e^-$$

Taking these definitions into account, when a redox half-reaction is written in the oxidation direction then the algebraic stoichiometric number of electrons corresponds to the positive number ($n > 0$) of the electrons exchanged:

▸ in the oxidation direction, the electrons are counted among the products of the balanced reaction:

$$v_{Ox} > 0 \quad v_e = n > 0 \quad v_{Red} < 0$$

▸ in the reduction direction, they are among the reactants:

$$v_{Ox} < 0 \quad v_e = -n < 0 \quad v_{Red} > 0$$

Some of the species involved are ionic, i.e., they have a charge which is a multiple of the absolute value of the charge of the electron, $|e| = 1.6 \times 10^{-19}$ C. The factor of proportionality, which is therefore a positive or negative integer, is called the charge number and will be denoted by z_i in this book.

---

do away with the written notation of the direction of the chemical reaction chosen in this context. In fact, this direction does not correspond to any physical reality in the equilibrium state (see sections 2.1.2.1 and 3.4.1).

[18] *Sometimes also named stoichiometric coefficients.*

THE ORIGINS OF THE VOLTA BATTERY

Document written with the kind collaboration of M. COMTAT,
Laboratoire de Génie Chimique de l'Université Paul Sabatier, in Toulouse, France and
D. DELABOUGLISE, Laboratoire des Matériaux et du Génie Physique, Phelma, Grenoble INP, in France

GALVANI or VOLTA, who was right, who was wrong?

Luigi GALVANI (1737-1798), an anatomy professor at the University of Bologna, showed great interest in the influence of electricity on the nerves and on muscle stimulation. Between 1780 and 1791, he focused his numerous experiments on frogs which he prepared by leaving only the lower limbs attached to the spinal cord. In his laboratory, he happened to notice that muscular contractions were caused when he touched a nerve with a metallic scalpel in his hand. However these contractions occurred only when an electric machine, switched on in the room generated a spark. GALVANI pursued the experiment on the terrace of his house, where he showed the effects of atmospheric electricity. He also noticed that when 'specially prepared frogs', with a copper hook stuck in the spinal cord, were laid on the iron bars of the balcony, the contractions occurred when he placed the copper hook in contact with the iron bar, even when the weather was good. He reproduced this experiment in his laboratory. Thus no external electricity source was necessary for stimulating the muscles, as the contact created between nerve and muscle via two different metals was sufficient. Herein lies the origin behind the theory of animal electricity, whereby a current is discharged when the nerves and the muscles are linked by metals.

At the end of his long, meticulous study, he published a memoir in latin in 1791 (the common medical language in use at that time), entitled *De viribus electricitatis in motu musculari commentarius* (Notes on the electric forces in muscular motion).

Example of a demonstration model of the VOLTA battery, built using zinc-copper pilings, used in practical experiments at Phelma (Grenoble-INP) by Professor D. DELABOUGLISE, who corrected in this demonstration battery a conception error made by VOLTA: in fact, only one metallic disc was needed at each end of the pile (copper at the bottom on the photographs). To keep true to VOLTA's historic conviction, an additional zinc disc should have been laid at the bottom of the pile below the copper disc.

Alessandro VOLTA (1745-1827), a physics professor in Como then in Pavia, was already a renowned scientist in 1792. Discovering methane, inventing the electrophorus and then travelling and working with his foreign colleagues all contributed to making him famous. During his travels from 1780 to 1783, he visited Switzerland where he discussed ideas with VOLTAIRE, France where he worked on atmospheric electricity with LAVOISIER and LAPLACE, Germany where he discussed ideas with LICHTENBERG, the Netherlands where he worked with VAN MARUM, and England where he worked with PRIESTLEY. In 1792 when he first came upon GALVANI's dissertation, he was initially skeptical, but then became enthusiastic and decided to follow up the research on his own. He quickly grasped the idea that the muscle contraction was triggered by metallic electricity being generated when two different metals were brought into contact. He tried out the experiments on himself (on his tongue, in his nostrils, in his ears and on his eyelids). He observed effects that he decided to explain using the laws of physics. He would readily use the Ag-Zn, Cu-Zn couples but also Ag-Sn, Cu-Sn, Pb-Zn. He presented a classification of the effects of these couples. In doing so, he contested the very existence of animal electricity.

GALVANI reacted to VOLTA's experiments and a series of counter-experiments thereafter ensued, each following the other, further feeding the rivalry. The controversy stretched out across Europe as far as London where Galvanist and Voltaic Societies were created. Napoléon BONAPARTE unwittingly put an end to this controversy in 1797 when he invaded Italy and demanded a plea of allegiance from the State employees. VOLTA accepted but GALVANI refused and was excluded from the university. He died a few months later.

VOLTA was trying to find how to increase the effect of the contact between two different metals. Inspired by GALVANI's study of the electric organ of the torpedo fish, and its description as a sequence of small flat hexagons, he decided to build his own generator by piling up elements. Each element was made out of a zinc disc with a silver disc laid on top, itself covered with a cardboard disc soaked in brine. The suffused cardboard disc was meant to represent the muscle. He managed to pile up as many as 20 elements and then connected several piles in series. In 1800, he sent his results via a letter addressed to the President of the Royal Society in London. When invited by BONAPARTE in 1801 to the *Académie des Sciences* in Paris to give a presentation of his battery, VOLTA was awarded a gold medal and given a pension. Later he was made senator of the French Empire. Subsequently, BONAPARTE created an award bearing the name of the italian scientist. The award was meant to encourage others to make 'a leap forward in the field of electricity and galvanism comparable to the contribution made to these sciences by FRANKLIN and VOLTA'. BONAPARTE later convinced GAY-LUSSAC and THÉNARD to build an enlarged version of the VOLTA battery at the *Ecole Polytechnique* in Paris. This same battery was at the heart of quite a significant number of theoretical and experimental developments.

Given the success of his battery, VOLTA appeared to be the clear winner in his competition against GALVANI. However, he was incorrect in his interpretation of the working principle of the battery since he thought that it was the mere contact between two different metals that produced electricity. As for GALVANI, he had been similarly mistaken in his interpretation, but was nevertheless the precursor of electrophysiology, a science that emerged a few decades later and that was to undergo considerable development before eventually becoming a major field as we know it today.

The species is an anion when its charge has the same sign as that of the electron ($z_i < 0$); it is a cation in the opposite case ($z_i > 0$). However, there is no link between the charge of the ion and its oxidizing or reducing properties. An anion or a cation can be, depending on the case, an oxidant or a reductant.

▶ The three examples below illustrate the fact that a reductant can be neutral, cationic as well as anionic:

$$Li \rightleftharpoons Li^+ + e^-$$
$$Fe^{2+} \rightleftharpoons Fe^{3+} + e^-$$
$$Fe(CN)_6^{4-} \rightleftharpoons Fe(CN)_6^{3-} + e^-$$

The charge numbers of these species are:

$$Li \qquad z_{Li} = 0$$
$$Li^+ \qquad z_{Li^+} = +1$$
$$Fe^{2+} \qquad z_{Fe^{2+}} = +2$$
$$Fe^{3+} \qquad z_{Fe^{3+}} = +3$$
$$Fe(CN)_6^{4-} \qquad z_{Fe(CN)_6^{4-}} = -4$$
$$Fe(CN)_6^{3-} \qquad z_{Fe(CN)_6^{3-}} = -3$$

Similarly, an oxidant may bear any charge.   ◢

In other more complicated situations, several species are involved in the redox half-reaction[19]. By convention, only those having actually exchanged electrons are mentioned in the name of the redox couple.

▶ For example, the redox half-reaction of the AgCl/Ag couple is:

$$AgCl + e^- \rightleftharpoons Cl^- + Ag$$

where the Cl^- species is neither oxidized nor reduced but plays a role in the reaction.   ◢

It may happen that a given species is the oxidant in one couple and the reductant in another couple: this case is similar to that of amphoteric species in acido-basic equilibria.

▶ For example, Fe^{2+} an oxidant in the Fe^{2+}/Fe couple and a reductant in the Fe^{3+}/Fe^{2+} couple:

$$Fe^{2+} + 2\,e^- \rightleftharpoons Fe$$
$$Fe^{2+} \rightleftharpoons Fe^{3+} + e^-$$
                    ◢

1.2.2 - OXIDATION NUMBER

The definition given by the IUPAC[20] for the oxidation number or oxidation degree of an element in a compound, denoted by o.n. is "*the charge that would be left on an atom*

---

[19] *The method for establishing the balanced expression of a redox half-reaction is described in section 1.2.3.*

[20] *IUPAC: International Union of Pure and Applied Chemistry.*

if all the electrons of each bond ending up at that atom were attributed to the most electronegative atom " [21].

This is a formal definition since the bonds in polyatomic structures are generally partially covalent. The electrons are therefore only partially shifted towards the most electronegative atom. Moreover, this notion becomes more difficult to grasp in the case of bonds between atoms with similar electronegativities.

For any compound, the method for writing how the overall charge is preserved is illustrated with the following equation:

$$\sum_{\text{various elements}} (\text{number of elements in the compound}) \times \text{o.n.} = \text{charge number of the compound}$$

Taking into account the electronegativity values of the elements, we should keep in mind that:

- the oxidation number of hydrogen in most compounds is equal to +I (excluding dihydrogen where it is equal to 0 and the hydride ion where it is equal to –I);
- the oxidation number of oxygen in most compounds is equal to –II (except in the case of dioxygen where it is 0, peroxides where it is –I, and fluorinated compounds with O-F bonds where it is +I);
- the oxidation number of a halogen atom X (i.e., fluorine F, chlorine Cl, bromine Br, iodine I and astatine At) in most halogenated compounds is equal to –I (except in pure substances such as X_2 where it is 0, and in compounds having at least one bond between X and a more electronegative element such as O in ClO^- or ClO_4^- where it is positive);
- the oxidation number of an alkali atom (i.e., lithium Li, sodium Na, potassium K, rubidium Rb, caesium Cs and francium Fr) in most compounds is equal to +I, except in the corresponding metals where it is 0.

This short list of observations can be used to determine the algebraic value of the oxidation numbers of a large number of usual compounds.

For instance, in the MnO_4^- anion, the oxidation number of manganese is equal to +VII [because, $-1 = +\text{VII} + 4\times(-\text{II})$].
In the IO_3^- anion, the oxidation number of iodine is not equal to –I since there is a O-I bond. It is equal to +V [because $-1 = +\text{V} + 3\times(-\text{II})$].

However, in certain cases it is also necessary to be familiar with the structure of the compound (e.g., the LEWIS structure) in order to determine the oxidation numbers.

For example, the peroxodisulfate ion $S_2O_8^{2-}$ has the following structure: $(O_3S\text{-}O\text{-}O\text{-}SO_3)^{2-}$.
It thus has an O-O peroxo bridge and each of the two sulphur atoms has an oxidation number of +VI [because, $-2 = 2\times(+\text{VI}) + 6\times(-\text{II}) + 2\times(-\text{I})$].

---

[21] *The electronegativity of atoms can be defined in different manners (MULLIKEN scale, PAULING scale, etc.). But it generally refers to properties which are not outside the electrochemical sphere since it uses the energy linked to the process of extracting or adding an electron from/to an atom in the gaseous phase.*

In other cases, this method yields a virtual oxidation number which only corresponds to a mean value, though it is still useful for writing redox half-reactions. A simple example is given by the iron oxide Fe_3O_4, which contains Fe^{2+} and Fe^{3+} ions in its crystal lattice, but which can be formally considered as an iron oxide with an iron oxidation number equal to +8/3. This is also the case with numerous compounds used in batteries [22], which are called insertion materials, such as Li_xMnO_2, H_xWO_3, and $Li_xV_2O_5$, etc. Their oxidation numbers still remain a subject of discussion today.

▼ Taking the example of Li_xMnO_2, we will suppose that the oxidation number of lithium is +I, that of oxygen is −II and therefore by way of deduction that of manganese is (+IV −x). ◢

1.2.3 - HOW TO WRITE A REDOX HALF-REACTION

When both of the compounds in a redox couple are known, then the balanced redox half-reaction can be found in different ways. Two of these methods are briefly described below.

▶▶ First method

The first step is to write both compounds of the redox couple and when necessary to adjust the stoichiometric numbers, in order to ensure that the element with the variable oxidation number is preserved. The number of electrons exchanged is then determined from the difference between the oxidation numbers of the element in its oxidized and reduced states, taking into account the stoichiometry. Protons [23] are added if necessary to ensure that the sum of the charges on both sides of the overall equation are kept the same. Finally, water molecules H_2O are added to balance the oxygen (or hydrogen) element. It is then possible to check if the hydrogen (or oxygen) element is also balanced in the overall equation.

▼ For example, for the Fe_2O_3/FeO couple, preserving the iron element on both sides of the equation is written in the following way:

$$2\,FeO \rightleftharpoons Fe_2O_3$$

The number of electrons exchanged can then be determined since the oxidation number of iron changes from +II in FeO to +III in Fe_2O_3:

$$2\,FeO \rightleftharpoons Fe_2O_3 \;+\; 2\,e^-$$

oxidation number: $2\times(+II)$ $2\times(+III)\;+\;2\times(-I)$

When the charge is preserved on both sides of the reaction by means of adding protons, then it gives the following:

$$2\,FeO \rightleftharpoons Fe_2O_3 + 2\,H^+ + 2\,e^-$$

Preserving the oxygen element by adding water molecules finally gives:

$$2\,FeO + H_2O \rightleftharpoons Fe_2O_3 + 2\,H^+ + 2\,e^-$$

The final way of writing the redox half-reaction also provides a means of checking that the hydrogen element is preserved in the equation. ◢

---

[22] See the illustrated board entitled 'Energy storage: the Li-Metal-Polymer (LMP) batteries'.

[23] Protons are usually written H_3O^+ or H^+ to simplify. The latter form will be used in this document.

▸▸ *Second method*

The first step is to write the two compounds of the redox couple and to adjust their stoichiometric numbers, if necessary, in order to ensure that the element with the variable oxidation number is preserved. Water molecules, H_2O, and protons, H^+, may then be added to ensure that the O and H elements are preserved respectively. The final step consists of adding electrons in order to balance the charge on both sides of the reaction equation. This method, which is more formal than the previous one, avoids calculating oxidation numbers, but may be tricky to carry out if the redox couples are not sufficiently well-identified.

▷ Back to the Fe_2O_3/FeO couple, let us first express the fact of preserving the iron element as follows:

$$2\,FeO \rightleftharpoons Fe_2O_3$$

Preserving the oxygen element by adding water molecules yields:

$$2\,FeO + \mathbf{H_2O} \rightleftharpoons Fe_2O_3$$

Preserving the hydrogen element by adding protons yields:

$$2\,FeO + H_2O \rightleftharpoons Fe_2O_3 + \mathbf{2\,H^+}$$

Preserving the charge by adding electrons yields:

$$2\,FeO + H_2O \rightleftharpoons Fe_2O_3 + 2\,H^+ + \mathbf{2\,e^-}$$ ◢

In the preceding examples, protons were added to preserve the H element. Such a formal choice has nothing to do with the pH of the medium in which the reaction occurs. One may prefer to use hydroxide ions rather than protons to ensure the hydrogen element is preserved in the reaction, e.g., in a reaction involving cyanide ions, CN^-, which are not stable in acidic media:

▸ for the first method, one can ensure the charge balance merely by adding hydroxide ions (OH^-) instead of protons,

▷ Let us return again to the Fe_2O_3/FeO couple. Once the preservation of the iron element has been taken into account, as well as the change in its oxidation number, then we have:

$$2\,FeO \rightleftharpoons Fe_2O_3 + 2\,e^-$$

The charge balance on both sides of the reaction is ensured by adding hydroxide ions instead of protons, which yields:

$$2\,FeO + \mathbf{2\,OH^-} \rightleftharpoons Fe_2O_3 + 2\,e^-$$

Preserving the oxygen element by adding water molecules yields:

$$2\,FeO + 2\,OH^- \rightleftharpoons Fe_2O_3 + \mathbf{H_2O} + 2\,e^-$$

We can check to see if the hydrogen element has also been preserved in the final redox half-reaction. ◢

▸ in the second method, one can add the water autoprotolysis (or autoionisation) equilibrium to the redox half-reaction simply by applying the correct multiplication coefficient, in order to eliminate the protons in the balanced reaction.

▷ In the preceding example, the water autoprotolysis equilibrium

$$H_2O \rightleftharpoons H^+ + OH^-$$

must be multiplied by -2 for the protons to be eliminated:

$$\begin{array}{llll} (\ 2\,FeO + H_2O & \rightleftharpoons & Fe_2O_3 + \mathbf{2\,H^+} + 2\,e^-\ ) & \times +1 \\ (\ H_2O & \rightleftharpoons & \mathbf{H^+} + OH^-\ ) & \times -2 \\ \hline 2\,FeO + 2\,OH^- & \rightleftharpoons & Fe_2O_3 + H_2O + 2\,e^- & \end{array}$$

In other cases, water and/or protons are not involved when writing the equilibrium: other simple ions, such as alkali M^+ or halide ions X^- must be involved.

This is the case for example with the redox half-reaction of the MnO_2/Li_xMnO_2 couple. When examining the oxidation numbers (the manganese oxidation number is what changes in this formal representation) one can see that the written form of the reaction initially gives:

$$Li_xMnO_2 \rightleftharpoons MnO_2 + \mathbf{x\,e^-}$$

oxidation number: (+IV -x) +IV + −x

For this half-reaction to be balanced, one needs to add a compound containing a lithium atom. Here we will choose the Li^+ ion so as to end up with the following balanced reaction:

$$Li_xMnO_2 \rightleftharpoons MnO_2 + x\,e^- + \mathbf{x\,Li^+}$$

In an electrochemical system, exactly as in homogeneous redox chemistry, any reduction reaction occurring is accompanied by at least one oxidation reaction with the same charge transferred. It is common in certain areas (e.g., in industrial mass balance for electrolytic processes or when describing how batteries work) to write a chemical reaction expressing the result of these two processes[24]. Writing the overall electrochemical reaction results from combining two half-reactions with the adequate multiplying coefficients so that the charges exchanged are identical. The electrons do not appear in the balanced overall redox reaction.

- For example, when writing the redox reaction between the Ag^+/Ag and Cu^{2+}/Cu couples, the half-reaction of the first couple must be multiplied by −2 whereas that of the second couple is multiplied by +1 in order to eliminate the electrons:

$$\begin{array}{llll} (\ Ag & \rightleftharpoons & Ag^+ + \mathbf{e^-}\ ) & \times -2 \\ (\ Cu & \rightleftharpoons & Cu^{2+} + \mathbf{2\,e^-}\ ) & \times +1 \\ \hline Cu + 2\,Ag^+ & \rightleftharpoons & 2\,Ag + Cu^{2+} & \end{array}$$

- In certain cases, a given species acts simultaneously as both an oxidant in one of the couples and a reductant in the other couple. This reaction is called dismutation (or disproportionation). To balance such a reaction, one must first spot the fact that it is a case of dismutation, i.e., a particular chemical reaction involving two different redox couples. For example, the following equation may indeed be formally balanced, yet it does not correctly represent the chemical dismutation of diiodine because electrons appear on the right side, as in a redox half-reaction:

$$\cancel{I_2 + 3\,H_2O \rightleftharpoons I^- + IO_3^- + 6\,H^+ + 4\,e^-}$$

One can spot dismutation by calculating the oxidation number of the three species involved. The oxidation degree of iodine is 0 in I_2, −I in I^- and +V in IO_3^-. This equilibrium is not a redox half-reaction, but an overall reaction which is the balance of the two redox half-reactions corresponding to the IO_3^-/I_2 and I_2/I^- couples.

[24] This notion of the overall reaction is of immediate use in homogeneous redox chemistry. However, in electrochemistry it must be handled with precautions as soon as several redox reactions occur simultaneously at the same interface. Those wishing to explore this question more deeply can refer to section 2.2.2.3 which deals with the faradic yield.

These two redox couples give the following equilibria:

$$I_2 + 6H_2O \rightleftharpoons 2IO_3^- + 12H^+ + 10e^-$$

$$I_2 + 2e^- \rightleftharpoons 2I^-$$

By combining these two equilibria (with a coefficient of +5 for the second), one can finally obtain the overall dismutation reaction:

$$6I_2 + 6H_2O \rightleftharpoons 10I^- + 2IO_3^- + 12H^+$$

1.3 - THE NOTION OF CURRENT

Electric current is a macroscopic notion which was given one of its first formal definitions by AMPÈRE in 1820: "*the overall movement of charges in a conductor*". This notion is linked to its microscopic origin: the movement of charged species, which corresponds to the charge flow rate or charge flux. Insofar as an elementary charged species has a mass, then the notion of current is consequently linked to mass transport [25].

1.3.1 - MACROSCOPIC QUANTITIES DEFINING THE CURRENT

1.3.1.1 - CURRENT DENSITY

Most of the time, several different kinds of charged species are in movement, and the overall current is created by the movement of various charge carriers. The current density, $\mathbf{j}$, which is a vector with a modulus expressed in $A\,m^{-2}$, represents the overall charge flux density, i.e., the sum of the charge flux densities of each charge carrier:

$$\mathbf{j} = \sum_i \mathbf{j}_i$$

1.3.1.2 - CURRENT

For any surface (S) oriented by the normal vector $\mathbf{n}$ and having an area S, the current intensity through that surface is equal to the quantity of charge moving across it per time unit. It is thus defined by:

$$I = \iint\limits_{(S)} \mathbf{j} \cdot \mathbf{n} \, dS = \frac{dq}{dt}$$

with:
I	the current intensity through the surface (S) (the sign of I being defined by the orientation of $\mathbf{n}$ [26])	[A]
$\mathbf{j}$	the local current density, with a modulus in	[A m^{-2}]
dq	a small element of charge crossing (S)	[C]
dt	a small element of time	[s]

---

[25] For instance the current flowing through a metal is the result of the overall movement of the electrons. Since the electrons have a mass, then mass movement also occurs. Mass transport is mainly characterized by mass flux, and the link between mass flux and current is studied in detail in section 4.1.1.

[26] The link between the sign of the current and the choice of the orientation of the normal to the surface is discussed in detail in section 1.4.1.3.

This precise definition refers to the current intensity which is a scalar corresponding to the overall charge flux or flow rate (1 A = 1 C s^{-1}). Naturally, it must be defined with reference to a surface. However, in certain conditions including those fulfilled in this document [27], the absolute value of the current intensity in a conductor can be defined by considering any section of the conductor. For these reasons, which is moreover usual practice in numerous documents, we will use 'current' to stand for 'current intensity across a section of area S'.

1.3.1.3 - ELECTRONEUTRALITY AND CONSERVATIVE CURRENT

Electroneutrality can be assumed in the bulk of all conducting media at the local level, on the macroscopic scale, only after a very short transient period, necessary in order for the charges to be rearranged (about 1 femtosecond in a metal [28]). Thanks to the property of volume electroneutrality in a conducting material, the current can be shown [29] to have the same value in any section of the conductor, whether it is normal to the current density or not. The current flowing through the conductor can then be clearly defined (except for the sign [30]), since it does not depend on the section selected, which can be any surface crossed by the overall current lines:

$$I = \iint\limits_{(S)} \mathbf{j} \cdot \mathbf{n} \, dS = \iint\limits_{(S')} \mathbf{j'} \cdot \mathbf{n'} \, dS$$

We can also say that the current is conservative.

One of the important consequences of this property is illustrated in figure 1.3 which focuses on a conductor with a variable section area.

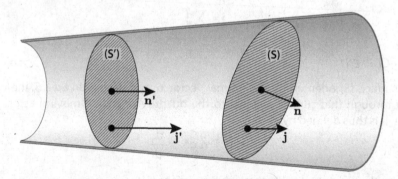

Figure 1.3 - Diagram of a conductor volume with a variable section crossed by a current

---

[27] *These are listed in the next section (1.3.1.3); the reasoning behind this result is given in section 4.1.2.*

[28] *The order of magnitude of the transient time needed for establishing electroneutrality in a conducting medium is calculated in section 3.1.1.1.*

[29] *The reasoning behind this result is presented in section 4.1.2.*

[30] *The question of which sign to attribute to the current, an important issue in electrochemistry, is addressed in detail in section 1.4.1.3.*

The current is the same in both sections. Consequently, the current mean densities at these levels cannot be the same. The mean current density is higher for the small section than that for the larger section, resulting in the following equation:

$$I = \langle j \rangle S = \langle j' \rangle S'$$

1.3.2 - CONDUCTING MEDIA

Electrochemistry makes use of various electricity conducting materials. The microscopic mechanisms linked to the current flow may differ from one material to another. The various conducting materials can be classified according to the nature of the charge carriers involved.

1.3.2.1 - DIFFERENT CHARGE CARRIERS

A current exists as a consequence of the overall movement of charged species. It is only since the beginning of the 20th century that science has managed to gain a clear understanding of the microscopic nature of the various charge carriers.

When considering the full scope of materials in question, there are different types of charge carriers that can be distinguished:

▸ electrons,
▸ holes (fictitious particles arising from an electronic deficiency in the valence band of a solid lattice),
▸ ions (anions or cations, simple or complex),
▸ vacancies or charged defects in solid structures.

Knowing the nature and quantities of the charge carriers in a conducting medium is an essential prerequisite for understanding conduction phenomena. However, this is not always obvious.

▸ For example, let us consider an aqueous solution of ferrous and ferric salts. In the presence of nitrate ions, the Fe-containing charge carriers are mainly cations (aqua-complexes of Fe^{3+} and Fe^{2+}) whereas in the presence of cyanide ions, they will mainly be anions ($Fe(CN)_6^{3-}$ and $Fe(CN)_6^{4-}$).

Determining the nature of charge carriers is still the goal of research in various study areas such as aqueous solutions, organic solutions (where ion pairs are frequently observed), molten salts, polymer and solid state media.

1.3.2.2 - DIFFERENT CLASSES OF CONDUCTORS

Starting with the main classes of charge carriers previously defined, conductors can be classified according to the nature of the major charge carriers [31]:

---

[31] The macroscopic quantities characteristic of conduction and the various conduction mechanisms at microscopic level are outlined in section 4.2.

▸▸ *Electronic conductors (ionic insulators)*

This category includes:

▸ metals (and superconductors)
  for example copper, with about one free electron per copper atom,

▸ semiconductors
  for example silicon, with a valence of IV. This is an n-type semiconductor using electrons as its main charge carriers provided it is doped with a controlled content of valence V impurities such as phosphorus atoms. It becomes a p-type semiconductor, with holes as major charge carriers, when it is doped with valence III impurities such as aluminium atoms.

▸▸ *Ionic conductors (electronic insulators) or electrolytes*

A wide variety of different examples exist:

▸ electrolytic solutions, such as an aqueous solution containing KCl,

▸ molten salts, such as NaCl at high temperature,

▸ solid oxides, such as $(ZrO_2)_{1-x}(Y_2O_3)_{x/2}$, used in solid oxide fuel cells SOFC[32],

▸ polymer electrolytes, such as $LiClO_4$ dissolved in poly(ethylene oxide) (PEO).

Today the term electrolyte refers to an ionic conducting medium, whereas in the case of solutions, it equally refers to the compound, also called the solute, which is dissolved in the solvent and which is what gives the medium its conducting properties. In this latter definition, there is a distinction to be made between strong electrolytes, whereby the quantity of charge carriers is proportional to the amount of solute introduced, and weak electrolytes, which are to be found in the other remaining cases, whereby the solute is partially dissociated.

> Let us take the example of an aqueous solution containing acetic acid. The variations observed in the electric conduction properties in relation to the amount of acid introduced show that the quantity of charge carriers (acetate anion and solvated protons) is not proportional to the amount of acid introduced. This is because acetic acid is a weak acid and undergoes partial dissociation. Consequently, it is also a weak electrolyte.
>
> Another less famous example can be found in many salts, which once dissolved in organic solvents or polymers, prove to be weak electrolytes. This can occur very frequently in such complex media, where the presence of undissociated ion pairs should generally not be ignored.

In certain applications, separating two electrolytes is a necessary precaution taken in order to avoid a rapid mixing process while ensuring the current flow by allowing the ions through[33]. To achieve this, separators must have good mechanical properties. Moreover, some of them have conduction properties that are particular to a given ion or to several types of ions, thus offering more or less selectivity.

---

[32] SOFC : Solid Oxide Fuel Cell, i.e., the generic term for the fuel cell devices using a solid oxide as electrolyte and working at high temperature.

[33] Section 4.4.2 describes certain aspects of mass transport in electrochemical systems where two compartments are separated by a membrane.

ON ELECTRODES

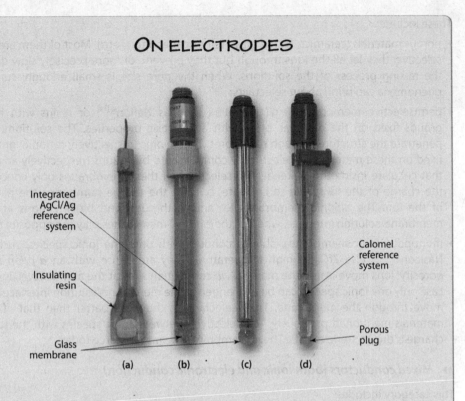

Integrated
AgCl/Ag
reference
system

Insulating
resin

Glass
membrane

Calomel
reference
system

Porous
plug

(a)   (b)   (c)   (d)

Strictly speaking, the term electrode should be restricted to defining the interface where the redox half-reaction occurs (see section 1.3.3). Generally, this term refers to the whole device: the electronic conducting material, its support and the electric connexion.

Device (a) pictured above is a metal electrode with its electric connexion embedded within an insulating resin, which moreover serves to delimit a well-defined surface area in contact with the electrolyte.

A reference electrode is usually a more complex device: it contains an electrochemical system which has a known and stable potential provided that the system is immersed in adequate solution. A porous material is what then ensures that contact is made with the cell's electrolyte. The calomel electrode (d) is an example of such a reference electrode (see section 1.5.1.2).

By extension, the term electrode usually refers to all half-cells ending in an electrical connexion, which includes specific electrodes that have an ionic membrane separating the electrochemical system from the solution. Such is the case with device (c) which uses a glass membrane specific to protons (membranes specific to other ions exist).

Also on the market can be found complete cells with an integrated reference electrode, generally coaxial, which is put in contact with the solution via a porous material placed on the lateral part. Such is the case with the combined pH electrode (b).

These include:

▸ porous materials (ceramics, fritted glass, felt or paper filters, etc.). Most of them are not selective: they let all the ions through but they prevent, or more precisely slow down the mixing process of the solutions. When the pore size is small enough, surface phenomena can bring about selectivity;

▸ permselective membranes, e.g., polymers such as Nafion®[34] or resins with ionic groups fixed on the material, often with acido-basic properties. The solutions can penetrate the structure through nanopores. The anionic (respectively cationic) groups fixed on these materials are electrically compensated by cations (respectively anions) that circulate inside the material. The selectivity of these membranes only concerns the charge of the ions able to circulate, but not the charge number or the nature of the ion. The anionic membranes let anions through and block cations at the membrane/solution interface, whereas the cationic membranes play the opposite role;

▸ monopolar ionic membranes, able to conduct with only one ionic species, such as Nasicon[35] or else ZrO_2 at high temperature. They are dense with an *a priori* zero porosity: ions move inside the material *via* conduction sites at the atomic level. In this case only one ionic species can be exchanged at the membrane/solution interface and move through the membrane. Their selectivity is therefore better than that of the materials mentioned previously. The selectivity between ionic species with the same charge is due to steric effects. These membranes are rather scarce today.

▸▸ *Mixed conductors (both ionic and electronic conduction)*

This category includes:

▸ certain oxides, such as the oxide film that builds up during the dry corrosion of metal in contact with dioxygen, or perovskites such as $LaSr_xCoO_3$ used in some fuel cells of the SOFC type,

▸ insertion materials such as derivatives of K_xC graphite or tungsten oxide bronzes,

▸ plasmas (ionised gases),

▸ molten salts containing an alkali metal in solution,

▸ liquid ammonia containing dissolved sodium which reveals solvated electrons and ions resulting from the autoprotolysis of ammonia.

1.3.3 - ELECTRODES AND INTERFACES

Electrochemistry involves the contact between different materials which conduct electricity. The two terminals in the electrochemical system linked to the external control device must be electronic conducting materials if the electric parameter is to be controlled, for instance using a direct current (DC) power supply. This system must also include at least one ionic conducting material. To illustrate, an electronic n/p junction

---

[34] Nafion is a perfluorinated polymer with numerous SO_3^- endings on the polymeric chains. It is a cationic conducting material.

[35] Nasicon (Natrium Super Ionic Conductor) is a ceramic material able to conduct via Na^+ ions at room temperature.

cannot be classified under the field of electrochemistry. Electrochemistry always deals with heterogeneous systems with the two ends made of electronic conducting materials. Such a heterogeneous system is sometimes called an electrochemical chain[36].

The term 'electrode'[37] is widely used in electrochemistry. However, it designates objects that can significantly vary depending on the situation. For the purposes of this document, in examples chosen to illustrate simple electrochemical systems, the term will most often refer to the metal which constitutes one of the terminals in the system in question. For instance, a platinum electrode or a copper rotating disc electrode [38]will be mentioned. When the system includes more than three materials, then the term electrode usually refers to the whole set of successive materials inserted between the metallic ending and the electrolyte material which makes up the core of the system. For instance, the term 'modified electrode' will be used to refer to a metal whose surface has been covered with a film of conducting material or the term 'positive electrode' in a battery will be used to refer to the composite material which is in contact with the electrolyte. In a third context, the term electrode will be used for an electrochemical half-cell [39]: this is the case with the 'pH electrode' or 'reference electrode'. In the final version of its meaning, the term electrode even stands for two half-cells combined to form the device, e.g., in the case of commercial systems for pH measurements by means of a 'combined electrode'[40].

Electrochemistry often focuses on the study of the high heterogeneity zone, which is generally a very narrow area (with a typical thickness of a few nanometres) that lies between two materials with different conduction modes. This zone is called the electrochemical interface. More generally speaking, an interface is the physical separation between two phases in a heterogeneous system. Such a separation cannot be described as a simple mathematical surface of discontinuity and therefore it is more accurate to use the term interface zone or interphase. However, as is customary in most electrochemistry books, the term interface will be used in the following work. Inside this zone, parameters such as concentrations or potential undergo large spatial gradients. Therefore the profiles, i.e., the spatial variation curves, generally show a discontinuity at the interface level, on a macroscopic scale [41].

The simplest example of an electrochemical interface is the contact zone between a metal (electrode) and a solution containing species likely to react or take part in equilibrium at this interface. The latter are called electroactive species. By contrast, a non-electroactive species does not take part in the redox half-reactions, but can play a

---

[36] The term galvanic chain, linked to the early stages of electrochemistry (see section 1.1.2), is also widely used. However, considering that its precise definition depends on the authors, we prefer to avoid this term in this document. For instance some authors restrict the use of the term 'galvanic chain' or 'galvanic cell' to systems working as power supplies, excluding electrolysis cells.

[37] See the illustrated board enttitled 'On electrodes'.

[38] See figure 1.17 in section 1.6.4.

[39] A half-cell designates one part of an electrochemical system, as defined in section 1.4.1.1.

[40] A combined electrode for pH measurement integrates the reference electrode in the body of the glass electrode (see the illustrated board entitled 'On electrodes').

[41] The potential and concentration profiles at an interface in usual cases are described in section 4.3.1.

role in carrying the current. The electrochemical interface can therefore be the highly specific zone where a redox half-reaction takes place involving electroactive species. This reaction is also called an electron transfer reaction[42], a charge transfer or indeed an electrode reaction, thus stressing the fact that it is a heterogeneous reaction localized in a very narrow zone.

By definition, when an interface sees a single redox half-reaction occurring, the electrode where oxidation takes place is called the anode, and the electrode where reduction takes place is called the cathode[43].

It is worth emphasizing that these terms are not defined in the case of open-circuit systems, and should not be used to designate electrodes, because the notions of anode or cathode are defined based only on the reactions occurring at the interface[44]. Interfaces in systems where the overall current value is zero behave in a heterogeneous way on a microscopic scale. This means that at any given moment certain zones behave as anodes while other zones behave as cathodes, resulting in an overall current equal to zero. This holds true for all systems in equilibrium: the notion of equilibrium is dynamic, and on a microscopic or even atomic scale it spans a range of oxidation and reduction events which cancel each other out completely. Another example of an open-circuit system, though this time not in equilibrium, is that of corrosion resulting from the galvanic coupling between two metals when put in contact with each other or resulting from the generalised or idiomorphic corrosion of a single metal. Here again, at a given instant, certain parts of the electrode behave as anodes while others behave as cathodes, the resulting overall current being equal to zero. But the overall chemical change is not zero[45].

These electrochemical interfaces still belong to an electrochemical chain comprising different conducting materials. Various types of interfaces can be seen and classified according to the nature of the conductors in contact:

▸ electronic junction for an interface between two electronic conductors,

▸ ionic junction for an interface between two ionic conductors,

▸ electrochemical interface for an interface between an electronic conducting medium and an ionic conducting medium,

▸ mixed junction in all the other cases.

Therefore, studying electrochemical systems requires one to understand the volume conduction of the conducting materials, as well as examine closely the interfaces

---

[42] *The term electron transfer may be ambiguous since it is also used for the shift of electrons from one medium to another in electrochemical systems where metals are in contact with mixed conductors. In such cases mobile electrons exist in both phases (e.g., solvated electrons in a molten salt) and the current flow corresponds partly to a simple electron transfer from one phase to the other, without any redox reaction occurring.*

[43] *As a suggestion, a mnemonic way of remembering these definitions is the following: oxidation and anode begin by a vowel whereas reduction and cathode begin by a consonant.*

[44] *Similarly, the use of the terms anode and cathode is not recommended for systems with blocking electrodes, where there is a transient current without any redox reaction occurring at the interfaces as represented qualitatively in section 2.2.1.2.*

[45] *Section 2.4.1 gives a simplified description of these phenomena using the current-potential curves.*

between two different conducting media. In the most general case which involves mixed conduction materials, the shift from an electronic conduction mode to an ionic conduction mode is not confined to a narrow zone. Such cases are frequently found in electrochemical applications, e.g., in the energy storage sector. In these instances the term 'volumic electrode' is used, meaning that the charge transfer reaction can take place in relatively large volumes, which is quite the opposite to usual electrochemical interfaces. This document, which presents the fundamental notions of electrochemistry, will limit its scope to localised interfaces, i.e., to systems where the interfacial zone is very thin.

In certain cases direct electron transfer can occur between chemical species in a homogeneous phase, such as aqueous solution. This homogeneous redox chemistry does not, *sensu stricto*, come under the title of electrochemistry, even though correlations can be made between homogeneous and heterogeneous systems.

1.4 - DESCRIPTION AND OPERATION OF AN ELECTROCHEMICAL CHAIN

1.4.1 - GENERAL FEATURES

1.4.1.1 - ELECTROCHEMICAL CELL AND CHAIN

The terms electrochemical cell or electrochemical chain are used for designating an electrochemical system. The first term is generally used for a tangible, concrete object which can be handled in the lab for example. The second term underlines the heterogeneous nature of the system which is made up of different successive materials which are in contact. By convention each interface is represented by a vertical stroke.

▶ The VOLTA battery is one of the simplest examples of an electrochemical chain, with each element made up of the following sequence of three conducting materials[46]:

Zn | NaCl aqueous solution | Cu   ◀

If the electrochemical chain includes several successive electrolyte media, then the abbreviated notation || is often used to denote the separation zone between two electrolytes. Such a notation represents either the porous material filled with a mixture of the two liquid electrolytes, or all of the various phases of a salt bridge (see example below). The nature of this intermediate zone is often such that the global zero current junction voltage can be neglected[47].

---

[46] *See the illustrated board entitled 'The origins of the VOLTA battery'.*

[47] *The junction voltage is very often mistakenly called junction potential in scientific literature. The elements needed for estimating the junction voltage between two solutions with different compositions are given in appendix A.1.1. A few numerical calculations show that the voltage is most often lower than 30 mV. When a salt bridge is implemented, the overall junction voltage can easily be made lower than 1 mV, which explains the approximation suggested here. However we must be careful and re-consider this approximation in cases where precise voltage values are required, which is notably the case in thermodynamic studies.*

To simplify matters, we will limit our scope to electrochemical cells having only two electrochemical interfaces and consequently with only two electrodes. Each system can be split into two half-cells where each of them has only one electrochemical interface.

▶ Let us take the example of the DANIELL cell presented in figure 1.4. It includes two compartments containing respectively a $ZnSO_4$ aqueous solution in contact with zinc metal and a $CuSO_4$ aqueous solution in contact with copper metal. These two compartments are electrically connected by a third aqueous solution, e.g., a concentrated KNO_3 solution, which is called a salt bridge.

The cell is represented below:

Zn | $ZnSO_4$ aqueous solution | KNO_3 aqueous solution | $CuSO_4$ aqueous solution | Cu

A very detailed description should also include the phases separating the various solutions. These separations are often composed of porous materials, e.g., porous plugs. It is difficult to give a precise definition of the composition of the solution inside the pores at the end of the salt bridge which is in contact with the $CuSO_4$ solution. It is a mixture of both the copper sulphate solution and the potassium nitrate solution. The same remark holds good for the solution filling the pores at the other end of the salt bridge. The brief representation is then:

Zn | $ZnSO_4$ aqueous solution | | $CuSO_4$ aqueous solution | Cu

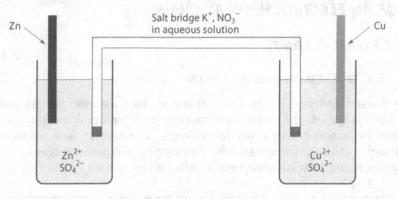

Figure 1.4 - Diagram of a DANIELL cell

1.4.1.2 - THE POLARITY OF THE ELECTRODES

The voltage of an electrochemical cell, denoted by U, is the potential difference between the two terminals of the cell; it is expressed in volts (V)[48]. The notation U is not systematically used in scientific literature, but here it has been chosen in preference to the notation E which is also frequently seen in the field of electrochemistry and also in electricity. In this document the symbol E will be kept for the potentials or voltages of

[48] A voltage measured by a voltmeter always represents the potential difference between two identical metals, in keeping with the measurement principle of a voltmeter. The electrochemical chain corresponding to a cell voltage should therefore be written with its electronic ending junctions included, although this thorough description of the electrochemical chain is often omitted. For instance, the voltage measurement of the DANIELL cell pictured above corresponds to the following electrochemical chain:

Cu'|Zn| $ZnSO_4$ aqueous solution || $CuSO_4$ aqueous solution |Cu

This point will be looked at again in the introduction of section 3.4 and an example will be studied in detail in section 3.4.1.1.

half-cells. The term emf (electromotive force) is also used in certain text-books for designating U. The term emf will be reserved here for the particular case of equilibrium [49], while the general term voltage will be kept for designating the parameter U.

When giving an algebraic value to the voltage of an electrochemical cell both of the two electrodes need to be distinguished. The privileged electrode is called the working electrode (most often noted WE or W). The other electrode is called the counter-electrode (CE) or auxiliary electrode. The algebraic voltage of the corresponding electrochemical cell is then the difference between the electric potentials of the two electrodes provided that the potential reference [50] is kept the same for both electrodes:

$$U = \varphi_{WE} - \varphi_{CE}$$

The polarity of the electrodes, which should not be confused with the polarisation [51], is defined according to the sign of the cell voltage. For example, if the voltage U is positive, the polarity + is attributed to the working electrode, which is sometimes also called the positive electrode. The polarity − is attributed to the counter-electrode, which is sometimes called the negative electrode. Conversely, if U is negative, then the polarity − is attributed to the working electrode.

Moreover, as soon as a current flows through the system, one of the electrodes becomes the anode and the other the cathode [52]. However there is no automatic link between the polarity of an electrode and its role as an anode or a cathode [53]. Unlike the notions of anode and cathode, the polarity of an electrode remains defined when the system is at open circuit. Finally, depending on the operating conditions of the electrochemical system in question, an electrode can be either the anode or cathode and also change its polarity [54].

1.4.1.3 - SIGN CONVENTION FOR THE CURRENT THROUGH AN INTERFACE

In all operating conditions, the direction of the current flow within the whole system, i.e., the orientation of the current density vector at each point, is always unambiguously defined [53]. It is only necessary to establish a sign convention for current through the

---

[49] The notion of thermodynamic equilibrium is addressed in chapter 3.

[50] The notion of reference for the potentials in electrochemistry is described in section 1.5.1. The notions of potentials are taken up again in a more detailed and rigorous way in chapter 3. In particular, the distinction between the VOLTA and GALVANI potentials is covered in section 3.1.1, and the precise description of the nature of the voltage of an electrochemical cell is given in section 3.4, in the particular case of thermodynamic equilibrium.

[51] The notion of polarisation, which has nothing to do with that used in physics (in electrostatics or in optics for instance), is defined in section 1.5.2.

[52] This sentence is not quite general but it holds for a very large majority of the systems considered in this document. It should however be handled cautiously, for example in the case of systems with so-called blocking electrodes which are crossed by a current despite the fact that no redox reaction occurs at the interfaces, as described qualitatively in section 2.2.1.2.

[53] The two possible cases (an electrochemical cell working as either a power source or as an electrolyser) are described in sections 1.4.2 and 1.4.3.

[54] The different operating conditions of an electrochemical system on the basis of current-potential curves are addressed in section 2.4.5.

working electrode if one needs to rapidly identify if the latter is an anode or a cathode. Presenting the experimental data (the current and potential of the WE) is made easier by distinguishing between the anodic and cathodic working points. We will adopt the following convention for defining an algebraic value of the current through an interface[55]:

The current is taken as positive if the electrode is on the overall an oxidation site
anode $I > 0$

The current is taken as negative if the electrode is on the overall a reduction site
cathode $I < 0$

Taking into account the definition of the current through a surface, the convention adopted here comes down to orienting the normal vector to the interface from metal towards the electrolyte (see figures 1.5 and 1.7). This convention corresponds moreover to another sign convention commonly found in thermodynamics: what leaves the system (or rather what is supplied by the system) is counted negatively; what enters the system (or rather what is supplied to the system) is counted positively.

This is also consistent with the sign convention used by electricians for an electrical load[56].

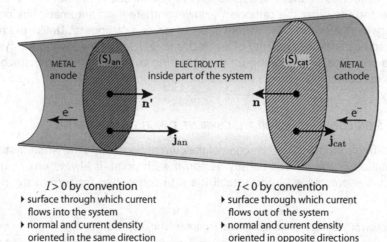

$I > 0$ by convention
▸ surface through which current flows into the system
▸ normal and current density oriented in the same direction

$I < 0$ by convention
▸ surface through which current flows out of the system
▸ normal and current density oriented in opposite directions

Figure 1.5 - Diagram of the sign convention for the current
in an electrochemical system

Due to the electroneutral nature of all conducting volumes, one can speak in terms of the current through the medium since the current is the same at whatever section[57].

---

[55] *This sign convention has also been widely adopted today by European electrochemists. However, the opposite convention has been in use for a long time for historical reasons: electrochemistry underwent significant development in the 1950's with studies carried out on the mercury electrode in aqueous electrolyte where only reduction reactions could be investigated. Hence, the working electrode was frequently a cathode.*

[56] *See the illustrated board entitled ' Sign convention for current '.*

[57] *This notion, presented in section 1.3.1.3, is laid out in section 4.1.2.*

This same result can be extended throughout the electrochemical cell. It remains linked to the overall electroneutrality of the volume in question, even when it includes one or several interfaces. In particular, the current crossing the anodic interface has the same absolute value as that of the current crossing the cathodic interface. This property is of practical importance because the respective surfaces of the two electrodes are not necessarily equal. This implies that the mean current densities at each interface do not generally have the same absolute value. Throughout the rest of this document, unless otherwise specified, j represents the projection of the vector **j** on the normal vector to the surface, respecting the sign convention shared by electrochemists.

When taking into account the sign convention previously outlined, we always end up with

$$|I| = \langle j_{an} \rangle S_{an} = -\langle j_{cat} \rangle S_{cat}$$

▶ ▶ To illustrate this property in the framework of an important case of analytical electrochemistry, let us consider the system in figure 1.6. The working electrode is a 4 mm in diameter copper disc (cross section of a metal wire embedded in an insulating material); the counter-electrode is a 1 mm in diameter platinum wire with a length ℓ immersed in the electrolyte.

Let us estimate the platinum wire length required for the mean current density at its surface to be 100 times smaller than the current density at the working electrode when both electrodes are connected to an external power supply.

The preserving property of the current at each electrode can be expressed with the following:

$$\left| \langle j_{copper} \rangle \right| \times \pi r^2 = \left| \langle j_{platinum} \rangle \right| \times (\pi r'^2 + 2 \pi r' \ell)$$

A platinum length of about 20 cm is therefore needed.

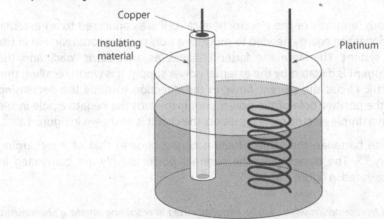

Figure 1.6 - Diagram of an electrochemical cell with a copper disc as working electrode and a counter-electrode made of a platinum wire with an immersed length ℓ

▶ Let us take a second example from galvanic corrosion, an area where this property is particularly important. Imagine an ordinary steel sheet (essentially composed of iron) with an area of $6\,m^2$, fixed to a structure by means of 60 steel rivets (16 mm in diameter) with a galvanized surface, i.e., covered by a zinc layer. In moist air conditions, significant corrosion can be observed in the rivets. The reason behind this is that dioxygen reduction occurs at the surface of the iron sheet with a current density of $100\,\mu A\,cm^{-2}$ (usual value for atmospheric conditions with no forced convection).

Moreover, coupled oxidation of the zinc rivets can be seen with the equation:

$$\left| \langle j_{sheet} \rangle \right| \times 6 \times 10^4 = \left| \langle j_{rivets} \rangle \right| \times 60 \times \pi \, (0.8)^2$$

The corrosion rate of the rivets is therefore about $50 \ mA \ cm^{-2}$ which, according to FARADAY's law (see section 2.2.2.2) corresponds to a dissolution of zinc with a rate of $85 \ \mu m \ h^{-1}$ (with the molar mass of zinc equal to $65 \ g \ mol^{-1}$ and its density equal to $7.1 \ g \ cm^{-3}$).

When studying a single interface, one generally chooses to represent it with a vertical interface, placing the metal to the left and the electrolyte to the right (figure 1.7)[58].

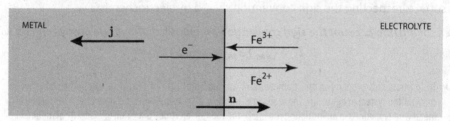

Figure 1.7 - Diagram of an electrochemical interface with the commonly used sign convention for current: in this case the reduction of Fe^{3+} ions to Fe^{2+} ions, $I < 0$

Due to the convention previously outlined, the normal vector to the surface on this diagram is always represented as a horizontal line oriented towards the right, for the anode as well as for the cathode.

1.4.2 - FORCED CURRENT FLOW: ELECTROLYSER MODE

In electrolysis, both terminals of the electrochemical cell are connected to an external source which supplies the energy needed to produce a non-spontaneous reaction in the electrochemical system. Therefore the latter behaves as an electric load, and the direction of the current is dictated by the external power supply. It is worth recalling that in a passive electric circuit the current flows in the direction towards the decreasing potentials: from the positive pole of the power supply towards the negative pole in the external circuit. In a simple electric representation, the circuit is as shown in figure 1.8[59].

When dealing with batteries, the same situation corresponds to that of a recharging secondary battery[60]. The battery and the external power supply are connected in opposition, as illustrated in figure 1.9.

---

[58] On this kind of representation which will be frequently used in describing interface phenomena, the arrows stand for the molar flux density vectors or for the velocities of the various species (here, Fe^{2+}, Fe^{3+} ions and electrons). The length of the arrows is most often arbitrary, i.e., without any link with the actual moduli of the corresponding vectors.

[59] The symbol selected for the external power supply (figures 1.8, 1.9 and 1.10) is that of an ideal voltage generator, where the point of the arrow indicates the highest potential. The symbol used in figure 1.8 for the electrochemical system working as an electrolysis cell should not be mixed up with that of an electrostatic capacitor.

[60] One refers to a 'secondary battery', or rechargeable battery, when the system can function easily in two modes, namely that of electrolyser as well as power source (e.g., in the case of a mobile phone battery). One uses the term 'primary battery', non-rechargeable battery, or indeed disposable battery, when it is impossible or difficult to recharge, i.e., to have it working as an electrolyser.

SIGN CONVENTION FOR CURRENT

A magnetic needle placed either close to the anode or close to the cathode would deviate in the same manner thus indicating that the direction of the current is actually the same in both situations (as demonstrated in OERSTED experiments on magnetism). Besides, both ammeters should display identical current readings if they are connected as indicated on the drawing. To remain in keeping with the convention agreed amongst electrochemists dictating that the current entering the system is positive, and the current leaving the system is negative, the two ammeters need simply to be connected in opposition.

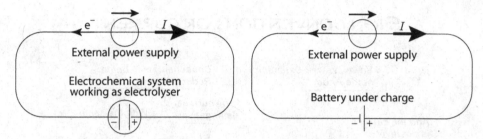

Figure 1.8 - Circuit diagram of an electrolysis cell with its power supply

Figure 1.9 - Circuit diagram of a battery being recharged and its external power supply

In an electrochemical system where electrolysis is taking place (or in the case of a battery being recharged) electrons enter at the negative electrode. Because there are no free electrons in the electrolyte and because electrons cannot durably accumulate at the interface, only a reduction reaction can use the electrons arriving at the interface. The negative electrode is therefore the cathode of the electrochemical cell. This reasoning is perfectly symmetrical at the positive electrode, which is the anode. The preceding diagram can therefore be completed as illustrated in figure 1.10.

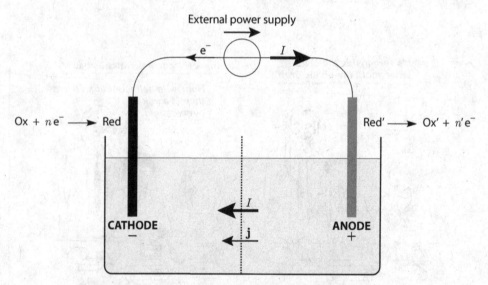

Figure 1.10 - Diagram of an electrolyser and its external power supply
The dotted line represents the membrane possibly implemented to prevent the electrolytes from mixing.

To sum up, in the case of an electrolyser (or charging battery):
the positive electrode is the anode,
the negative electrode is the cathode

1.4.3 - SPONTANEOUS CURRENT FLOW: POWER SOURCE MODE

For an electrochemical system working as a power source, (i.e., supplying energy to an external circuit) both of the two terminals of the electrochemical cell are connected, for instance by a resistance. The electrochemical system, which is where the spontaneous reaction occurs, is what imposes the direction of the current. In an external circuit, the current always flows from the positive pole of the power supply towards its negative pole. As far as batteries are concerned, it corresponds to the discharging mode. In a simple electric representation the circuit is as shown in figure 1.11.

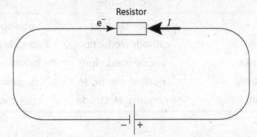

Figure 1.11 - Circuit diagram of an electrochemical system working as a power source

In this electrochemical system which is set in power source mode, electrons enter at the positive electrode. Because there are no free electrons in the electrolyte and because electrons cannot durably accumulate at the interface, only a reduction reaction can use the electrons arriving at the interface. The positive electrode is therefore the cathode of the electrochemical cell. The preceding diagram can be completed as illustrated in figure 1.12.

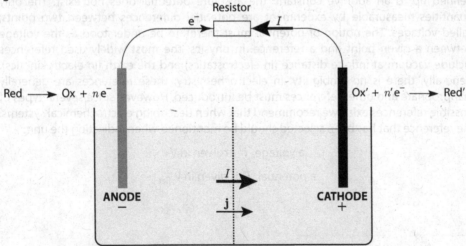

Figure 1.12 - Diagram of an electrochemical system working as a power source
The dotted line represents the membrane possibly implemented to prevent the electrolytes from mixing.

To sum up, in the power source mode:
the positive electrode is the cathode,
the negative electrode is the anode

1.4.4 - SPONTANEOUS OR FORCED CURRENT FLOW

Table 1.1 summarizes the two possible situations pertaining to an electrochemical system when crossed by a current.

Table 1.1 – Comparing the characteristics of an electrochemical system with forced or spontaneous current flow

Operating mode For a battery[61]	power source discharge	electrolyser charge
reactions	spontaneous	non-spontaneous
positive electrode	cathode (reduction)	anode (oxidation)
negative electrode	anode (oxidation)	cathode (reduction)
cathode (reduction)	positive electrode	negative electrode
anode (oxidation)	negative electrode	positive electrode

1.5 - NOTIONS OF POTENTIAL - VOLTAGE - POLARISATION

1.5.1 - VOLTAGES AND POTENTIALS IN AN ELECTROCHEMICAL CELL

The word potential should be understood as an abbreviation of electric potential *vs* a reference[62]. As in physics (electrostatics or electrokinetics) the potential is always defined up to an additive constant: the absolute potential does not exist. The only quantities measurable by experiment are potential differences between two points, called voltages. The notion of potential must therefore be understood as the voltage between a given point and a reference. In physics, the most widely used references include vacuum at infinite distance (in electrostatics) and the earth (in electrokinetics). Generally, there is no ambiguity. In electrochemistry, these references are generally inappropriate and other references must be introduced. However, since several types of possible references exist, we recommend that when describing electrochemical systems, the reference that has been selected should be mentioned when indicating the unit:

a voltage, U, is given in V

a potential, E, is given in $V_{/Ref}$

---

[61] It is increasingly common in current industrial and scientific literature on the subject of batteries to come across the misuse of the words anode and cathode for the negative and positive electrodes respectively. When referring to a primary battery it can be justified. However, for a secondary battery, the misuse of these words should be banned because it leads to significant confusion.

[62] The different notions of electric potential are laid out in section 3.1.1 (VOLTA and GALVANI potentials). Other notions making use of this term potential will also be defined in section 3.1.2 (chemical and electrochemical potentials).

1.5.1.1 - STANDARD HYDROGEN ELECTRODE

In order to define a potential reference suitable for electrochemistry, a reference redox couple must firstly be chosen: the H^+/H_2 couple in its thermodynamic standard state. This reference system is called Standard Hydrogen Electrode (SHE). It is well-defined even though it is theoretical, and it is the reference used in all contemporary data tables in thermodynamics and electrochemistry[63].

In certain electrochemical applications (namely when semiconductors as opposed to metals are involved) it may be important to link the scale of the relative potentials *vs* SHE used in electrochemistry to the scale of the potentials commonly used in physics, where the zero reference potential is vacuum at infinite distance. Establishing that link quantitatively is no easy task because it is not simply the result of experiments: what is required is hypotheses and a model. The commonly accepted value for H^+/H_2 in aqueous solution is [64]:

$$E_{/vacuum} = E_{/SHE} - 4.5 \text{ V}$$

1.5.1.2 - REFERENCE ELECTRODES

Given that the SHE electrode is an ideal system, it cannot be created through any possible experiment. However, the availability of experimental reference systems are of the utmost importance when studying and characterizing electrochemical systems. Various electrochemical half-cells have therefore been developed as references for potential measurements in electrochemical systems. Some examples which are frequently used in electrochemical devices are briefly described below [65].

▸ *Hydrogen electrode (HE)*

A hydrogen electrode, as represented in figure 1.13, is obtained by bubbling dihydrogen into a solution with a known pH, on a platinum (or platinised platinum [66]) electrode. The corresponding half-cell is:

Pt platinised, H_2 (1 bar) | H^+ with a concentration C |

---

[63] *A few thermodynamic notions are given in section 3.1.2, including in particular the definition of the standard state for various systems.*

[64] *Comparing these quantities, here in relation to the H^+/H_2 couple, in different media (for example changing the solvent) involves difficult reasoning. This aspect will be briefly addressed in section 3.1.2.3.*

[65] *Section 3.4.2 goes into greater detail on certain aspects of these systems in connection with their thermodynamic characteristics. Also listed in this section are the main characteristics of redox systems that make them good candidates for being reference systems.*

[66] *Platinised platinum is obtained by forming this metal by means of electrolysis, which allows for a highly powdery deposit to be built up, and therefore creating an extremely rough surface. Photons penetrating the open porosity are trapped, resulting in a black colour. The contact area with the solution is significantly increased, which consequently decreases the current density flowing through it, as compared to a smooth platinum electrode with the same geometric area (see section 1.4.1.3). The stability of the potential of this half-cell is therefore improved because the influence of the kinetic characteristics of the redox half-reaction is decreased. The fact of obtaining a very rough metallic surface also leads to interesting properties in the context of conductivity measurements, as mentioned in section 4.2.2.2.*

It involves the H^+/H_2 couple:

$$H_2 \rightleftharpoons 2H^+ + 2e^-$$

If an acid is chosen with a concentration of 1 mol L^{-1} then the result is a NHE (with N for 'normal', i.e., molar for a monoacid). However, it does not quite constitute a SHE because the real compounds are not close to their standard state. For example, the proton activity of an acid with a concentration equal to 1 mol L^{-1} is not equal to 1. The value of the NHE potential is about 6 mV$_{/SHE}$ at room temperature for hydrogen chloride. In practice, if one wishes to have a hydrogen electrode with a potential as close as possible to that of the SHE, then an acidic solution is implemented with a concentration slightly higher than 1 mol L^{-1} [67]. Using this type of electrode is a difficult task and must be reserved to particular applications: indeed, many particular experimental precautions are necessary for ensuring an equilibrium state in such a system.

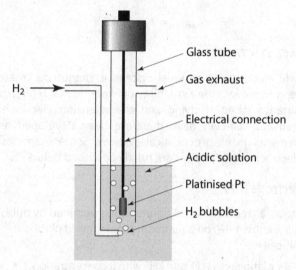

Glass tube

Gas exhaust

H_2

Electrical connection

Acidic solution

Platinised Pt

H_2 bubbles

Figure 1.13 - Diagram of a hydrogen electrode

⇥ Silver chloride electrode

A silver chloride electrode is made up of a silver wire coated with a solid silver chloride deposit (AgCl, scarcely soluble in water) immersed in a KCl solution. The electrochemical chain of this half-cell is therefore of the following type:

Ag | AgCl | KCl aqueous solution with a concentration C |

The overall equilibrium in this half-cell uses the following AgCl/Ag couple:

$$Ag + Cl^- \rightleftharpoons AgCl + e^-$$

The KCl concentration is generally set at a high value (from 1 to 3 mol L^{-1}), though it remains lower than the value for a KCl saturated solution [67]. This gives for example:

$$E_{AgCl/Ag}(KCl, 3\ mol\ L^{-1}) = +0.21\ V_{/SHE}$$

---

[67] *The reasons behind this choice are given in section 3.4.2.2.*

⇢ Calomel electrode (Hg_2Cl_2)

A calomel electrode is obtained by putting mercury with calomel (mercury(I) chloride, scarcely soluble in water) in contact with an aqueous solution containing potassium chloride. The electrochemical chain of this half-cell is therefore of the following type:

$$Hg \mid Hg_2Cl_2 \mid KCl \text{ aqueous solution with a concentration } C \mid$$

The relevant redox couple is Hg_2Cl_2/Hg:

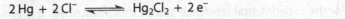

$$2\,Hg + 2\,Cl^- \rightleftharpoons Hg_2Cl_2 + 2\,e^-$$

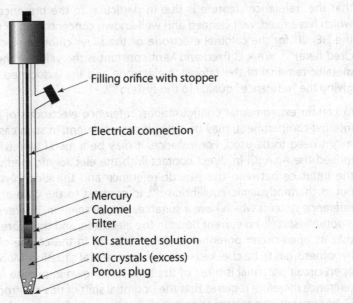

Filling orifice with stopper

Electrical connection

Mercury
Calomel
Filter
KCl saturated solution
KCl crystals (excess)
Porous plug

Figure 1.14 - Drawing of a saturated calomel electrode, SCE

A saturated KCl aqueous solution (which means about 5 mol L^{-1} at room temperature) is most often used. In this instance one then refers to a saturated calomel electrode, SCE (figure 1.14). Nowadays, it is the most widespread commercial system used for potential measurements in electrochemistry. The value of its potential at 25°C is:

$$E_{SCE} = +0.24 \text{ V}_{/SHE}$$

Changing from one potential scale to another one is done *via* a translation. It is not a difficult operation. One must however be cautious about the direction of the translation. A simple procedure is to spot, on the first scale, the new reference (0 in the new scale), e.g., +0.24 V for a change from SHE to SCE, as illustrated in figure 1.15.

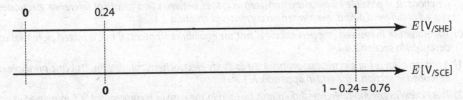

Figure 1.15 - Illustration of the change in reference on a potential scale

All the potentials must therefore be translated by −0.24 V to shift to the new scale: +1 V in the SHE scale corresponds to +0.76 V in the SCE scale.

This can also be expressed using the following equation to illustrate the preceding example of the shift from the SHE scale to the SCE scale:

$$\varphi - \varphi_{SCE} = \varphi - \varphi_{SHE} + \varphi_{SHE} - \varphi_{SCE} = \varphi - \varphi_{SHE} - (\varphi_{SCE} - \varphi_{SHE})$$
$$= 1 - 0.24 = 0.76 \text{ V}$$

For the experimental reference systems just described, it is important to stress the fact that the 'reference' feature is due in particular to the presence of the inner solution which has a fixed, well-defined and well-known concentration of the relevant ion (H^+ for the HE, Cl^- for the calomel electrode or the silver chloride electrode in the examples cited here)[68]. What is fixed and kept constant is the value of the voltage between the metallic terminal of the reference electrode and the associated internal solution, thus giving the 'reference' quality to the system[69].

In certain experimental configurations, reference electrodes of this type, i.e., with an internal compartment, may be difficult to implement. In such cases a pseudo-reference might need to be used. For instance, it may be a metal wire (silver, platinum, etc.) or indeed the Ag,AgCl in direct contact with the electrolytic medium. In these examples the interface between the pseudo-reference and the electrolyte studied is generally not in thermodynamic equilibrium[70], in contrast to the case of the interface in usual reference systems which have a suitable internal solution. However, thanks to the use of a potentiostat[71], no current flows in the electrode, and therefore it is correct to assume that its open-circuit potential remains constant in the course of the experiment. This hypothesis has to be checked in each experimental situation. Moreover, the value of this open-circuit potential is most of the time not known in precise terms. Using a pseudo-reference therefore requires that the potential shift of this electrode be determined, e.g., by implementing a reference compound at the end of the experiment[72].

---

[68] *The internal solution is in contact with the electrolyte of the system studied via a porous material. The latter is essential because it ensures contact is made with a solution having a fixed, well known concentration during the experiment. Nevertheless, it introduces an ionic junction between the internal solution and the electrolyte of the system being studied. This can be the cause of errors on the potentials measured, which are generally difficult to estimate. However, when certain experimental precautions are taken, these errors are considered insignificant most of the time (see section 3.4.2.2 and appendix A.1.1).*

[69] *As an example to shed light on this property, which may look strange at first sight, one can measure a non-zero voltage between two strictly identical reference electrodes placed in different positions in a cell. This is the case when there is a current flowing in the cell. The voltage between two points of the electrolyte, called the ohmic drop, can be measured by means of two identical reference electrodes (see figures 2.11 and 2.12, section 2.2.3). In cells with ionic and/or electronic junctions, it is possible to measure non-zero voltages between two identical reference electrodes placed at different points, even when no current is circulating.*

[70] *The potential measured at open circuit is not an equilibrium potential but a mixed potential as described in section 2.4.1.*

[71] *A potentiostat is the electric apparatus used to study electrochemical systems. Its basic principle is described in section 1.6.2 and in appendix A.1.2.*

[72] *This notion of pseudo-reference should also be used in the case of a commercial SCE immersed in an organic electrolyte, with an internal compartment filled with a saturated KCl aqueous solution. It is*

1.5.1.3 - THE POLARITY OF THE ELECTRODES

The polarity of the electrodes merely indicates the sign of the electrochemical cell voltage [73]. Thus, when applying a usual sign convention for voltages in electrochemical cells, which writes the voltage as the difference between the working (WE) and the counter-electrode (CE), it gives the following voltage:

$$U = \varphi_{WE} - \varphi_{CE} = E_{WE/Ref} - E_{CE/Ref}$$

The polarity of the electrodes has no relation to the sign of the voltage of the corresponding half-cells, which itself depends on the reference that has been selected. Depending on the nature of the reference electrode, the potential of the positive electrode can very well be negative. The potential of the negative electrode, in relation to the same reference electrode, will therefore also be negative, with a larger absolute value. Conversely, the potential of the negative electrode can be positive, however its value will be smaller than that of the positive electrode. Another practical consequence of these definitions is that it does not make any sense to attribute a relative precision (e.g., in percentage) to the value of an electrode potential, which depends on the selected reference.

▶ Let us consider a copper electrode immersed in an aqueous solution containing Cu^{2+} ions with a concentration of 10^{-3} mol L^{-1}. Its open-circuit potential is +0.25 V$_{/SHE}$ (close to the thermodynamic value). If this open-circuit potential is measured by means of an SCE, the value found is +0.01 V$_{/SCE}$, whereas choosing a mercurous sulphate electrode Hg_2SO_4/Hg with a saturated K_2SO_4 solution would yield the value of −0.39 V$_{/MSE}$. A measurement error of about 1 mV on these voltages would thus yield a relative error of 10% in the first case, and of 0.3% in the second case. But this is meaningless. ◀

1.5.2 - POLARISATIONS AND OVERPOTENTIALS IN AN ELECTROCHEMICAL CELL

When a current flows in an electrochemical cell, by definition the system is not in equilibrium. On the other hand, if no current flows (as in an open-circuit system), then the system could be either in thermodynamic equilibrium or not. This is because although the equilibrium state can always be defined, it is not necessarily always possible to observe. In fact the time scale of the experiment may be too short to reach the equilibrium state.

For a system which initially begins in thermodynamic equilibrium, the magnitude of the perturbation to its equilibrium state is an important parameter for describing the phenomena: it is the driving force behind the evolution. This is why the voltage between the terminals of an electrochemical cell with current flow is often compared to the open-circuit voltage [74].

---

impossible to know the precise value of the junction voltage between both electrolytes (aqueous and organic). Therefore the exact value of the potential of the reference electrode cannot be known, though it can nevertheless be used as a reference point provided it is constant.

[73] These notions have previously been defined in section 1.4.1.2.

[74] The voltage in thermodynamic equilibrium is also called emf. Its links with the thermodynamic reaction quantities are laid out in section 3.4.1.1.

This parameter is called overpotential or overvoltage and is denoted by η:

$$\eta_{overall} = U - U(I = 0) = U - emf$$

Whatever the open-circuit state of the system, and in particular when not in equilibrium, the term polarisation [75], π, is used:

$$\pi_{overall} = U - U(I = 0)$$

This term overall overpotential or overall polarisation is often used in the industrial world because the process of cell design generally makes it impossible to differentiate the contribution of each electrode to the overpotential. This is not the case in laboratory cells where it is possible to discriminate the contribution of each electrode by implementing one or even two reference electrodes [76]:

$$\pi_+ = E_{+/Ref} - E_{+/Ref}(I = 0) \qquad \pi_- = E_{-/Ref} - E_{-/Ref}(I = 0)$$

$$\pi_{cat} = E_{cat/Ref} - E_{cat/Ref}(I = 0) \qquad \pi_{an} = E_{an/Ref} - E_{an/Ref}(I = 0)$$

These values, which do not depend on the nature of the reference electrode, may depend on the spatial position of the reference electrode in the electrochemical system [77].

Such a splitting of the overall voltage in an electrochemical system leads to a better understanding of the phenomena being analysed. It is an important tool which will be developed in the following work using current-potential curves [78].

1.6 - EXPERIMENTATION IN ELECTROCHEMISTRY

1.6.1 - MEASUREMENT DEVICES

Three parameters play an important role in each electrochemical experiment: time, current and voltage. The latter parameter can be measured between different points. It is therefore necessary to have high performance voltmeters and ammeters.

Measuring a voltage can be based on various physical principles. To minimize any disturbance that may be caused to the phenomenon being studied by the measurement process itself, the current generated by that measurement must be kept extremely low. Therefore, when taking electrochemical measurements, notably at open circuit, voltmeters are used with field effect transistors with extremely high input impedance [79].

---

[75] This term was introduced by BECQUEREL.

[76] It would make sense to use overvoltage for the whole cell and overpotential for a single electrode. However, the authors mostly use both terms as synonyms and not as a means of discriminating two different quantities. In the following work, we will keep the term 'overpotential'.

[77] Each bench scientist must keep in mind the question of the position of the reference electrode, whenever ohmic drop in the electrolyte cannot be neglected. A schematic illustration of such a situation is detailed in section 2.2.3 and figure 2.12.

[78] For a detailed presentation of these aspects, refer to sections 2.3 and 2.4.

[79] The order of magnitude of the input impedances of voltmeters used in electrochemistry is $10^{12}\,\Omega$.

The characteristics of the ammeters required depend on the types of systems being studied. The current densities may vary from one system to another, however the surfaces are what constitute the parameter which varies to the largest degree. The current scale must therefore be very wide.

The following examples give various orders of magnitude.

▶ in an industrial cell, such as in aluminium plants, the electrolysis cells use currents about 100 000 A,

▶ in the starting battery of a car, the current requirements for the phase where the engine starts are about 100 A,

▶ in a mobile phone battery, when a call is taken the current is about 100 mA,

▶ in the button battery of a watch, the current is about 10 µA,

▶ in analytical electrochemical experiments with microelectrodes, the usual currents used are about 1 nA and can be lowered to 1 pA and even less[80].

1.6.2 - POWER SUPPLY AND CONTROL DEVICES

It is necessary to supply energy to the electrochemical system whenever it is working as an electric load (electrolyser mode). The most frequently used supply devices are electric current sources and voltage sources. A mere variable charge resistance can suffice to study an electrochemical system in the power supply mode.

When controlling the current in the system, we use the terms intensiostatic or galvanostatic setups. This applies even when the imposed current is time-dependent (see below), and the corresponding experiments are said to be potentiometry: in this case, the current is imposed and the voltages in the system are measured. When the voltage is imposed at the system's terminals then one refers to the term potentiostatic setup, even if the imposed voltage is time-dependent, and the corresponding experiments are called amperometry: in this case, the voltage between the system's terminals is imposed and the current passing through is measured.

Table 1.2 summarizes the terms for the main types of electrochemical experiments.

Table 1.2 - Main categories of electrochemical experiments

	Potentiometry	Amperometry
The setup uses a source of ...	controlled current (intensiostatic or galvanostatic setup)	controlled voltage (potentiostatic setup or a potentiostat)
The controlled parameter is...	the current $I(t)$	the voltage $U(t)$ or the potential $E(t)$
The response is in the measure of...	the voltage $U(t)$ or potential $E(t)$	the current $I(t)$

[80] *See the illustrated board entitled 'Scanning electrochemical microscope'*

Opting in favour of one of these two command modes is not always straightforward.

In a number of amperometry experiments, an apparatus is used that is more sophisticated than a mere voltage generator in order to gain precise control of the voltage, E, between the working electrode and a reference electrode. The use of such a setup is more high-performance than the method of merely controlling the voltage U between the system's two terminals: working and counter-electrodes. In fact, in the 2-electrode setup, only the overall voltage is monitored and the distribution of the voltage between the two interfaces cannot be controlled. The device essentially needed for any quantitative research study to be carried out in electrochemistry is therefore the 3-electrode setup with a potentiostat. The three electrodes used are then the working electrode (WE), the counter-electrode (CE) and a reference (Ref), through which no current flows. Figure 1.16 is a simplified diagram of an experimental electrochemical setup making use of a potentiostat[81]. The working point illustrated corresponds to a case in which the working electrode is the anode ($I > 0$ with the usual convention and the electrochemical cell works in the electrolyser mode.

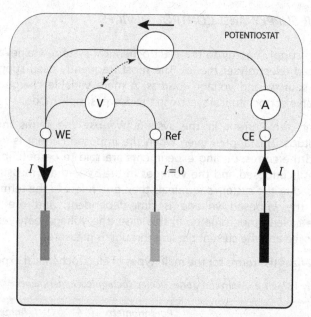

Figure 1.16 - Simplified diagram of an electrochemical setup using a potentiostat
WE: working electrode; Ref: reference electrode; CE: counter-electrode

The dotted arrow symbolizes the existing link between the voltage indicated by the voltmeter (controlled voltage between WE and Ref) and the actual voltage delivered by the power supply between WE and CE.

---

[81] *The implementation principles of these electrochemical setups are explained in appendix A.1.2. In particular, a simplified electronic circuit of a potentiostat is described (see figure A.1), underlining the use of an operational amplifier to ensure the control of E and U and to maintain a zero current in the portion of circuit containing the reference electrode (see also the illustrated board entitled 'Electrochemical devices').*

1.6.3 - DIFFERENT TYPES OF ELECTRIC CONTROL

There is a wide range of possible experiments that can be chosen for a given system depending on the type of command used and on their time dependence. The simplest example is the response given by a system where a constant voltage or current is imposed. Here one refers to the term potentiostatic or intensiostatic control. In general cases the response is time-dependent and the term transient state is then used. The response given by a system to a chronoamperomety experiment (potentiostatic control) is a curve $I(t)$; the response to a chronopotentiometry experiment (intensiostatic control) is a curve $U(t)$ or $E(t)$ [82]. It may also be of interest to note the evolution over time when the value of the command quantity applied is changed after a fixed time. In this type of experiment, depending on the kind of signal imposed, one refers to the terms single step, double step methods and other more sophisticated techniques with fixed duration pulses.

In another important category of electrochemical experiments, there is a command signal with a linear variation with time that is imposed. Included in this category is voltamperometry, frequently abbreviated to voltammetry, in which a linear potential sweep is imposed. The slope of the $U(t)$ or $E(t)$ curve is called the scan rate (in V s^{-1}). The response given by the system is usually represented by a curve $I = f(U)$ or $I = f(E)$, in which U or E, and consequently I vary with time. The direction of the potential sweep can be reversed when a given value of the potential is reached. In this case, the experiment is then called cyclic voltammetry.

Even though our main objective here is not to present all the usual techniques, it is nonetheless essential to mention the techniques where the command signal varies periodically with time. The most widely used example is electrochemical impedance spectroscopy (or abbreviated to impedance spectroscopy or impedancemetry) where the applied signal is sinusoidal with a low amplitude and a controlled frequency f. After a transient response, a steady-state response has to be reached whose amplitude must vary linearly with the input signal so that usual analysis can be carried out. This linearity can be obtained if the amplitude of the input signal is sufficiently low. Those performing the experiment must firstly check that the system fulfils both these criteria, regarding the steady character and the linearity before validating the data obtained, because the valid conditions can vary from one system to another. In these conditions, the response

---

[82] *Etymologically speaking, the term suggests that chronoamperometry (respectively chronopotenti-ometry) methods encompass all experiments in which one can expect information to emerge as an outcome of the variation with the time of the current (respectively potential) being measured. However, these terms were historically introduced for single or double-step pulses with a constant signal during each pulse. Moreover, they were restricted to experiments in which diffusion was the only transport mode for the electroactive species (such examples are presented in section 4.3.1.3). More generally speaking, the terms chronoamperometry or chronopotentiometry are used today for experiments with a constant input signal, at least for the duration of one pulse, where the mass transport modes are immaterial. For example, diffusion, migration or even forced convection, can be studied in the transient period preceding the steady state. On the other hand, experiments where the open-circuit potential is recorded, which are very useful in corrosion or titration studies, are not designated by the term chronopotentiometry, even though it would seem appropriate in etymological terms.*

is sinusoidal with the same frequency as that of the input signal. It can therefore be expressed in an impedance form which is a function of the frequency. The quantities measured can be represented on a BODE plot [83] (frequently used by electricians). However, in electrochemistry, they are more frequently presented in the form of a NYQUIST plot: the opposite of the imaginary part of the complex impedance, $-\text{Im}\,(Z)$, is plotted as a function of the real part $\text{Re}\,(Z)$ for various values of the frequency.

1.6.4 - STEADY STATE

After letting the experiment last for a reasonable duration, it is sometimes possible to reach a current/voltage point that does not change with time. In this case one refers to a steady or stationary state. The steady (U,I) or (E,I) curves that are plotted in these conditions are useful tools for analysing the behaviour of electrochemical systems [84].

This notion must not be confused with the notion of equilibrium: in other words, the terms 'steady' and 'constant' are not interchangeable. For a steady state to be established, it is merely necessary that the relevant parameters, at any point in space, do not vary with time. Strictly speaking, an electrochemical system in a steady state can therefore present spatial gradients, but the latter must not vary with time. For instance, as far as the concentrations are concerned, the steady state is expressed by the following:

$$\forall i \quad \frac{\partial C_i}{\partial t} = 0 \quad \text{but possibly} \quad \frac{\partial C_i}{\partial x} \neq 0$$

Equilibrium states are particular cases of steady states in which the overall current is zero.

From an experimental point of view, what is frequently observed are (U,I) or (E,I) curves which are steady throughout the duration of the experiment. But the steady character of the various parameters, such as the concentrations in the electrolyte, is not necessarily strictly verified. Therefore one should use the more accurate term of quasi-steady state. It is very exceptional in electrochemistry to find strict steady states differing from equilibrium states. This is because no time evolution of the mean composition of each phase must then occur [85]. The systems showing steady (U,I) or (E,I) characteristics throughout the duration of the experiment are thus therefore most of the time in a quasi-steady state.

The first precondition for securing a quasi-steady state is to ensure that the system's overall chemical composition be kept practically unchanged from start to end of the experiment.

---

[83] A BODE plot is commonly used in electronics and represents the modulus or the phase of the complex quantity as a function of the frequency on a logarithmic scale.

[84] Sections 2.3 and 2.4 deal with the shape of the current-potential curves and show how they can be used to understand and describe electrochemical systems.

[85] Examples of this type are described in appendix A.4.1 for a system where the anodic and cathodic reactions are exactly opposite. The electrolyte may then reach a steady state in the strictest sense.

Electrochemical devices

Document written with the kind collaboration of B. Petrescu,
R&D engineer for the company Bio-Logic, based in Claix in France

All electrochemical studies require an apparatus that is capable of controlling and measuring voltages and currents between the terminals of an electrochemical cell. These instruments are often called potentiostats or galvanostats. Today manufacturers tend to provide a comprehensive response to users' needs by offering better integration and modularity on both hardware and software levels. Depending on the specific application, intelligent modules can be added to the basic potentiostat, often using the same chassis. Specific high current amplification modules (up to 100 A) meet the needs of storage or energy supply (batteries, fuel cells, photovoltaic devices), while low current devices (resolution below 1 pA) meet the needs of research on materials, biochips, sensors, etc. In addition, the number of measurement channels can be modulated and the range extends from single channel to multichannel.

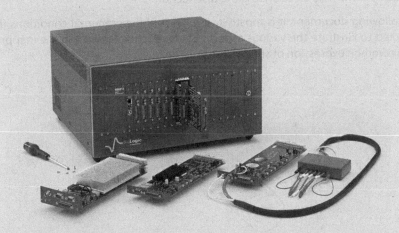

Modular multichannel potentiostat VMP3 manufactured by Bio-Logic (© Biologic SAS)

The huge technological advances that have been made in electronics and the increasingly common use of digital technology seen in recent years have collectively had a significant impact on miniaturizing and simplifying the setup around electrochemical cells. It has also opened up the door to new functions: potentiostat, galvanostat and impedance analyser configurations, intelligent recording and the possibility of having unlimited numbers of experimental recorded data, chaining techniques, analysing the peak or parametric identification in impedance, data protection, USB and Ethernet connections are just some of the possibilities that a modern potentiostat can offer. Remember that only until recently, a cyclic voltammetry experiment required the use of an analogic potentiostat coupled with a signal generator and a plotter. In order to carry out an electrochemical impedance measurement, the user often had to handle several instruments and different types of software at the same time. Today, only one instrument and only one software package are required for monitoring the experiment, recording the data and analysing the results. All this aims towards saving time for those carrying out the experiment, while improving the performance and reliability of the measurements.

This may occur in two rather different cases:

▶ a typical situation in analytical chemistry: the electrolyte's volume is sufficient to allow an extremely large amount of substance to be contained, compared to that produced or consumed during the experiment. Thus, the measurement causes no perturbation within the system being analysed;

▶ a macro-electrolysis situation in a system with a circulating electrolyte, which ensures the electrolytic cell is continuously fed with an electrolyte with a constant composition while the reaction products are being simultaneously evacuated.

This condition in terms of the system's chemical composition does not suffice to obtain quasi-steady states that are distinct from equilibrium states and that can be observed moreover under experiment for a reasonable time length. The rotating disc electrode (RDE), illustrated in figure 1.17, is an example of a device that can be used easily to obtain a quasi-steady state in the lab. The cylindrical shaft of this electrode is attached to an electric motor ensuring the rotation of the electrode around its axis. The only metallic part in contact with the electrolyte is the disc section.

In the following document, it is mostly these kinds of experimental conditions that will be selected to illustrate the various aspects of electrochemistry using the inappropriate though common expression of steady state.

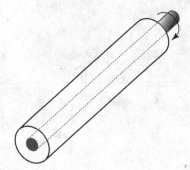

Figure 1.17 - Schematic drawing of a rotating disc electrode, RDE

1.6.5 - MAIN ELECTROCHEMICAL METHODS

Usual amperometry experiments can be briefly classified under four main categories which are shown in table 1.3. A similar table could be drawn up for usual potentiometry techniques, apart from voltammetry because its equivalent in potentiometry is not used.

Table 1.3 - Main types of amperometry experiments found in electrochemistry

	Imposed voltage	Response
Steady state (for some experimental conditions)	U, E vs t	I vs t — Steady state
Chronoamperometry single step	U, E vs t	I vs t — Transient state
double step	U, E vs t	I vs t
Voltamperometry	U, E vs t — v_{sweep}	I vs U
Impedancemetry	U, E vs t	$-\mathrm{Im}(Z)$ vs $\mathrm{Re}(Z)$ — f

QUESTIONS ON CHAPTER 1

1 - An anion is always negatively charged   true   false

2 - An oxidant is always a cation   true   false

3 - In the following half-reaction:

$$Co + 4\,Cl^- \longrightarrow CoCl_4^{2-} + 2\,e^-$$

indicate:

- ▸ the redox couple involved   ....
- ▸ the oxidized species of the couple   ....
- ▸ the (algebraic) charge number of the oxidant   ....
- ▸ the (algebraic) stoichiometric number
  of the reducing agent   ....
- ▸ the element undergoing oxidation   ....
- ▸ the oxidation number of the oxidized element   ....
- ▸ the direction of the reaction   oxidation   reduction

4 - An anion can be reduced at the cathode   true   false

5 - What is the usual oxidation number of oxygen in a compound?   ....

    Among the following compounds, circle where oxygen features:

- ▸ at its usual oxidation number

H_2O FeO H_2O_2 OH^- ClO_4^- F_2O CO_2 CO

- ▸ at a higher oxidation number

H_2O FeO H_2O_2 OH^- ClO_4^- F_2O CO_2 CO

6 - What is the oxidation number of oxygen in O_3?

         −III   −I   0   +I   +III

7 - Write the redox half-reaction of the SiO_2/Si couple in an acidic medium

    ....

8 - An electrolyte is:

- ▸ an ionically conducting medium   true   false
- ▸ a vessel used for performing electrolysis   true   false
- ▸ a compound that dissolves in a solvent giving rise to ions   true   false
- ▸ a man performing electrolysis   true   false
- ▸ an electrocuted person   true   false

9 - Molten salts are media with mainly electronic conduction   true   false

10 - Semiconductors are media with an electronic conduction type   true   false

11 - An electrolyte can exist in a .... state:
  ▶ solid   true   false
  ▶ liquid   true   false
  ▶ gas   true   false

12 - The usual order of magnitude for the width of a metal│aqueous
electrolytic solution interfacial zone is a few micrometres   true   false

13 - In electrochemistry, the cathode:
  ▶ is always the negative electrode of the system   true   false
  ▶ always has a negative potential *vs* SHE   true   false
  ▶ is always a reduction site   true   false

14 - An electrolysis process is carried out between an electrode with a surface of $1\,m^2$
where the current density is equal to $1\,mA\,cm^{-2}$ and an electrode whose active
surface is a $10\,cm \times 10\,cm$ square. The absolute value of the current density at this
second electrode is

$$10^{-3}\,A\,m^{-2} \qquad 1\,A\,m^{-2} \qquad 10^3\,A\,m^{-2}$$

15 - Considering the electrode reactions given below, complete the following diagram,
by specifying:
  ▶ the positions of the anode and the cathode
  ▶ the direction of the current (or of the current density)
  ▶ the type of the external circuit component
    (indicate your answers by replacing the question marks on the diagram)

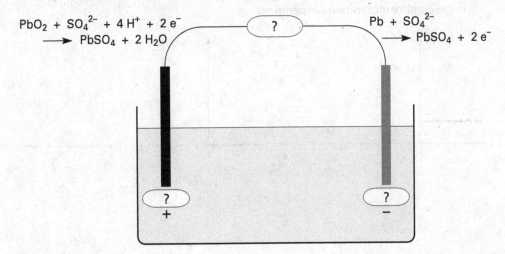

$$PbO_2 + SO_4^{2-} + 4\,H^+ + 2\,e^-$$
$$\longrightarrow PbSO_4 + 2\,H_2O$$

$$Pb + SO_4^{2-}$$
$$\longrightarrow PbSO_4 + 2\,e^-$$

16 - In a 3-electrode setup, these electrodes are called:
  ▸ . . . .
  ▸ . . . .
  ▸ . . . .

  What is the name of the electronic device generally
  used in the lab in this case?   . . . .

17 - In electrochemistry, an electrode playing a specific role is the SHE.
  ▸ What is this specific role?   . . . .
  ▸ What do the initials stand for?   . . . .
  ▸ What is the redox couple involved?   . . . .
  ▸ A potential difference exists between SHE and NHE   true   false

18 - Cite two types of reference electrodes of experimental use, and specify the redox
  couple involved.
  ▸ . . . .
  ▸ . . . .

19 - A silver wire coated with silver chloride is dipped into an
  aqueous solution containing copper nitrate. This electrode
  can be used as reference electrode for measuring potentials
  that can be spotted in the potential scale   true   false

20 - When a system, not equilibrium at open circuit,
  is crossed by a current, then one must exclu-
  sively use the term   polarisation   overpotential

21 - Complete the diagram by showing the appropriate shape of the curves that would
  indicate the variations of the voltage and the current as a function of time, in a
  simple chronoamperometry experiment.

2 - SIMPLIFIED DESCRIPTION OF ELECTROCHEMICAL SYSTEMS

When studying an electrochemical system, the first vital step is to understand the rules that define its current/voltage working point at each instant. This chapter focuses on simple systems to illustrate the thermodynamic and kinetic aspects of the phenomena occurring in those heterogeneous systems. For these purposes, an explanation, though without demonstration, will be given of the two main quantitative laws of basic electrochemistry: NERNST and FARADAY's laws. Then, on a qualitative level, we will explain how to understand the working points of an electrochemical system thanks to the current-potential curves [1].

2.1 - CHARACTERISTICS OF SYSTEMS IN THERMODYNAMIC EQUILIBRIUM

In thermodynamics it is important to establish the precise definition of a system, i.e., its boundaries and exchanges with its surroundings. In the description adopted here, the electrochemical system is confined to the metallic materials that make up the two electrodes. If heat exchanges are not taken into account, then this open-circuit system is isolated, with neither energy nor mass exchanged with the surroundings. The aim of this section is to define the thermodynamic equilibrium state of this isolated system. However, excluded from the current issue will be the description of a system that supplies energy to a device, as with a discharging battery. In the latter case, given that electrons and electric energy exchanges occur with the surroundings, then these exchanges trigger a transformation, that is to say a progression towards a new equilibrium state. Here an analogy can be drawn with a gas system: the isolated system in equilibrium is equivalent to the gas when contained in a cylinder under pressure, with fixed temperature and pressure. If a hole is drilled allowing the gas to expand in an initially empty volume, then the transformation towards a new equilibrium state can be seen.

The thermodynamic equilibrium of an electrochemical chain, which itself is rigorously defined by the equilibria within all the phases and at all the interfaces, cannot necessarily be observed on a usual time scale.

2.1.1 - DISTRIBUTION OF THE ELECTRIC POTENTIALS AT EQUILIBRIUM

In equilibrium, no current flows through the electrochemical chain and each conducting volume is equipotential. The voltage between the terminals is the algebraic sum of the

[1] Other authors also use the term polarisation curve, as will be seen in section 2.3.

junction voltages and/or of the voltages of the electrochemical interfaces. This is illustrated in figure 2.1 which presents a simplified example of a plausible potential profile inside a DANIELL cell in open-circuit conditions[2].

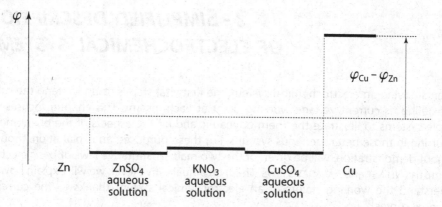

Figure 2.1 - Potential profile inside an open-circuit DANIELL cell

Each of the five conducting volumes inside the electrochemical chain is equipotential. However, there is a potential difference at each interface (in this simplified representation the thickness is assumed to be insignificant), including for ionic junctions.

In numerous electrochemical systems, especially where compartments need to be separated (see example above), ionic junctions appear in the electrochemical chain between solutions with different compositions. Strictly speaking, the thermodynamic equilibrium of such a system amounts to the perfect mixing of all the solutions[3]. Therefore all junction voltages are strictly equal to zero in thermodynamic equilibrium. However, the experimental setup is designed in such a way that actually establishing the equilibrium state is considerably slowed down by the fact of implementing a porous frit, a membrane or a gel[4]. Therefore from an experimental point of view, we are interested in the quasi-steady state, which can be reached rapidly by such a system and thus observed on a usual time scale. However this state differs from the equilibrium state. Moreover, we will admit that, when a salt bridge is used, as illustrated in the example above, the fact of opting for an intermediate electrolyte with a much higher concentration than that of the ion concentrations in each of the two other compartments, leads to the ionic junction voltages being minimised in the quasi-steady state.

---

[2] *The actual potential profile is not measurable in an experiment. To draw this diagram, we therefore had to make arbitrary choices for the signs of the various junction or interface voltages. On the other hand, the overall voltage is easily measurable. Its sign, which is positive if copper is chosen as the working electrode, is therefore correctly indicated in figure 2.1. As for the voltages at the electrochemical interfaces, their signs can be determined only if one knows the relative positions of the open-circuit potential of the electrode and that of its zero charge potential. However this discussion stretches well beyond the scope of this document.*

[3] *Detailed study of the thermodynamic equilibrium of the interfaces is given in section 3.3. The result stated here will be demonstrated in section 3.3.5.*

[4] *Section 4.4.2 will describe some aspects of mass transport phenomena in electrochemical systems with two compartments separated by a membrane, a gel or a porous frit.*

Let us recall the convention stating that the voltage of the electrochemical chain is defined as the difference between the potentials of the working electrode and the counter-electrode:

$$U = \varphi_{WE} - \varphi_{CE}$$

When the system is in thermodynamic equilibrium, this voltage is called the emf of the electrochemical cell [5]. It is an algebraic quantity expressed in volts (V).

2.1.2 - POTENTIOMETRY AT EQUILIBRIUM

2.1.2.1 - NERNST'S LAW

If you begin by expressing the equilibrium potential difference across an interface where both elements of a redox couple are present, then you can derive the NERNST law [6] which expresses the potential of the corresponding electrode *vs* a reference electrode.

This book will use the following expression of the NERNST law:

$$E_{/Ref} = E°_{/Ref} + \frac{RT}{v_e \mathscr{F}} \ln \prod_i a_i^{v_i}$$

where i represents each of the species involved in the redox half-reaction apart from the electron,

with : $E_{/Ref}$ the potential of the electrode *vs* the selected reference electrode,
when the relevant redox couple is present   [$V_{/Ref}$]

$E°_{/Ref}$ the standard potential of the relevant redox couple,
vs the selected reference,   [$V_{/Ref}$]

R   the ideal gas constant   [J K^{-1} mol^{-1}]

T the temperature   [K]

v_e the algebraic stoichiometric number of the electron

a_i the activity of species i

v_i the algebraic stoichiometric number of species i

$\mathscr{F}$ FARADAY's constant [7]   [C mol^{-1}]
1 $\mathscr{F}$ = 96 485 C mol^{-1} ≈ 96 500 C mol^{-1}

In most documents, the expression of the NERNST law that is commonly chosen uses the number n of electrons exchanged. This expression then implicitly corresponds to a half-reaction written in the direction of oxidation (the activity of the oxidant is in the numerator). We prefer a more general expression involving the algebraic stoichiometric

---

[5]   *Section 3.4.1.1 will describe the thermodynamic equilibrium of the electrochemical systems and interfaces and explain the link to be made between the emf and the standard GIBBS energy of reaction. Strictly speaking, U is the potential difference between the metallic connections made of the same metal inside the voltmeter on the WE side and on the CE side.*

[6]   *The description of the thermodynamic equilibrium of electrochemical interfaces leading to the demonstration of the NERNST law is given in sections 3.3.4.2 and 3.4.1.2.*

[7] $\mathscr{F}$ *represents the absolute value of the molar electron charge:* 1 $\mathscr{F}$ = $\mathscr{N}$ |e| *with* $\mathscr{N}$ *the AVOGADRO constant,* $\mathscr{N}$ = 6.02×10^{23} mol^{-1} *and e the elementary charge of the electron, e = − 1.6×10^{-19} C.*

number of the electrons in the half-reaction. In fact, with this form of expression the NERNST law can be presented without having to consider the direction in which the half-reaction is written, as illustrated in the examples commented below:

▸ in the direction of oxidation : $v_{Ox} > 0$ $v_e = n > 0$ $v_{Red} < 0$,

▸ in the direction of reduction : $v_{Ox} < 0$ $v_e = -n < 0$ $v_{Red} > 0$.

At 25 °C, using decimal logarithms in place of natural logarithms, one can also write the following with $(RT/\mathscr{F})\ln 10 \approx 0.06$ V:

$$E_{/Ref} = E^{\circ}_{/Ref} + \frac{0.06}{v_e} \log \prod_i a_i^{v_i} .$$

Here it is worth noting a few elementary notions of thermochemistry. The thermodynamic quantities depend on the specific state of the system. The thermodynamic state of a given system, i.e., usually a mixture of several components, is defined by the activity of each component. Let us recall the definition of the activity, which is a dimensionless number, in the case of ideal systems [8]:

▸ for an ideal mixture of ideal gases $a_i = P_i / P^{\circ}$
  (ratio of the partial pressure to the standard pressure P° equal to 1 bar)

▸ for a pure liquid $a = 1$

▸ for the solvent of a solution (major component of the medium)   . $a = 1$

▸ for a solute (whether ionic or not) of an ideal solution $a_i = C_i / C^{\circ}$
  (ratio of the concentration to the standard concentration C° equal to 1 mol L^{-1})

▸ for an element or a compound alone in its solid phase $a = 1$

▸ for ideal solutions or ideal solid mixtures $a_i = x_i$
  (molar fraction x_i of the constituent under study). In electrochemistry the most frequently found examples of this type are metallic alloys or solid insertion materials. In fact they are rarely ideal systems, however their thermodynamic description always uses parameters that are close to the molar fractions, such as the insertion rate in an insertion material (see example below).

▎ The use of NERNST's law is illustrated below in some examples of ideal systems:

▸ Fe^{3+}/Fe^{2+}:   Fe^{2+} ⇌ Fe^{3+} + e$^-$

   The stoichiometric numbers are: $v_e = 1$ $v_{Fe^{2+}} = -1$ $v_{Fe^{3+}} = +1$

   The Fe^{3+} and Fe^{2+} ions are solutes in an aqueous solution, hence:

$$E = E^{\circ}_{Fe^{3+}/Fe^{2+}} + 0.06 \log \frac{[Fe^{3+}]}{[Fe^{2+}]}$$

▸ Cu^{2+}/Cu :   Cu^{2+} + 2 e$^-$ ⇌ Cu

   The stoichiometric numbers are: $v_e = -2$ $v_{Cu^{2+}} = -1$ $v_{Cu} = +1$

   The Cu^{2+} ions are solutes in an aqueous solution, while Cu is alone in its own phase, hence:

$$E = E^{\circ}_{Cu^{2+}/Cu} - 0.03 \log \frac{1}{[Cu^{2+}]} = E^{\circ}_{Cu^{2+}/Cu} + 0.03 \log [Cu^{2+}]$$

---

[8] These basic notions of thermochemistry are presented in more detail in section 3.1.2.1.

▸ $Zn(OH)_2/Zn$: $\quad$ $Zn(OH)_2 + 2\,e^- \; \rightleftharpoons \; Zn + 2\,OH^-$

The stoichiometric numbers are: $\quad$ $\nu_e = -2 \quad \nu_{Zn(OH)_2} = -1 \quad \nu_{Zn} = +1 \quad \nu_{OH^-} = +2$

The OH^- ions are solutes in an aqueous solution, Zn and $Zn(OH)_2$ are solid single components belonging to two distinct phases[9]:

$$E = E°_{Zn(OH)_2/Zn} - 0.03 \log[OH^-]^2 = E°_{Zn(OH)_2/Zn} - 0.06 \log[OH^-]$$

▸ Let us consider the interface between an organic electrolyte containing Li^+ ions and the cobalt oxide Li_xCoO_2, which is a solid insertion material of lithium. The redox half-reaction can be written in various different forms, such as:

$$Li_xCoO_2 \; \rightleftharpoons \; CoO_2 + x\,e^- + x\,Li^+,$$
$$LiCoO_2 \; \rightleftharpoons \; Li_{1-x'}CoO_2 + x'\,e^- + x'\,Li^+$$

Ideally, the two extreme states of this material should correspond to the $CoO_2/LiCoO_2$ couple. To express the NERNST law in this material in a simplified manner, one can consider only the equilibrium of lithium ions which keep their oxidation number +I. This implies that the insertion material is a sufficiently good electronic conductor for the insertion limit to be ruled by the ionic insertion sites. The redox half-reaction can then be written:

$$\langle Li \rangle \; \rightleftharpoons \; \langle \; \rangle + e^- + Li^+_{electrolyte}$$

where the symbol $\langle \; \rangle$ represents an insertion site of the host material: $\langle \; \rangle$ for a free site and $\langle Li \rangle$ for an occupied site. The NERNST law can then be written as:

$$E = E°_{CoO_2/LiCoO_2} + 0.06 \log \frac{a_{Li^+}\, a_{\langle \; \rangle}}{a_{\langle Li \rangle}} = E°_{CoO_2/LiCoO_2} + 0.06 \log a_{Li^+} + 0.06 \log \frac{1-y}{y}$$

where y is the ratio between the insertion rate of the material and its maximal value, in this simplified expression.

▸ For the electrochemical chain corresponding to the DANIELL cell (see figure 2.1 in section 2.1.1):

$$Zn \mid aqueous\ solution\ of\ ZnSO_4 \mid\mid aqueous\ solution\ of\ CuSO_4 \mid Cu$$

neglecting junction voltages and assuming that both interfaces are at thermodynamic equilibrium lead to the following equation[10]:

$$U = E°_{Cu^{2+}/Cu} + 0.03 \log[Cu^{2+}] - (E°_{Zn^{2+}/Zn} + 0.03 \log[Zn^{2+}])$$

---

[9] Given the half-reaction written in this example, the standard state that is of particular relevance is that of a solution with an activity for OH^- equal to 1: in other words with $pH = 14$. The standard potential for the half-reaction written with H^+ ions would be different, in that, although it corresponds to the same couple, the standard sate corresponds to a solution with a unit activity for H^+ ($pH = 0$). Therefore, the values of $E°$ in both equations written with OH^- or with H^+ must not be mixed up. Section 3.4.1.1 gives the same recommendation in the case of the standard GIBBS energy data. It also applies to the particular case of the apparent standard potential, which is developed in the next section.

[10] When a voltmeter is used for measuring a voltage, the measured quantity is always that of the potential difference between two metals of the same nature (following the measuring principle of a voltmeter). The electrochemical chain corresponding to the measure of the voltage of the DANIELL cell is:

$$Cu' \mid Zn \mid ZnSO_4\ aqueous\ solution \mid\mid CuSO_4\ aqueous\ solution \mid Cu$$

This point will be taken up again in the introduction of section 3.4 and set out with an example in section 3.4.1.1.

Industrial production of aluminium in France

Document prepared with the kind collaboration of Y. Bertaud and P. Palau,
engineers at the Rio Tinto Alcan Company, based in Voreppe, France

Aluminium was first produced between 1859 and 1887 in Salindres in the Gard region of France using the Sainte-Claire Deville process. This process involved the chemical reduction of a molten mixture of chloroaluminate $NaAlCl_4$ and cryolite Na_3AlF_6 by metallic sodium Na. In 1886, P. Héroult filed a patent for an electrolysing process for Al_2O_3. His patent was bought in 1888 by the French *Société Electrométallurgique Française*: the first plant was built in Froges (Isère). The first cells were made up of a rotating crucible to limit interfacial depletion in alumina and to encourage liquid aluminium to gather at the bottom of the crucible. With an imposed voltage of $U = 10$ V (for $U_{eq} = 1.18$ V), the current intensity was 4000 A. Later on, with immobile crucibles and an increased anode area, the voltage could be lowered to 8 V for the same current value.

First electrolysis cells in Froges in 1889

Subsequently, the electrolysis processes of alumina underwent major technological changes (with the automation of machines, recycling fumes, overcoming the electrolyte and metal stirring problems due to magnetic field, which made a considerable current increase possible). However the principle of the process remained the same.

The electrolyte is a liquid phase made up of molten cryolite Na_3AlF_6 with additional AlF_3 in which alumina is dissolved at a continuous rate. This electrolyte also contains a small % of CaF_2 and possibly LiF and MgF_2. Alumina dissolves based on the following overall reaction:

$$2 Al_2O_3 + 2 AlF_6^{3-} \rightleftharpoons 3 Al_2O_2F_4^{2-}$$

During the electrolysis process, the electrode reactions are:

at the anode $\quad\quad Al_2O_2F_4^{2-} + 4 F^- + C \longrightarrow CO_2 + 2 AlF_4^- + 4 e^-$

at the cathode $\quad\quad\quad\quad AlF_6^{3-} + 3 e^- \longrightarrow Al + 6 F^-$

Therefore, when taking into account the equilibrium between AlF_6^{3-} and AlF_4^-, the following overall reaction for the main electrolysis reactions is:

$$2 Al_2O_3 + 3 C \longrightarrow 4 Al + 3 CO_2$$

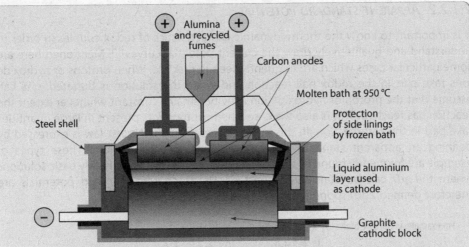

Block diagram of a modern cell for producing aluminium (© Rio Tinto Alcan, with permission)

The anode therefore sees a decrease in the $Al_2O_2F_4^{2-}$ ions content. If the solubility of the alumina is too low, or if its dissociation kinetics is too slow, the interfacial concentration may reach zero, depending on the anodic current density. Here, an 'anode effect' occurs and causes unwanted production of CF_4.

In modern cells, the supplied electric power between the cell terminals and the fact of regulating the inter-electrode distance both lead to a current intensity of about 300 000 A and to a voltage of about 4 V for each cell, including a 1.7 V ohmic drop within the electrolyte, with an energy yield value of about 0.5 and a faradic yield close to 0.95. The difference to 1 comes mainly from a ' chemical shuttle ', which is caused by the process of metal aluminium being dissolved in the electrolyte, and then followed by a re-oxidation process by dissolved CO_2 in the vicinity of the anode.

Series of 'AP 30' cells (300 000 A) (© Rio Tinto Alcan, with permission)

The key parameters that govern the technological developments are: current, faradic yield, energy consumption, the kinetics of alumina dissolution and, from more recent analysis, temperature. Moreover, in a bid to avoid CO_2 emissions, researchers are looking into replacing carbon anodes with dimensionally stable anodes which should produce dioxygen. However this would cause the energetic yield to decrease.

2.1.2.2 - APPARENT STANDARD POTENTIAL

It is important to know the thermodynamic characteristics of redox couples in order to understand and qualitatively draw the current-potential curves[11]. Mentioned here are some particular cases which are frequently seen in practice. When protons or hydroxide ions take part in the redox half-reaction and when the solution is buffered, one can assume that the proton or hydroxide ion activity remains constant whatever extent the reaction has reached. This is also the case when components present in large quantities take part in the half-reaction. In this instance therefore, the NERNST law is expressed by defining an apparent standard potential for this specific medium. These types of example are frequently found in systems involving highly acidic or highly basic solutions where the pH can be considered constant. Here apparent standard potentials are therefore defined, corresponding to aqueous solutions with fixed pH.

▶ For example, in the case of the IO_3^-/I^- couple, the redox half-reaction is:

$$I^- + 3 H_2O \rightleftharpoons IO_3^- + 6 H^+ + 6 e^-$$

We therefore have: $\nu_{I^-} = -1$ $\nu_{H_2O} = -3$ $\nu_{IO_3^-} = +1$ $\nu_{H^+} = +6$ $\nu_e = +6$

NERNST's law at 25 °C is: $$E = E°_{IO_3^-/I^-} + \frac{0.06}{6} \log \frac{(a_{H^+})^6 a'_{IO_3^-}}{a_{I^-}(a_{H_2O})^3}$$

Since H_2O is the solvent, its activity is taken to be equal to 1. The end result is:

$$E = E°_{IO_3^-/I^-} + 0.01 \log \frac{a_{IO_3^-}}{a_{I^-}} - 0.06 \, pH$$

If the solution under study is highly acidic (or highly basic) or if the solution is buffered, the NERNST law can be written as follows:

$$E = E°_{app_{IO_3^-/I^-}} + 0.01 \log \frac{a_{IO_3^-}}{a_{I^-}}$$

with $$E°_{app_{IO_3^-/I^-}} = E°_{IO_3^-/I^-} - 0.06 \, pH$$

Therefore, when numerical values are ascribed to apparent standard potentials, the relevant medium must necessarily be specified (for instance, in the preceding example, the pH value has to be given).   ◢

A similar situation is also found in systems using insertion materials where there is a high concentration of the ion being exchanged. The notion of apparent standard potential is again used in writing the NERNST law, in which only the insertion rate appears. One also refers to insertion isotherm.

▶ Let us consider the interface between an organic electrolyte containing Li^+ ions and the Li_xMnO_2 solid oxide which is an insertion material for lithium:

$$Li_xMnO_2 \rightleftharpoons MnO_2 + x e^- + x Li^+$$

The NERNST law can be written as a function of the insertion rate, y, of manganese oxide:

---

[11] Section 2.3 describes the various shapes and characteristics of current-potential curves. The thermodynamic standard (possibly apparent) potential of each relevant couple is always spotted on these curves.

$$E = E°_{MnO_2/LiMnO_2} + 0.06 \log a_{Li^+} + 0.06 \log \frac{1-y}{y}$$

In the case of ideal materials, and if the variations of the Li^+ concentration in the solution can be disregarded, the result is the following insertion isotherm, i.e., the relationship between the insertion rate and the equilibrium potential:

$$E = E°_{appMnO_2/LiMnO_2} + 0.06 \log \frac{1-y}{y}$$

with $$E°_{appMnO_2/LiMnO_2} = E°_{MnO_2/LiMnO_2} + 0.06 \log a_{Li^+}$$

2.1.2.3 - THE WATER REDOX COUPLES

Water, as a solvent or as a solute, can be oxidised or reduced [12]. It takes part in the two following couples:

$$O_2/H_2O \quad \text{and} \quad H_2O/H_2$$

One can write and balance the two corresponding redox half-reactions either with protons or with hydroxide ions:

Table 2.1 - Various ways of writing the two water redox couples

Medium	O_2/H_2O couple	H_2O/H_2 couple
Acidic	$2 H_2O \rightleftharpoons O_2 + 4 H^+ + 4 e^-$	$2 H^+ + 2 e^- \rightleftharpoons H_2$
Basic	$4 OH^- \rightleftharpoons O_2 + 2 H_2O + 4 e^-$	$2 H_2O + 2 e^- \rightleftharpoons 2 OH^- + H_2$

However, whichever of the forms is selected, the elements which see their oxidation numbers change are respectively oxygen and hydrogen:

▶ for the O_2/H_2O couple, it is the oxidation number of oxygen that changes: 0 in dioxygen and $-II$ in H_2O or in OH^-;

▶ for the H_2O/H_2 couple, it is the oxidation number of hydrogen that changes: 0 in dihydrogen and $+I$ in H_2O or in H^+.

Depending on the particular case, one writing mode is preferable to the other, though the choice will not in any way change the result when applying the NERNST law. The redox couple will remain the same and equally the equilibrium potential will not be affected by the writing mode selected. Figure 2.2 represents the thermodynamic stability domain of water as a function of pH, in a diagram of the E/pH type. This diagram is calculated by applying the NERNST law, and by using the convention which states that for gases each partial pressure is equal to 1 bar [13]:

[12] *The same water couples often intervene in non-aqueous media (e.g., when imperfectly dehydrated organic solvents are in question). But in this case thermodynamic data adapted to these media must be used (particular attention must be paid to the reference states chosen for water).*

[13] *This type of E/pH diagram, also called POURBAIX diagram, is frequently used for redox equilibria in solutions, and also in the field of corrosion studies. It is a graphic illustration of thermodynamic data which results from applying the NERNST law. Section 3.4.1.4 focuses on the example of a particular part of the POURBAIX diagram which deals with the chlorine element in aqueous solution. Numerous other examples can easily be found in scientific literature.*

▸ for the couple in which water undergoes oxidation:

$$2\,H_2O \rightleftharpoons O_2 + 4\,H^+ + 4\,e^-$$

$$E = E^\circ_{O_2/H_2O} + \frac{0.06}{4}\log\left(a_{H^+}\right)^4 \times \frac{1}{1^2} = 1.23 - 0.06\ pH$$

▸ for the couple in which water undergoes reduction:

$$2\,H^+ + 2\,e^- \rightleftharpoons H_2$$

$$E = E^\circ_{H^+/H_2} - \frac{0.06}{-2}\log\left(a_{H^+}\right)^2 \times \frac{1}{1} = 0 - 0.06\ pH$$

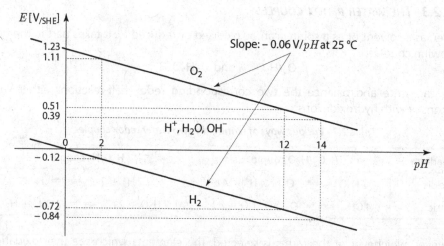

Figure 2.2 - E/pH diagram of the water couples in aqueous medium

Also noted will be the enormous impact of the pH on thermodynamic characteristics, especially in this example, since the fact of shifting from an acidic medium to a basic medium causes the potential to decrease by more than 600 mV.

When focusing on the redox reactions occurring, as opposed to when focusing on the equilibrium states, one must always take into account the two redox couples involving water[14]. This is particularly true when trying to predict and understand the reactions occurring in electrochemical systems with aqueous solutions. Here, the choice that needs to be made between the two possibilities for writing each redox couple becomes of key importance. It is an obvious choice for acidic or basic media. However it deserves particular attention when the medium is neither strongly acidic nor strongly basic. When making the choice, one needs to take into consideration the direction of the reaction in question and put the protons or hydroxide ions on the products side only and not on the reactants side. For example, keep in mind that when water is present, dihydrogen can be produced even in non-acidic media:

$$2\,H^+ + 2\,e^- \longrightarrow H_2 \qquad\qquad 2\,H_2O + 2\,e^- \longrightarrow 2\,OH^- + H_2$$

---

[14] In fact, as set out in sections 2.3.5 and 2.3.6, the two redox water couples play a part in the shape of the current-potential curves of the systems with aqueous electrolytes.

or that with a neutral electrolyte, reduction of dioxygen has to be written as the following:

$$O_2 + 2\,H_2O + 4\,e^- \longrightarrow 4\,OH^-$$

To help remember this element, in this document we suggest the following writing mode for these two couples:

$$O_2/H_2O,OH^- \qquad \text{for } O_2/H_2O \text{ and,or } OH^-$$

$$H^+,H_2O/H_2 \qquad \text{for } H^+ \text{ and,or } H_2O/H_2$$

Being able to gain a quantitative understanding of the influence of thermodynamic data of these two couples in the behaviour of systems that are not in equilibrium is a complex process [15]. This is especially true when drawing current-potential curves. We can remind ourselves that in qualitative terms, when dealing with very acidic or very basic media, it is the notion of apparent standard potential that is of key relevance. The latter is calculated by means of applying the NERNST law for a partial pressure of O_2 or H_2 equal to the standard pressure (see figure 2.2).

2.2 - CHARACTERISTICS OF SYSTEMS WITH A CURRENT FLOWING

2.2.1 - PHENOMENA OCCURRING WHEN A CURRENT IS FLOWING

In order to be able to describe the phenomena that occur when a current flows through electrochemical systems, one must first be able to describe what factor enables the current to circulate through the volumes of the various conductors and also through the various interfaces of the system. Incidentally, let us recall that current circulation, which corresponds to macroscopic charge movements, always corresponds to macroscopic mass movements.

2.2.1.1 - VOLUME CONDUCTION

The phenomena linked to the flow of current through a conducting medium can be approached at the microscopic level or from their macroscopic description. Here, we will only give a brief overview of the macroscopic description of conduction phenomena [16].

The various processes of mass transport are generally classified in three categories which are described below in a simplified manner:

▶ migration: the movement of charged species submitted to an electric field. Migration is generally characterised by the concentration and the molar conductivity of the charge carriers. The SI unit of molar conductivity (usually denoted by λ_i) is S m^2 mol^{-1}, but S cm^2 mol^{-1} is also used [17];

---

[15] Appendix A.2.1 gives a detailed explanation of the shape of the current-potential curve of the $H^+,H_2O/H_2$ couple that is assumed to be fast, validating the result given here.

[16] A quantitative macroscopic description of volume conduction is given in section 4.2.1. Some aspects of associated microscopic phenomena are described in section 4.2.2.

[17] Other notations different from the Siemens, S, can be found in other works (Ω^{-1} or else mho) but they are not recommended.

▸ diffusion: the movement of species submitted to a concentration gradient (rigorously speaking to an activity gradient). The species diffuse from the most concentrated zone towards the less concentrated zone. Diffusion is generally characterised by the diffusion coefficient (or diffusivity) of each species. The SI unit of a diffusion coefficient (usually denoted by D_i) is m^2 s^{-1}, but cm^2 s^{-1} is also used;

▸ convection: the overall movement of the medium when it is a fluid (liquid or gaseous). This movement, characterised by a velocity in m s^{-1}, carries the species i with a concentration C_i (in mol m^{-3}). It can take the form of natural convection, caused by density gradients, or forced convection corresponding to a homogenisation process caused by mechanical agitation.

In certain experimental conditions (see examples below), the electrolyte may have zones where diffusion plays a negligible role compared to the other mass transport modes (migration and convection). These are often called ' bulk electrolyte '. The zone(s) where diffusion cannot be disregarded is (are) then defined as the diffusion layer(s) [18]. Defining the diffusion layer volume in quantitative terms, and in particular defining its thickness, wholly depends on how precisely one chooses to observe and define the phenomena [19]. A diffusion layer, which is generally next to an interface, may see migration and convection also occur.

This can be illustrated in the example of a mechanical stirring device (forced convection) when used for homogenising the liquid electrolyte, mainly away from the zones that are next to the interfaces. The same case applies to analytical chemistry when a rotating disc electrode is involved or when a system is installed in industrial electrolysers in order to force the circulation of the electrolyte. By following a simplified model called the NERNST model [20], one can define the thickness of the diffusion layer (often denoted by δ) which is not time-dependent in this case. The order of magnitude of δ is a few µm. It increases with the viscosity of the medium and with the diffusion coefficient of the species concerned; it decreases when the stirring becomes more effective. The NERNST model is based on the simplifying hypothesis that the role played by convection in diffusion layers is insignificant, as indicated in figure 2.3, which shows the conduction modes involved in the various zones of the electrochemical system. In contrast, outside the two diffusion layers (thicknesses δ_1 and δ_2), i.e., in the bulk electrolyte, the very presence of forced convection means that the phenomenon of diffusion can be

---

[18] Care should be taken not to confuse this notion with another term also used in electrochemistry when describing the interfacial zone: the diffuse layer of the double layer (see sections 3.3.1 and 4.3.1.1).

[19] Section 4.3.1 focuses on the concentration profiles of several experiments that are commonly encountered in electrochemistry, serving to give a better visualization of this notion of the diffusion layer. This more detailed examination found in sections 4.3.1.3 and 4.3.1.4 helps one make a clear distinction between this general notion of a diffusion layer thickness, where diffusion represents 90, 95, 99 or 99.99%… of mass transport, and the different thickness values which result from various models (semi-infinite diffusion, NERNST's model, etc.). The latter models give precise thickness values (defined thanks to referring to the slope of the concentration profiles at the interface) which represent the right order of magnitude for the diffusion layer thickness (which itself is slightly undervalued, see for example figure 4.20 in section 4.3.1.4).

[20] Section 4.3.1.4 describes in more detail the results produced by the NERNST model, which uses hydrodynamic calculations as laid out by LEVICH.

disregarded. This simplified model creates a discontinuity in the transport modes at the boundary of the diffusion layers. This discontinuity has no physical reality, but has no impact provided the aim is not to gain a precise definition of the transition zones around δ_1 and δ_2.[21]

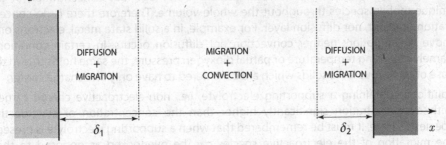

Figure 2.3 - Distribution of the conduction modes in an electrolyte
when submitted to forced convection, according to the NERNST model

Another example is a system, initially in equilibrium, where the electrolyte is not stirred, and moreover natural convection can be disregarded. Interfacial perturbations are then imposed to this system. Figure 2.4 illustrates the conduction modes involved in these conditions. The change in concentration does not immediately reach the electrolyte zones located far from the interfaces. In between, the electrolyte has a homogeneous zone where only migration occurs. As the experiment progresses with time, the size of this non-perturbed zone decreases, while the thicknesses of the diffusion layers increase.

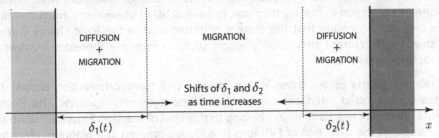

Figure 2.4 - Distribution of the conduction modes in an unstirred electrolyte
The thicknesses of diffusion layers are time-dependent.

In usual experimental conditions involving liquid electrolytes, it is impossible to prevent natural convection from occurring. In aqueous solutions at room temperature, natural convection occurs at distances of about 300 µm from the interface. Therefore the only way to keep convection negligible, is to carry out experiments for short periods, typically a few minutes[22].

---

[21] Figure 4.20 in section 4.3.1.4 illustrates the differences between the actual concentration profiles and those obtained when using the NERNST model.

[22] The order of magnitude of the diffusion layer thickness in a semi-infinite diffusion experiment is given by $\sqrt{Dt}$. This is discussed in more detail in section 4.3.1.3. For a diffusion coefficient of about 10^{-5} cm^2 s^{-1}, about 2 minutes are needed for the diffusion layer thickness to reach 300 µm, namely when natural convection phenomena begin to reach a significant scale.

Two other particular cases are frequently found to have an effective impact on the conduction modes that occur:

▸ namely media which conduct using a single mobile charged species. In these media, the electroneutrality determines that there is a homogenous concentration of the unique mobile species throughout the whole volume. Therefore there is no concentration gradient, nor diffusion layer. For example, in a solid state metal, electrons only move by migration: neither convection nor diffusion occurs. In certain conditions, namely regarding temperature or partial dioxygen pressure, the same holds true in the case of numerous ionic solids which are considered to have only one charge carrier;

▸ solutions containing a supporting electrolyte, i.e., non-electroactive charge carriers with a concentration significantly higher than the concentration of electroactive species[23]. Here, it must be remembered that when a supporting electrolyte is present, the migration of the electroactive species can be overlooked as opposed to their convection and diffusion. For instance, in the case of forced convection with a supporting electrolyte (see figure 2.3), the conduction modes of the electroactive species are diffusion within the diffusion layers and convection outside these layers.

2.2.1.2 - PHENOMENA OCCURRING AT INTERFACES

▸▸ Reactive phenomena

Let us consider the common case of the electrochemical interface between a metal and a solution. Given that a solution has no free electrons, which are what ensures the current circulation in metals, then one or several other phenomena must therefore come into play to ensure that the current is transmitted to the ionic charge carriers. The steady-state current circulation through such an interface therefore involves an electrochemical reaction.

The reactive interfaces are those where at least one chemical reaction occurs. The reactions occurring at interfaces are by nature heterogeneous reactions. The electrochemical electrode reactions always belong to this kind of reaction. Figure 2.5 illustrates the example of the reduction of Fe^{3+} ions in aqueous solution at a platinum electrode, specifying the directions in which the species move in relation to the reaction progression[24].

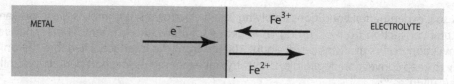

Figure 2.5 - Diagram of the reduction of Fe^{3+} at an interface with a platinum electrode

---

[23] Section 2.2.4.4 and appendix A.4.1 deal with several properties which are linked to the presence or absence of a supporting electrolyte.

[24] In this type of diagram, the arrows represent the molar flux density vectors, generally with arbitrary lengths.

When a gas is produced at a metal|electrolyte interface, the phenomenon generally occurs in several steps, involving adsorbed species and the dissolution of the molecular species produced into the electrolyte (figure 2.6).

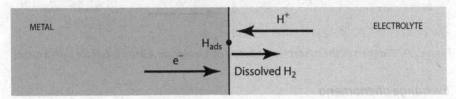

Figure 2.6 - Diagram of the reduction of H^+ at an interface with a platinum electrode [25]

As soon as the activity of the dissolved gas exceeds the critical solubility value, gas bubbles are formed.

This notion of a reactive electrode is used much more widely in modern science. For instance it can be found in phase transfer chemistry or in the study of liquid-liquid interfaces. Figure 2.7 represents the phase transfer reaction between a potassium chloride aqueous solution and a solution of a crown-ether (denoted by L) in an organic solvent [26].

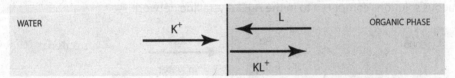

Figure 2.7 - Diagram of the complexation reaction of K^+ by the ligand L
at an interface between an aqueous solvent and an organic solvent

Cases exist among these interfacial reactions where some of the species involved are not mobile in both phases. Figures 2.8 and 2.9 illustrate two cases in which the species produced is immobile: it gathers in the interfacial zone.

▶ In the case of a metal (silver) deposition reaction by reducing cations (Ag^+) in solution, the atoms formed remain on the surface (figure 2.8) and the interface continuously develops on an atomic scale as the reaction progresses.

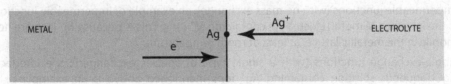

Figure 2.8 - Diagram of the reduction of an Ag^+ ion at an interface with a silver electrode

---

[25] *Quantitatively speaking, according to FARADAY's law (see section 4.1.4), the modulus of the mass flux density of dissolved dihydrogen is equal to half that of the protons.*

[26] *Here the organic solvent is supposed to be sufficiently dissociating. If that is not the case, then this diagram should be completed by adding anions which form ion pairs (KL^+, X^-) in the organic phase.*

▸ In the case of an adsorption reaction of a molecule A, the adsorbed species occupies a vacant site on the electrode surface, thus partly blocking it (figure 2.9).

Figure 2.9 - Diagram of the adsorption reaction of a molecule A at an interface with a metal

▸▸ Exchange phenomena

Let us consider two phases which are in contact with each other, each containing species of the same chemical nature, though with different interactions with the surroundings (figure 2.10). If these species are mobile in both phases, they can be exchanged through the interface. The interface is then said to be permeable to these species.

For instance, the ionic junction between solid silver chloride, which is an ionic solid where conduction is only due to Ag^+ ions, and an aqueous solution containing Ag^+ ions, is permeable to these ions. This is the case despite the fact that there is a difference between the interactions of these Ag^+ ions with the surrounding medium in the aqueous solutions compared to in the AgCl crystalline network.

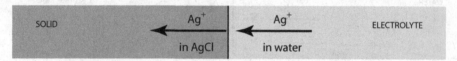

Figure 2.10 - Diagram of the exchange reaction of a silver ion
at an interface between AgCl and an aqueous solution containing Ag^+ ions

These exchange phenomena constitute a particular case of reactive phenomena since the interfacial exchange can be represented here by the following overall equation:

$$Ag^+_{water} \longrightarrow Ag^+_{AgCl}$$

Depending on the number of species that can be exchanged, one can distinguish the following:

▸ impermeable junctions (with no mass exchange): an example of impermeable junction is the interface M metal|electrolyte containing M^+ ions. This is because M^+ ions are not mobile in the metallic lattice, at least at room temperature;

▸ single-exchange junctions (with a single type of species exchanged): the interface AgCl|aqueous solution containing Ag^+ ions mentioned above is an example of a single-exchange ionic junction since chloride ions are not mobile in the ionic solid. Electronic junctions are also single-exchange junctions at room temperature: electrons are the only species that are mobile in both phases;

▸ multiple junctions (with at least two species of different natures exchanged): junctions between two electrolyte solutions belong to this kind of junction.

Although this terminology (permeable/impermeable) is frequently used in the field of junction thermodynamics, it will not be used in the remainder of this book. In fact, it is easier to apply a common treatment for electrochemical chains where different kinds of interfaces are involved. This is notably true when treating kinetic phenomena occurring at these interfaces. If you consider all these interfaces to be impermeable interfaces, whether reactive or not, then this unification is made possible (see next example). In the case of a permeable junction, the two exchanging species (e.g., Ag^+ in solid AgCl and Ag^+ in solution) will simply be considered as two different species: the first one is transformed into the second one by a reaction at the interface.

▶▶ *Accumulation phenomena*

It is important to distinguish between a ' non-reactive interface ' and an ' interface where nothing happens '. A non-reactive interface can transiently be a site of mass accumulation and consequently a site where charge accumulates if ions or electrons are concerned.

A key example of a non-reactive interface is the interface between a metal and an electrolyte which contains no electroactive species. Such a case is generally confined to a certain potential window where the electrode is said to be ideally polarisable or blocking. In such systems, applying a potential variation leads to the charge carriers moving transiently in each of the two phases. In fact, although no reaction occurs, the structure of the interfacial zone, and consequently the distribution of charge carriers, are adapted to fit the potential distribution variations. A simplified diagram of this interfacial zone would represent a capacitor where charge carriers of different signs accumulate on two planes which are placed very close to one another. This is called the electrochemical double layer[27]. It is worth mentioning that this particular feature is used in certain energy storage systems called supercapacitors[28].

From an experimental point of view it may be important to spot the particular case, notably in terms of potential, where there is no charge accumulated on both sides of the interface. This is known as the zero-charge potential.

Mass accumulation in the interfacial zone may or may not be a reactive phenomenon based on the chemical nature of the species in that particular zone. If chemical interactions do occur at the interface, then they are called adsorption reactions or chemisorptions. As in the case of catalysis, the notion of adsorption (or chemisorption) is only defined in relation to the order of magnitude of the interaction energy. The limit between an adsorption reaction and simple accumulation is not very clear. Electrosorption is an adsorption reaction involving an ion at the interface. In general cases, it is hard to determine the exact nature of an ion which has been specifically adsorbed, that is the effective charge of this new adsorbed species, which is generally not an integer multiple of the elementary charge.

---

[27] *Section 3.3.1 describes the potential profile observed in this interfacial zone which is linked to the particular charge distribution on both sides of the interface.*

[28] *See the illustrated board entitled ' Energy storage: supercapacitors '.*

2.2.2 - THE FARADIC PHENOMENA

2.2.2.1 - FARADIC CURRENT AND CAPACITIVE CURRENT

In most cases, current flow corresponds to reactive phenomena (including possibly exchange) and/or to accumulation phenomena at the electrodes. Though it may be a complicated task, notably if adsorption [29] occurs, the current can be separated into two additive contributions:

▸ the faradic current (named so because this component follows the FARADAY law) which is linked to the reactive phenomena with species consumed and produced at the electrodes [30];

▸ the capacitive current (named so because of the simplified model of a plane capacitor on the molecular scale) which is linked to the charge accumulation phenomena at the interfaces.

In numerous cases the capacitive component of the current can be disregarded, and this being so in particular in the quasi-totality of industrial applications involving current circulation in electrochemical systems.

Since this work is focused on describing simple electrochemical phenomena, we will only consider the faradic part of the current.

2.2.2.2 - FARADAY'S LAW

The faradic part of the current is a direct measure of the reaction extent rate in the case of a single reaction. Namely, this corresponds to how the system has evolved compared to the initial system, meaning that it characterises the amount of substance that has been transformed. The amount of electric charge is a measure of the reaction extent rate. This is what FARADAY's law represents in quantitative terms.

The FARADAY law is most often used in its integrated form which gives the following equation:

$$\Delta n_i^{\text{farad}} = \frac{v_i}{v_e \mathscr{F}} Q^{\text{farad}} = \frac{v_i}{v_e \mathscr{F}} \int\limits_{t}^{t+\Delta t} I^{\text{farad}}(t)\, dt$$

and when the current is constant, it gives:

$$\Delta n_i^{\text{farad}} = \frac{v_i}{v_e \mathscr{F}} I^{\text{farad}} \Delta t = \frac{v_i}{v_e \mathscr{F}} j^{\text{farad}} S\, \Delta t$$

with: $\Delta n_i^{\text{farad}}$ the algebraic variation of the amount of substance of   [mol]
         species i during the time interval Δt
     v_i the algebraic stoichiometric number of species i
     v_e the algebraic stoichiometric number of the electron
     $\mathscr{F}$ the FARADAY constant   [C mol^{-1}]

---

[29] Section 4.1.3 illustrates the consequences of adsorption phenomena at interfaces.

[30] The corresponding heterogeneous reaction is called a redox half-reaction or charge transfer or electronic transfer, as defined in section 1.3.3.

Q^{farad} the algebraic faradic charge amount exchanged   [C]
during the time interval Δt

I^{farad} the algebraic faradic current flowing through the interface   [A]

j^{farad} the algebraic faradic current density   [A m^{-2}]

Δt the time interval during which the current flows   [s]

S the surface area of the electrode in question   [m^2]

FARADAY's law, as well as its expression as a local vector, will be demonstrated in greater detail later in this book[31]. This law has numerous applications, notably in industry. For instance, it forms the basis of the theoretical calculation for battery capacity, for homogeneous corrosion rates of metals, and for the amount of substance produced or consumed in electrosynthesis processes, etc.

▼ In a fuel cell element with an area $S = 100$ cm^2 where current flows with $j = 0.20$ A cm^{-2} at room temperature, FARADAY's law enables one to determine the molar flux density of the dihydrogen consumed as well as the volume flow rate of the water produced.

The overall equation of the fuel cell is:

$$H_2 + \tfrac{1}{2}O_2 \xrightarrow{2\mathscr{F}} H_2O$$

When using FARADAY's law, it is vital to know the respective stoichiometries in the redox half-reactions including that of the electrons in particular. This is what the number written above the reaction arrow means. It can be determined by considering the anodic half-reaction:

$$H_2 \longrightarrow 2H^+ + 2e^-$$

We therefore have: $v_e = +2$ $v_{H_2} = -1$

The dihydrogen molar flow rate can be calculated from the following equation:

$$\frac{\Delta n_{H_2}}{\Delta t} = \frac{v_{H_2}}{v_e \, \mathscr{F}} j^{farad} S = -\frac{1}{2\mathscr{F}} j^{farad} S$$

1.0×10^{-4} mol s^{-1} or 0.37 mol h^{-1} of dihydrogen, i.e., 9.0 L h^{-1} of dihydrogen gas at room temperature and pressure ($V_{molar} = 24$ L mol^{-1}) are consumed.

The volume flow rate of water can be calculated by taking into account the stoichiometric numbers in the cathodic half-reaction:

$$2H^+ + 2e^- + \tfrac{1}{2}O_2 \longrightarrow H_2O$$

we therefore have : $v_e = -2$ $v_{H_2O} = +1$

$$\frac{\Delta V_{H_2O}}{\Delta t} = \frac{v_{H_2O}}{v_e \, \hat{}} j^{farad} S V_{molar} = -\frac{1}{2 \hat{}} j^{farad} S V_{molar} = \frac{1}{2 \hat{}} \left| j^{farad} \right| S V_{molar}$$

Take the molar volume of liquid water at room temperature, that is to say 18 cm^3 mol^{-1}, then the calculated water production is 6.7 cm^3 h^{-1}. ◢

2.2.2.3 - FARADIC YIELD

If several simultaneous, concurrent reactions occur at an electrochemical interface, the overall faradic current is the sum of the faradic currents of each redox half-reaction[32].

---

[31] Section 4.1.4 shows how the FARADAY law can be demonstrated by writing interfacial mass balances.

[32] Section 4.1.4 demonstrates this additive property of faradic currents in the case of an example.

The faradic yield of a given redox half-reaction is the ratio between the variation of the amount of a given species in the medium used and the corresponding variation that should be expected from FARADAY's law if the totality of the current was used for the half-reaction to progress:

$$r_{farad} = \frac{\Delta n^{effective}}{\Delta n^{farad}}$$

This notion of faradic yield is most often used for quantifying the competition between several reactions occurring simultaneously at a given interface. It represents the fraction of the actual current used for the half-reaction in question. The faradic yield (either anodic or cathodic) of a particular half-reaction is therefore inferior to 100%.

▶ For instance, when nickel is deposited using an acidic aqueous solution containing Ni^{2+} ions, one can observe a weak level of dihydrogen production, whilst simultaneously observing a reduction of the Ni^{2+} ions.

The thickness of the nickel layer deposited in an experiment where the current density is 50 mA cm^{-2}, and for the total duration of one hour, is equal to 60 μm.

The faraday yield of this electrochemical deposition can then be determined once the metallic deposit's molar mass, M, and the density, ρ_{vol}, are both known. A homogeneous thickness variation $\Delta \ell$ over a surface area S is linked to the variation of the amount of substance Δn by the following equation:

$$\Delta n = S \, \Delta \ell \, \frac{\rho_{vol}}{M}$$

The redox half-reaction of nickel deposition is:

$$Ni^{2+} + 2\,e^- \longrightarrow Ni$$

We therefore have: $\quad v_e = -2 \quad v_{Ni} = +1$

The actual current density used for the nickel deposition reaction can be calculated using FARADAY's law:

$$j_{deposition} = \frac{\Delta \ell}{\Delta t} \frac{v_e \, \mathscr{F}}{v_{Ni}} \frac{\rho_{vol}}{M} = -2 \mathscr{F} \frac{\Delta \ell}{\Delta t} \frac{\rho_{vol}}{M}$$

The current density calculated is negative, which is to be expected for a reduction reaction, following the algebraic form of FARADAY's law.

The molar mass of nickel is equal to 59 g mol^{-1}, its density is 8.9 g cm^{-3}. The calculated value of the current density is:

$$j_{deposition} = -2 \times 96\,500 \, \frac{60 \times 10^{-4}}{3600} \frac{8.9}{59} = -49 \text{ mA cm}^{-2}$$

In the end, the faradic yield obtained for nickel deposition is therefore 97%.   ◢

If the faradic yield for a given working point of the electrochemical system is close to 100% at both electrodes, then the two redox half-reactions are called main reactions. In such a case, writing the balanced results of these two half-reactions corresponds to the overall conversion of reactants to products within the system.

On the other hand, as soon as several half-reactions occur simultaneously at one of the two electrodes (or *a fortiori* at both electrodes), it is still possible to write the various overall reactions combining oxidation and reduction. However, this is of little interest since it does not represent a real overall result. A relevant example in the field of batteries is laid out below.

THE FIRST ELECTRIC VEHICLES

In the nineteenth century, there were three means of achieving propulsion competing to replace the old hackney cabs:

▶ steam following the invention of the steam carriage (4.8 km/h) by CUGNOT in 1769, and the construction of l'OBÉISSANTE by A. BOLLÉE in 1873, the first ever steam vehicle, designed for 12 passengers;

▶ electricity supplied by batteries (DAVENPORT's small-scale model in 1834, and a two-seater vehicle developed by M. FARMER in 1847);

▶ the internal combustion engine (the two-stroke engine developed by E. LENOIR in 1859, the four-stroke developed by E. DELAMARE-DEBOUTTEVILLE in 1883, and R. DIESEL's diesel engine in 1889).

To give a general idea, at the end of the eighteenth century, vehicles without horses were divided into three types: 40% steam vehicles, 38% electric traction vehicles and only 22% combustion engine vehicles.

La JAMAIS CONTENTE
(The electric automobile. France 1899. Manufactured by *Compagnie internationale des transports automobiles électriques* - JENATZY - 56, rue de la Victoire - Paris. Bodywork made out of partinium alloy, designed by RHEIMS and AUSCHER, with FULMEN accumulators. Photo by P. FABRY, with the kind permission of the *Musée national de la voiture et du tourisme* - Compiègne, France)

The struggle to penetrate the market manifested itself in competitions. The most legendary race against the clock was carried out in a matter of only a few months between Gaston DE CHASSELOUP and Camille JENATZY (1898-1899). JENATZY, the son of a Belgian manufacturer in the rubber industry (including tyres), won the race reaching up to a speed of 105.9 km/h over a distance of 1 km. He held this record for three years. La JAMAIS CONTENTE, which was shaped like a shell, had two rear engines and held over half a ton of lead acid batteries. Given the road conditions and the rustic state of mechanics at the time (steering, shock absorbers), this feat certainly merits applause. C. JENATZY continued to be an experienced racing driver while still managing his factory (he achieved a new speed record of 200 km/h in 1909, but with a combustion engine car).

Electric vehicles had clear benefits (silent, clean, basic maintenance, easy steering and long engine lifespan) compared to gasoline vehicles (dangerous crank start, need for a clutch, frequent maintenance, unpleasant smell, etc.). Although the first vehicles had low endurance levels in terms of autonomy, some of them could keep going for up to 300 km at a go. Large car fleets began to be developed in big cities (for taxis, the postal service, milk deliveries, garbage collection, etc.). Having eventually been taken over by gasoline or diesel (following the arrival of starters and FORD's assembly line), electric vehicles are now currently in the process of finding new life in light of prospected oil shortages.

▶ In a nearly fully recharged lead acid battery, two half-reactions are in competition at the positive electrode (anode):

$$PbSO_4 + 2H_2O \longrightarrow PbO_2 + 2e^- + HSO_4^- + 3H^+$$

$$H_2O \longrightarrow \tfrac{1}{2}O_2 + 2H^+ + 2e^-$$

At the negative electrode (cathode), in a battery which enables gases to escape (called a vented or flooded battery), competition between the following half-reactions can also occur:

$$PbSO_4 + H^+ + 2e^- \longrightarrow Pb + HSO_4^-$$

$$2H^+ + 2e^- \longrightarrow H_2$$

In most cases however, since there is an excess of active material at the negative electrode, the faradic yield of the positive electrode drops whereas that of the negative electrode remains very close to 100%. If we take an extreme case where the faradic yield at the cathode is 100%, then we can write the two overall reactions that are in competition at the end of the recharging operation:

$$2PbSO_4 + 2H_2O \longrightarrow PbO_2 + Pb + 2HSO_4^- + 2H^+$$

$$PbSO_4 + H_2O \longrightarrow Pb + \tfrac{1}{2}O_2 + H^+ + HSO_4^-$$

When the recharging operation is prolonged, the second reaction gradually becomes the only one to be considered, and this corresponds to an operating mode called floating [33]. In any case it makes no sense to write the balance between the reduction and oxidation of water. However there is a lot of literature about thoses batteries where one can frequently read about how the parasitic reaction is the electrolysis of water. ◀

2.2.3 - CELL VOLTAGE DISTRIBUTION

Generally speaking, in systems with localised interfaces, the voltage between the terminals in an electrochemical chain can be split into the sum of the interfacial voltages and the ohmic drops [34]. Let us recall that an interfacial voltage is the difference between the potentials of both sides of the interface, whose thickness is about a few nanometres most of the time. It is therefore either an ionic or electronic junction voltage or an electrochemical interface voltage. The ohmic drop is the potential difference between

---

[33] In floating mode, it is better to use VRLA (Valve Regulated Lead Acid) batteries where the reactions are water oxidation at the positive electrode and dioxygen reduction at the negative electrode. When a steady state is reached in an ideal VRLA battery, there is no mass consumption, which is not the case for a vented battery for which the floating mode leads to a continuous loss of water.

[34] When a voltage is measured by means of a voltmeter, the measurement represents the potential difference between two metals of the same nature (based on the measurement principle of a voltmeter). Rigorously speaking, one should add to the electrochemical chain the electronic junction(s) of the electrode connections in order to describe the distribution of the potentials which yield the voltage U that is measured between the terminals of the cell. Figure 2.11 (as well as figures 2.15 and 2.16 in section 2.2.4.2) should therefore include a first additional interface, namely an electronic junction between the copper (Cu') and the zinc on the negative electrode side. This interface is generally omitted. But rigorously speaking, the voltage between the two extreme points that is represented in this figure is different from the voltage U measured by a voltmeter. This discrepancy corresponds to the voltage of the Cu'|Zn electronic junction. Since this value is constant whatever the cell's operating mode, this precision is not essential provided that only voltage variations are studied. In this chapter, the voltage value, shifted by the electronic junction voltage, will be simply denoted by «U»:

$$\langle\!\langle U \rangle\!\rangle = \varphi_+ - \varphi_- = U - U_{connections\ electronic\ junction}$$

This point will be taken up again in the introduction to section 3.4 and laid out in detail in section 3.4.1.1.

the terminals of a given conductor volume that is part of a chain with a current flowing through it.

For each of the two extreme interfacial voltages, if we opt for the sign convention which writes the potential difference between the metal and the electrolyte, $\Delta\varphi = \varphi_{metal} - \varphi_{electrolyte}$, then it gives the following:

$$E_+ - E_- = \Delta\varphi_+ - \Delta\varphi_- + \sum U_{ohmic\ drop} + \sum U_{junction}$$

The ohmic drop terms can either be positive or negative depending on the situation[35]. In simple systems, the ohmic drop across a conducting volume is proportional to the current flowing through it[36]. It corresponds to the OHM law at the macroscopic level,

$$U = RI \qquad\qquad \text{(absolute value[35])}$$

Figure 2.11 gives a diagram representing a potential profile in a DANIELL cell:

Zn | ZnSO$_4$ aqueous sol. | KNO$_3$ aqueous sol. | CuSO$_4$ aqueous sol. | Cu

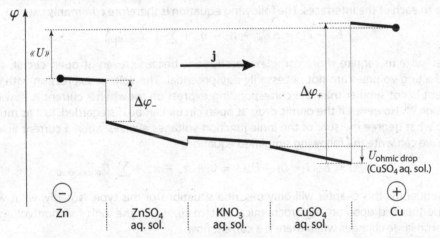

Figure 2.11 - Potential profile in a DANIELL cell working as a power source[37]
To avoid overloading the diagram, only the ohmic drop across
the CuSO$_4$ aqueous solution is represented by an arrow.

The diagram shows the profile on the scale of the system with a current flowing through it, whereby copper is the cathode and zinc the anode. In other words, the cell is working as a power source[38]. The voltage between the terminals in this electrochemical cell can

---

[35] Section 2.2.4.2 shows the potential profile in an electrochemical system working either as a power source or as an electrolyser. The differences between the ohmic drops are also highlighted (see figures 2.15 and 2.16).

[36] Appendix A.4.1 deals with a more complicated case, in which this relationship no longer holds

[37] For a precise definition of «U», refer to note [34].

[38] Note (see section 1.4.3) that the current always flows from the positive electrode towards the negative electrode in the circuit part which is out of the power supply device. Figure 2.11 describes the situation inside the electrochemical cell working as a power source where current flows from the negative electrode to the positive electrode. On the other hand, the current flow out of the cell is indeed directed from the positive towards the negative electrode.

be separated into two, firstly in ohmic drop terms (in the electronic conductors and in the three different ionic conductors) and secondly in interfacial voltage terms (for the electrochemical interfaces, $\Delta\varphi_+$ and $\Delta\varphi_-$, and the two ionic junctions between the solutions)[34]. It will be noted in this diagram that, as in figure 2.1 at open circuit, the various interfacial voltage signs, $\Delta\varphi_+$ and $\Delta\varphi_-$, are arbitrary (shown as positive in this diagram). On the other hand, the sign for potential variations in the conducting volumes (represented in the diagram by negative slopes) tallies logically with the direction of the current, and therefore with the sign for the various ohmic drops, which are here negative. It will also be noted that in power source mode, the direction of the electric field is opposite to what one would expect from the system's polarities.

For a system which starts in equilibrium at open circuit and with negligible ionic junction voltages, the word overpotential usually refers to the difference between the interfacial voltages observed with and without current flow. In connection with the general definition[39], the ohmic drop terms are therefore excluded from overpotential terms. It all comes down to being able to imagine that one can place two references infinitely close to each of the interfaces. The following equation is therefore commonly used:

$$E_+ - E_- = \Delta E_{eq} + \eta_+ - \eta_- + \sum U_{ohmic\ drop}$$

Other systems require more complex descriptions because, even at open circuit, the conducting volumes are not necessarily equipotential. The voltage expression with no current is not simpler than the corresponding expression for when a current is flowing through[40]. However, if the ohmic drop at open circuit can be disregarded, just as much as to what degree the sum of the ionic junction voltages evolves when a current flows, then we can write the following simplified equation:

$$E_+ - E_- = E_+(I = 0) - E_-(I = 0) + \pi_+ - \pi_- + \sum U_{ohmic\ drop}$$

Subsequently, this chapter will only describe situations of this type. Notably, what will not be touched upon are electrochemical systems in which the ionic junction voltages undergo large changes when there is a current flow.

From a practical point of view, it is often useful to bring in a reference electrode to separate the overall voltage between the system's terminals. The reference electrode splits the cell into two parts, allowing one to distinguish between the contributions at both extremities. However, one must keep in mind the fact that if the various ohmic drops are significant in value, then the spatial position of the reference electrode has an impact on the experimental values of the voltages being measured.

▶ Returning to the example of the DANIELL cell (figure 2.11), two types of potential measurements can be compared, when the reference electrode is dipped either into the $CuSO_4$ aqueous solution or in $ZnSO_4$ aqueous solution. As shown in the diagram in figure 2.12, the first case gives a measurement for the potential of the positive electrode ($E_{Cu} - E_{Ref\,1}$) with an ohmic drop contribution that is smaller than that of the negative

---

[39] Refer to section 1.5.2.

[40] For instance, in an open-circuit system not in equilibrium, such as a metal undergoing corrosion, the current density is not strictly zero at any point in the electrolyte. Even if the overall current is zero, the electrolyte and the electrodes are not necessarily equipotential and therefore the ohmic drop terms may not necessarily be zero in open-circuit conditions.

electrode ($E_{Zn} - E_{Ref\ 1}$). In the second case (with Ref 2), the measure of the potential of the negative electrode will be closest to its interfacial voltage.

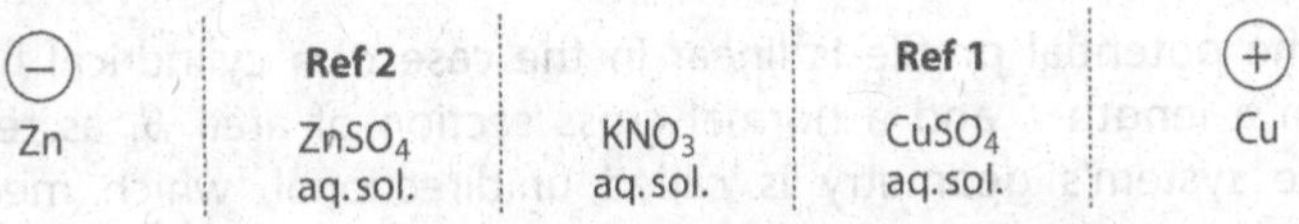

$\ominus$	**Ref 2**		**Ref 1**	$\oplus$
Zn	ZnSO$_4$ aq. sol.	KNO$_3$ aq. sol.	CuSO$_4$ aq. sol.	Cu

Figure 2.12 - Different positions for a reference electrode in a DANIELL cell

2.2.4 - OHMIC DROP IN A CONDUCTING MEDIUM

The sum of ohmic drops across each conducting medium represents a part of the overall voltage between the terminals in an electrochemical cell. The precise size of that share depends on the system in question. The aim here is to evaluate these terms by focusing on simple cases.

2.2.4.1 - OHM'S LAW AND THE OHMIC DROP

In certain experimental conditions the diffusion layers occupy only a small part of the total volume of each conducting medium. Therefore, as far as conduction modes are concerned, the quasi-totality of the volume of each conducting medium presents a case of pure migration [41]. In other instances, when a supporting electrolyte [42] is present and if the overall current and migration current can be considered as equal, then once again it is simple to define the potential and calculate the ohmic drop, even if the diffusion layers are thick. We can then consider that the current density of each charge carrier is proportional to the electric field [43].

It is the local expression of OHM's law:

$$\mathbf{j} = -\left(\sum_i \sigma_i \right) \mathbf{grad}\ \varphi = \sigma\, \mathbf{E} = \frac{1}{\rho}\, \mathbf{E}$$

with:
$\mathbf{j}$	the current density, with the modulus in	[A m^{-2}]
σ_i	the electric conductivity due to species i	[S m^{-1}]
σ	the electric conductivity of the medium	[S m^{-1}]
$\mathbf{E}$	the electric field, with the modulus in	[V m^{-1}]
ρ	the electric resistivity of the medium	[Ω m]

In any conducting medium, outside the diffusion layers, there is a constant electric conductivity for each charge carrier [44]. By integrating the local expression of OHM's law

---

[41] *When convection occurs, the overall convection current density is zero (see section 4.2.1.5). Therefore, when expressing the overall current density in a zone where diffusion can be disregarded, only the migration terms are to be taken into consideration.*

[42] *Sections 2.2.1.1 and 2.2.4.4 as well as appendix A.4.1 all give a definition of a supporting electrolyte, and explore some of the properties which are directly linked to its presence.*

[43] *Refer to section 4.2.1 which describes the macroscopic laws of conduction, and in particular, migration terms.*

[44] *The various concentrations are homogeneous outside the diffusion layers. The conductivity is also constant in the corresponding volume, since it is only dependent on these concentrations (for an isothermal system, see sections 2.2.4.3 and 4.2.2).*

taking into account the system's geometry, one obtains the ohmic drop across the medium ($U = RI$, in absolute value).

For instance, the potential profile is linear in the case of a cylindrical homogeneous conductor with a length ℓ and a normal cross section of area S, as represented in figure 2.13. The system's geometry is called unidirectional, which means that the equipotential surfaces are planes and the current lines are straight [45].

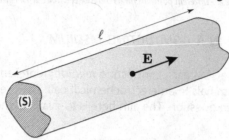

Figure 2.13 - Diagram of a conducting volume with a unidirectional geometry

The resistance of the electrolyte volume, R, and the potential difference, U, between two sections positioned at a distance equal to ℓ are:

$$R = \frac{\rho \ell}{S} = \frac{\ell}{\sigma S} \quad \text{and} \quad U = \frac{\rho \ell}{S} I = \frac{\ell}{\sigma} j$$

with: R the resistance of the electrolyte volume   [Ω]
         ρ the electric resistivity   [Ω m]
         σ the electric conductivity   [S m^{-1}]
         ℓ the length of the cylinder   [m]
         S the surface area   [m^2]
         U the potential difference   [V]
         I the current   [A]
         j the current density modulus   [A m^{-2}]

▶ When fixing an order of magnitude for the ohmic drop in an electrochemical cell, let us consider a cylindrical electrochemical cell where two electrodes with identical geometry are positioned 5 cm apart. It is assumed that the thickness of the diffusion layers will be small enough to guarantee that the only significant quantity is the ohmic drop in the bulk electrolyte. During electrolysis, a current with a density of 7 mA cm^{-2} flows through the cell (same area for the electrodes, therefore the same current density). The electrolyte contains sulphuric acid with a concentration of 0.5 mol L^{-1}, which acts as a supporting electrolyte and therefore sets the solution's conductivity at a value of $\sigma = 43$ S m$^{-1} = 0.43$ S cm^{-1}.

In this cell with unidirectional geometry the ohmic drop is given by the following equation:

$$U = \frac{\rho \ell}{S} I = \frac{\ell}{\sigma} j \qquad\qquad \text{hence the ohmic drop is 81 mV.}$$

---

[45] The first step when studying electrochemical engineering often involves determining how the current lines and equipotential surfaces are distributed. In this first step, the only phenomenon taken into account is migration in the electrolyte given the cell geometry. In this case we commonly refer to the primary distribution of the current. The geometry in question here, which is an example of unidirectional geometry, was selected on the grounds that it is very simple. Obviously, it does not fit all the different types of experimental device.

This value is rather high, although the solution used is highly concentrated. One can spot the great influence of the distance between the electrodes.

To determine the various orders of magnitude, one can calculate that the ohmic drop across a metallic sample with identical geometry, and with a conductivity of $\sigma = 5 \times 10^7 \, S \, m^{-1} = 5 \times 10^5 \, S \, cm^{-1}$, would be equal to $0.1 \, \mu V$. In most electrochemical systems the ohmic drop is found mainly in the electrolyte. However, this is not a hard and fast rule. For example, in certain industrial plants where aluminium is produced[46], the current density is about $1 \, A \, cm^{-2}$, the mean voltage is about 4.3 V, including an ohmic drop of 2.4 V. This ohmic drop is divided up into 1.7 V in the electrolyte and 0.7 V in the electrodes (aluminium, carbon, etc.), as well as the current leads and the connections between adjacent cells (an industrial installation is made up of many cells connected in series).

2.2.4.2 - MOVEMENT DIRECTION VIA MIGRATION

In the case of migration, negatively charged species (electrons, anions) and positively charged species (electron holes, cations) move in opposite directions: positive charges migrate in the same direction as that of the electric field and negative charges move in the opposite direction. Yet the corresponding current densities always have the same direction as that of the electric field, as illustrated in figure 2.14.

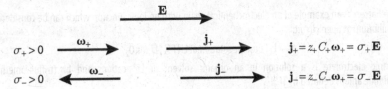

Figure 2.14 - Diagram showing the movement of charge carriers in an electric field: ω_i, velocity, and $\mathbf{j}_i$, current density of species i

The conductivity of each type of charge carrier (as well as the overall electric conductivity) is always positive. Therefore when considering any conducting material, including mixed conductors, one has:

$$\sigma = \underbrace{\sigma_{electrons} + \sigma_{holes}}_{\sigma_{electronic}} + \underbrace{\sigma_{anions} + \sigma_{cations}}_{\sigma_{ionic}}$$

It is important to underline the fact that in such a simple example of pure migration, the direction in which the ions move is unequivocally linked to the direction of the current density. Cations move in the same direction as the current density and anions move in the opposite direction. The following sentence expresses these characteristics, with the merit of being easy to remember and applicable for both a power source and an electrolyser.

Anions migrate towards the anode, cations migrate towards the cathode.

This result can be generalised to fit most electrochemical systems: the sufficient condition is that the migration current densities are in the same direction as the total current density[47].

---

[46] See the illustrated board entitled 'Industrial production of aluminium in France'.

[47] Section 4.2.1.5 gives an example which lays out in detail the link between the overall current density and migration current density.

One should avoid oversimplifying and focusing exclusively on an electrostatic viewpoint by only looking at the polarities of the electrodes. Nor should one jump to the conclusion that cations are attracted and therefore migrate towards the electrode with negative polarity. In a cell which works as a power source, the cations migrate towards the cathode, which is the positive electrode in the system. The great error here would be to take into consideration the overall potential difference between the two electrodes (merely focusing on the polarities), whereas the electric field that governs the migration process is the one found in the electrolyte, that is having already excluded the two interfacial zones. This aspect has been illustrated previously in figure 2.11, showing the potential profile in a DANIELL cell in the power source mode, as well as by the following simpler example where power source and electrolyser modes are compared in a system with no ionic junction.

In other respects, one must carefully keep in mind the fact that migration is nothing more than just one amongst other transport modes. Just because a cation migrates towards the cathode, this does not mean that it cannot react at the anode[48]. In the latter case, the effect of the convection and diffusion of the cation pushes it in the opposite way to that of migration[49].

▶ A lithium battery is an example of an electrochemical chain with no ionic junction which can be considered to be in equilibrium at open circuit:

Li | organic solution of LiTFSI | Li$_x$MnO$_2$

The organic electrolyte is a solution in an organic solvent of Li$^+$ cations and bis-(trifluoromethane-sulfonyl)imide anions denoted by TFSI$^-$.

▸ The diagrams in figure 2.15 represent the potential profile on the scale of the system, either when it is at open circuit or when a current is flowing through it in the power source mode (or discharge mode): here lithium metal, which constitutes the negative electrode, is the anode.

As in figures 2.1 and 2.11, the signs for $\Delta\varphi_+(I=0)$ and $\Delta\varphi_-(I=0)$ are chosen at random. Each of the diagrams are set with the same given potential at a random point in the electrolyte, whether it be in the system at open circuit or in the same system with a current flowing through it. To give an example, this random point might be where a reference electrode could be placed. On the other hand, the sign of the potential variation in the conducting volumes (corresponding to the various ohmic drops and represented in the diagram by straight-lined slopes) follows the direction of the current flow.

The diagrams highlight the following general relationships[50]:

$$\Delta\varphi_- - \Delta\varphi_-(I=0) = \eta_- = \eta_{an} > 0$$
$$\Delta\varphi_+ - \Delta\varphi_+(I=0) = \eta_+ = \eta_{cat} < 0$$

We therefore end up with a decrease in the overall voltage, $U < U(I=0)$, for a system working as a power source.

---

[48] Section 1.2.1 has already shown that a cation can act as either oxidant or reductor in a redox couple.

[49] Systems where the migration and diffusion fluxes directions are different are presented in qualitative terms in section 4.3.1.5.

[50] These general relationships can be illustrated in a different way by using the current-potential curves to spot the working point of an electrochemical system, as described in sections 2.4.3 and 2.4.4.

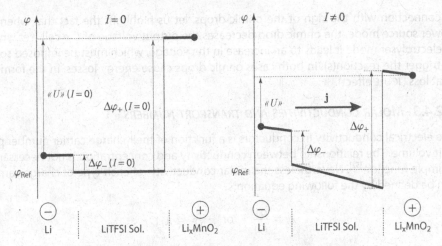

Figure 2.15 - Comparison of the potential profiles in a system in equilibrium with the same system working as a power source [51]

▶ The second diagram in figure 2.16 corresponds to the same system, which is however in electrolyser mode (or recharging battery mode): the lithium negative electrode is the cathode.

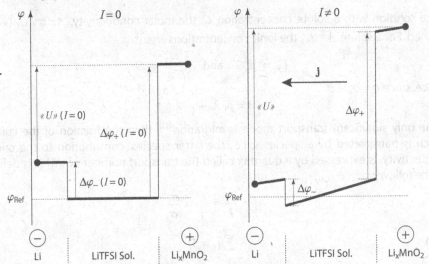

Figure 2.16 - Comparison of the potential profiles in a system in equilibrium with the same system working as an electrolyser [51]

In this second case, the diagrams highlight the following general relationships [48].

$$\Delta\varphi_- - \Delta\varphi_-(I = 0) = \eta_- = \eta_{cat} < 0$$

$$\Delta\varphi_+ - \Delta\varphi_+(I = 0) = \eta_+ = \eta_{an} < 0$$

We therefore end up with an increase in the overall voltage, $U > U(I=0)$, for a device working as an electrolyser.

---

[51] The voltage «U» is the voltage between the cell terminals, up to a constant: see note [34] of section 2.2.3.

In connection with the sign of the ohmic drops, let us highlight the fact that when in power source mode, the ohmic drop decreases the output voltage of the cell. Yet when in electrolyser mode it leads to an increase in the voltage, which must be imposed so as to trigger the reaction(s). In both cases ohmic drops cause energy losses, in the form of heat loss (JOULE effect).

2.2.4.3 - MOLAR CONDUCTIVITIES AND TRANSPORT NUMBERS

The electrical conductivity in conductors is a function of their charge carrier number per unit volume. The relationship between conductivity and concentration is not necessarily a simple function[52]. Nevertheless the molar conductivity of each type of charge carrier can be defined by the following equation:

$$\lambda_i = \frac{\sigma_i}{C_i} \quad \text{or} \quad \sigma_i = \lambda_i C_i$$

with: λ_i the molar conductivity of charge carrier i [S m^2 mol^{-1}]
       σ_i the electric conductivity of charge carrier i [S m^{-1}]
       C_i the concentration in charge carrier i [mol m^{-3}]

λ_i is, as σ_i, a positive quantity (see figure 2.14).

For a solution with a solute concentration C, the molar conductivity, $\Lambda = \sigma / C$, is also defined. For a solute $A_{p_+}B_{p_-}$, the ionic concentrations are:

$$C_+ = p_+ C \quad \text{and} \quad C_- = p_- C$$

hence, since $\sigma = \sigma_+ + \sigma_-$,

$$\Lambda = p_+ \lambda_+ + p_- \lambda_-$$

If the only significant transport mode is migration[53], then the fraction of the current which is transported by a species i, i.e., the latter species' contribution to the overall conductivity, is expressed by a quantity called the transport number of i. This is defined by the following:

$$t_i = \frac{I_i}{I} = \frac{j_i}{j} = \frac{\sigma_i}{\sigma}$$

with

$$\sum_i t_i = 1$$

Generally, this notion of transport number is used when the only transport mode that is to be considered is migration. Since all the current densities are parallel and have the same direction (that of the electric field, see figure 2.14), then the transport number values are between 0 and 1.

---

[52] Section 4.2.2.5 describes how the relationship between conductivity and concentration is, for instance, rather complicated for concentrated solutions

[53] The general definition for transport numbers in the case where diffusion, migration and convection are all simultaneously involved in mass transport is given in section 4.1.1.3 and illustrated in appendix A.4.1.

When dealing with all the various types of conducting media, the total conductivity is the quantity that is easiest to measure. If one wants to distinguish between the different contributions of the various charge carriers then this requires a set of complementary transport number measurements which are often more complex in experimental terms.

▼ The level of conductivity in a decimolar aqueous solution of calcium nitrate, $Ca(NO_3)_2$, which is a strong electrolyte, is measured as $\sigma = 26.2$ mS cm^{-1} at 25 °C. Assuming that the molar conductivity is unrelated to the concentrations, we are able to calculate the molar conductivity of the electrolyte, that of the calcium ions and the transport numbers of the two types of ions present in the solution (with data provided in scientific literature, see table 4.2 in section 4.2.2.4):

$$\lambda_{NO_3^-} = \lambda^\circ_{NO_3^-} = 7.14 \times 10^{-3}\ S\ m^2\ mol^{-1}$$
$$\sigma = 26.2\ mS\ cm^{-1} = 2.62\ S\ m^{-1}$$

$$\begin{cases} [NO_3^-] = 2\,C = 0.2\ mol\ L^{-1} = 200\ mol\ m^{-3} \\ [Ca^{2+}] = C = 0.1\ mol\ L^{-1} = 100\ mol\ m^{-3} \end{cases}$$

$$\sigma = \lambda_{NO_3^-}[NO_3^-] + \lambda_{Ca^{2+}}[Ca^{2+}] = (2\,\lambda_{NO_3^-} + \lambda_{Ca^{2+}})\,C = \Lambda\,C$$

hence 　　　　　　　　　　$\Lambda = 26.2 \times 10^{-3}\ S\ m^2\ mol^{-1}$

and 　　　　　　　　　　$\lambda_{Ca^{2+}} = 11.9 \times 10^{-3}\ S\ m^2\ mol^{-1}$

Often, data tables which are drawn up for solutions still take into account the values of the so-called equivalent conductivity λ_{eq}. For example, in the case of Ca^{2+} one finds:

$$\lambda_{\frac{1}{2}Ca^{2+}} = \frac{1}{2}\,\lambda_{Ca^{2+}} = \lambda_{eq} = 5.96 \times 10^{-3}\ S\ m^2\ mol^{-1}$$

However, the notion of equivalent, which tends to disappear, will not be used here, because it will only confuse the issue.

As for transport numbers, the following ratio is determined:

$$t_{NO_3^-} = \frac{\sigma_{NO_3^-}}{\sigma} = \frac{\lambda_{NO_3^-}}{\lambda_{NO_3^-} + \frac{1}{2}\lambda_{Ca^{2+}}} \quad \text{and} \quad t_{Ca^{2+}} = 1 - t_{NO_3^-}$$

$$t_{NO_3^-} = 0.545 \quad \text{and} \quad t_{Ca^{2+}} = 0.455$$　　　◢

2.2.4.4 - THE SUPPORTING ELECTROLYTE

Now let's imagine that one introduces into an ionically conducting medium a strong electrolyte in high concentration in comparison to the concentration level of other ions in the medium (a factor of 100 is generally considered to be sufficient). Here we observe how the quasi-totality of the migration current is due to this electrolyte. This is what is called a supporting electrolyte. Usually, the supporting electrolyte is made up of non-electroactive ions, and plays no part in the faradic phenomena (redox reactions) at the interfaces. On the other hand, it does contribute to the current flow in the electrolyte volume and possibly to the capacitive current at the interfaces.

▼ The effects of introducing a supporting electrolyte can be highlighted by numerical calculations involving transport numbers and conductivities in two different solutions:
  ▸ an aqueous solution containing silver nitrate with a concentration of 10^{-3} mol L^{-1},

▶ an aqueous solution containing silver nitrate with a concentration of 10^{-3} mol L^{-1} and potassium nitrate with a concentration of 0.1 mol L^{-1}.

To find the different values of molar conductivity at infinite dilution one can refer to the relevant literature (see table 4.2, section 4.2.2.4). For the purposes of simplicity, we will assume here that the molar conductivities are unrelated to the concentrations:

$$\lambda_{NO_3^-} = \lambda^\circ_{NO_3^-} = 7.14 \times 10^{-3} \text{ S m}^2 \text{ mol}^{-1}$$
$$\lambda_{Ag^+} = \lambda^\circ_{Ag^+} = 6.19 \times 10^{-3} \text{ S m}^2 \text{ mol}^{-1}$$
$$\lambda_{K^+} = \lambda^\circ_{K^+} = 7.35 \times 10^{-3} \text{ S m}^2 \text{ mol}^{-1}$$

▶ silver nitrate solution:

$$[Ag^+] = [NO_3^-] = 10^{-3} \text{ mol L}^{-1} = 1 \text{ mol m}^{-3}$$

The solution's conductivity is calculated using the following equation:

$$\sigma = \lambda_{NO_3^-} [NO_3^-] + \lambda_{Ag^+} [Ag^+] = (\lambda_{NO_3^-} + \lambda_{Ag^+}) [Ag^+]$$

hence $\sigma = 13.3 \times 10^{-3}$ S m^{-1} = 1.33×10^{-4} S cm^{-1}

The transport numbers of the two ions are equal to:

$$\begin{cases} t_{Ag^+} = \dfrac{\sigma_{Ag^+}}{\sigma} = \dfrac{\lambda_{Ag^+}}{\lambda_{NO_3^-} + \lambda_{Ag^+}} = 0.46 \\[3mm] t_{NO_3^-} = \dfrac{\sigma_{NO_3^-}}{\sigma} = \dfrac{\lambda_{NO_3^-}}{\lambda_{NO_3^-} + \lambda_{Ag^+}} = 0.54 \end{cases}$$

▶ silver and potassium nitrate solution:

$$\begin{cases} [Ag^+] = 10^{-3} \text{ mol L}^{-1} = 1 \text{ mol m}^{-3} \\ [K^+] = 0.1 \text{ mol L}^{-1} = 100 \text{ mol m}^{-3} \\ [NO_3^-] = 0.1 + 10^{-3} \approx 0.101 \text{ mol L}^{-1} = 101 \text{ mol m}^{-3} \end{cases}$$

The solution's conductivity is calculated using the following equation:

$$\sigma = \lambda_{NO_3^-} [NO_3^-] + \lambda_{Ag^+} [Ag^+] + \lambda_{K^+} [K^+] \approx (\lambda_{NO_3^-} + \lambda_{K^+}) [K^+]$$

hence $\sigma = 14.6 \times 10^{-1}$ S m^{-1} = 1.46×10^{-2} S cm^{-1}

This value is about 100 times higher than in the last solution. Moreover, it gives the following transport numbers:

$$\begin{cases} t_{Ag^+} = \dfrac{\sigma_{Ag^+}}{\sigma} = \dfrac{10^{-3} \lambda_{Ag^+}}{10^{-3} \lambda_{Ag^+} + 0.1 \lambda_{NO_3^-} + 0.1 \lambda_{K^+}} \approx 0.004 \\[3mm] t_{K^+} = \dfrac{\sigma_{K^+}}{\sigma} = \dfrac{0.1 \lambda_{K^+}}{10^{-3} \lambda_{Ag^+} + 0.1 \lambda_{NO_3^-} + 0.1 \lambda_{K^+}} \approx 0.49 \\[3mm] t_{NO_3^-} = \dfrac{\sigma_{NO_3^-}}{\sigma} = \dfrac{0.1 \lambda_{NO_3^-}}{10^{-3} \lambda_{Ag^+} + 0.1 \lambda_{NO_3^-} + 0.1 \lambda_{K^+}} \approx 0.51 \end{cases}$$

We will keep in mind that here in this second case, when a supporting electrolyte KNO_3 is present, then the solution's conductivity is much higher, and that:

$$t_{Ag^+} \ll t_{K^+} \quad \text{and} \quad t_{Ag^+} \ll t_{NO_3^-}$$

Adding a supporting electrolyte, whenever possible (that is to say, in liquid electrolytes), only serves a purpose in a certain range of applications. For example, being able to disregard the contribution made by electroactive species in the migration transport, often makes easier the use of experimental data, for instance in analytical applications. Adding a supporting electrolyte also increases the electrolyte's overall conductivity. However, in many industrial applications, there is a sufficiently large amount of electroactive species to ensure that the electrolyte has a good level of conductivity, all of which generally does away with the need for an additional compound. This is particularly key for example when the weight of the overall system is seen as an important criterion as in applications for batteries in portable electronic devices.

2.3 - THE SHAPE OF THE CURRENT-POTENTIAL CURVES

In the description given outlining electrochemical systems in which a current flows, key parameters include the variations of the anodic and cathodic polarisations (or overpotentials if applicable) as a function of current and time. These relationships are generally represented in the form of current-potential curves of an electrode, $I = f(E)$, where E is the voltage between the electrode in question and a reference electrode[54]. The experimental results can also be presented in the form of current density-potential curves. However, when the study concerns the whole electrochemical system and is not just focused on the working electrode, it is best to keep the current-potential representation[55].

Generally speaking, these characteristics are time-dependent. Here, we will only look at the steady-state current-potential curves, so called because they are obtained in steady-state conditions. However the shapes presented remain valid in qualitative terms at all times. Moreover, the voltage that is accessible in experimental conditions includes an ohmic drop term between the electrode in question and the reference electrode. However, this simplified description will not take into account this ohmic drop. Yet this type of curve, once corrected for the ohmic drop, can be determined in most cases in experimental conditions. Here we will discuss their shape in qualitative terms but not the detail of the curves[56].

The current-potential curves reflect the kinetic behaviour of redox systems since they indicate the changes in current, and therefore also indicate the changes in the reaction rate for different values of the potential. The word kinetic is very broad and includes the

---

[54] Some books also call these curves polarisation curves, especially those which choose to represent the potential as a function of the current. The EVANS diagrams, depicting the use of currents in corrosion, provide an example of such a type of representation. In this document, we will only choose to use the current-potential representation which is quite a natural choice if the data are recorded using potentiostatic techniques.

[55] Sections 2.4.2 to 2.4.5 describe various examples that illustrate the importance of using current-potential curves when discussing the operating conditions in electrochemical systems. This choice is explained by the fact that, contrary to current density, the current remains constant in all sections of the system, as presented in section 1.4.1.3.

[56] Quantitative analytic expressions of these curves are given for simple redox systems in section 4.3.3.

whole phenomena leading to the current flow. In electrochemistry, the two main types of phenomena involved are frequently distinguished using the expressions 'mass transport kinetics' and 'redox reaction kinetics'[57]. Sections 2.3.2 and 2.3.3 describe the main impact of both types of kinetic phenomena on the shape of the current-potential curves.

2.3.1 - GENERAL CHARACTERISTICS

2.3.1.1 - POLARISATION SIGN

The polarisation sign depends on each particular situation encountered. Drawing an analogy with mechanical systems can help one understand certain rules about this sign. Whilst an object moves spontaneously from an initial velocity in an environment where it is only subject to friction force, then the mechanical energy decreases. Therefore, to maintain a given speed, a supplement is needed to compensate for the energy that is dissipated by the friction force. To cause a current to flow in an electrochemical chain operating as an electrolyser, that is to say, in order to produce a given reaction rate, an additional electric energy supply is needed due to electric polarisations. One can argue that electric polarisations play a role that matches that of the force in brakes, namely friction. They must be overcome in order to impose a given rate, in other words a given current. This extra energy supply is reduced to heat in the electrochemical system, as in the JOULE effect.

To sum up, given the chosen convention for the current sign[58], the following relationships are generally true:

$$\pi_+ \, I_+ > 0 \quad \text{and} \quad \pi_- \, I_- > 0$$

and similarly

$$\pi_{\text{an}} \, I_{\text{an}} > 0 \quad \text{and} \quad \pi_{\text{cat}} \, I_{\text{cat}} > 0$$

Remember that anode polarisation is positive whilst cathode polarisation is negative.

The current-potential curves are therefore often monotonous and with a positive slope (following the sign conventions chosen in this document), however this is not an absolute rule[59]. In this book we will limit ourselves to monotonous curves.

It is important not to confuse the concepts of the polarity and the polarisation of an electrode. There is no necessary connection between the signs of these two quantities. Thus during electrolysis, the positive electrode is the anode and its polarisation is positive. In other words, its potential, when a current is flowing, is higher than when it is measured without a current. Yet when the system is working as a power source, then the

---

[57] Section 4.3.2.1 gives a thorough, qualitative description of the phenomena that are involved when current flows through an electrochemical system

[58] The current sign convention used in electrochemistry is presented in detail in section 1.4.1.3.

[59] For example, when a metal surface is passivated due to an insulating layer forming (oxide, hydroxide, etc.) then there is a blockage in the redox reaction. The current then decreases sharply. This phenomenon can be observed when drawing a curve $I(E)$ in the potentiostatic mode. In intensiostatic mode, the potential suddenly jumps towards a value where a new redox reaction occurs.

positive electrode is the cathode and its polarisation is negative. In the latter situation, the fact that there is a current flowing implies that one is only able to retrieve a reduced part of the electric energy, due to the electric polarisations in the system.

If both two species of a redox couple Ox/Red are present, then the current-potential curve takes on the shape as shown in figure 2.17. In this type of representation, the use of an arrow enables us to identify the polarisation at a given working point. The sign for polarisation is positive if the arrow points in the same direction as the potential axis, as shown in figure 2.17, and negative if the arrow points in the opposite direction. By convention, on a current-potential curve the two elements of the redox couple which are responsible for the current, are indicated as well as an arrow specifying the direction of the reaction (oxidation direction for a positive current). Given the direction of the potentials axis (increasing potential to the right), the reduced species (Red) of the couple is placed on the left and the oxidised species (Ox) on the right. This corresponds to a similar representation in predominance diagrams, which are used for instance for acido-basic equilibria[60]. When following common electrochemical conventions, one must therefore always represent the reactions by using an arrow to the right for positive currents (oxidation) and an arrow to the left for negative currents (reduction).

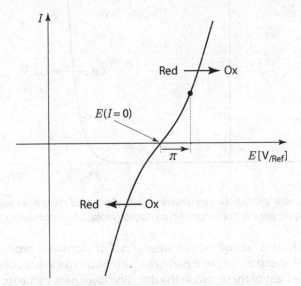

*Figure 2.17 - Current-potential curve of an electrode
when both elements of a redox couple are present*

In the particular case in point here, the two elements of a single redox couple are present and therefore the open-circuit potential corresponds to the equilibrium potential given by the NERNST law that is applied to that couple. In this case, we can use the term of overpotential instead of polarisation.

---

[60] In fact an analogy can be made between the NERNST law for the potential of a redox couple and the law which expresses the pH of a solution based on the pK_a and the respective concentrations of the two forms of an acido-basic couple.

2.3.1.2 - STEADY-STATE CURVES

If only one species of the couple is present in the system studied, then the shape of the current-potential curve is different from that shown in figure 2.17 and the open-circuit potential cannot be defined simply by this species alone. In particular, the NERNST law can no longer be applied. The activity of one of the species of the redox couple is zero, therefore giving no value for the equilibrium potential. On the other hand, insofar as we are only representing steady-state current-potential curves here, if a species is absent in the initial system, then this remains the case throughout the experiment. Therefore this species cannot contribute to the faradic current. Only one part of the curve is seen in this case, and can be either anodic or cathodic depending on the situation. For example, if we look at the interface between a copper electrode and a large volume of aqueous electrolyte devoid of Cu^{2+} ions, then the steady reduction of the Cu^{2+}/Cu redox couple is not possible. The corresponding current-potential curve is therefore situated entirely in the anodic zone corresponding to copper oxidation, as shown in figure 2.18.

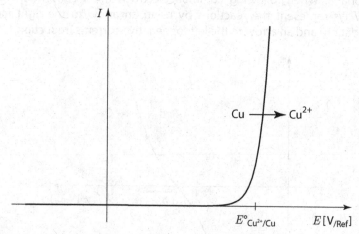

Figure 2.18 - Current-potential curve when only one element of the redox couple is present
Example of a copper electrode dipped in an aqueous solution containing no Cu^{2+} ions.

In this system, which is actually quasi-steady[61], Cu^{2+} ions are produced when one measures a point corresponding to a non-zero current on the anodic branch. There can be a high concentration of these ions in the diffusion layer next to the working electrode (copper). However, given how large the volume of the electrolyte, the amount of Cu^{2+} ions produced during this measuring process is very low. Therefore in practice, when examining the operating conditions at lower potentials, where a reduction of Cu^{2+} ions would be expected, no steady-state current can be found. This is because the Cu^{2+} ions previously produced are extremely diluted in the overall volume of the electrolyte and are no longer in the diffusion layer[62].

---

[61] Section 1.6.4 focuses on the notion of strictly steady systems.

[62] This particular characteristic of current-potential curves is linked to the steady character, which is chosen for this description. In transient experiments such as voltammetry, using a copper electrode with no Cu^{2+} ions in the solution, one would see a reduction current during the reverse scan of the

Another important point, as far as the steady-state current-potential curves are concerned, is the position on the potential axis of the branch plotted, which in the example above is anodic. As previously underlined, no equilibrium potential can be spotted because the open-circuit potential cannot be defined by the NERNST law. However, it is quite useful to have the thermodynamic constant associated with the redox couple in question, which in this case is the standard potential[63] $E°(Cu^{2+}/Cu)$, because this enables one to position the current-potential curve, i.e., the potential zone where the current undergoes sharp variations (see figures 2.18 above and 2.21 in section 2.3.2.2)[64]. Our aim here is to plot and use the shapes of the curves rather than to estimate their precise mathematical expression or any potential values to the mV.

If a current flows through a system for a long time, then one generally witnesses a variation in the chemical composition of the medium. Such a system is no longer steady, unlike the system presented in the example above. Then one observes a current which corresponds to a species that is produced at the electrode but which was not present in the electrolyte from the start. This is illustrated in figure 2.20 in section 2.3.2.1.

2.3.2 - ROLE OF MASS TRANSPORT KINETICS

2.3.2.1 - LIMITING CURRENT

When mass transport kinetics interferes, the steady-state current-potential curves show a limiting current for the redox process, which brings into play this mass transport. The limiting current is shown on a current-potential curve as a horizontal plateau. It is often also called a limiting diffusion current. It is dependent on the mass transport parameters (migration, diffusion and convection) and is, with a few exceptions[65], proportional to the concentration of the consumed species, because their mass transport limits the reaction rate and therefore the current flow[66].

Let us return to the example of the interface between a copper electrode and an aqueous electrolyte now containing Cu^{2+} ions. With both elements of the Cu^{2+}/Cu couple present, the two branches of the current-potential curve are then observed. For a cathodic working point, Cu^{2+} ions must be carried to the cathode to be reduced. As shown in figure 2.19, a reduction limiting current is observed, with a value proportional to the Cu^{2+} concentration in the bulk solution.

---

voltammogram (which is also a current-potential curve). The Cu^{2+} ions produced by the oxidation of the copper electrode during the direct scan are still partly present at the interface during the reverse scan.

[63] *Generally, the value to consider here is the apparent standard potential when it is relevant (see section 2.1.2.2 with, for instance, the IO_3^-/I^- couple in a buffered medium).*

[64] *This property applies for rapid systems. The changes of current-potential curves are discussed in section 2.3.3, when the kinetics of the redox reactions interferes.*

[65] *Such exceptions are found for example in the case of electrolytes with both members of a given redox couple in conditions where their migration cannot be disregarded (an absence or too small a quantity of supporting electrolyte).*

[66] *Section 4.3.3.1 outlines the quantitative laws linking the limiting current to the concentration in consumed species in simple systems.*

On the other hand, the shape of the current-potential curve is different in the oxidation branch because metallic copper is not concerned by any mass transport phenomenon, since copper is always present at the interface. The zone where the current undergoes large variations is close to the open-circuit potential, which is, in this case, equal to the equilibrium potential of the system. The latter, which can be calculated using the NERNST law, is shifted slightly from the standard potential. To give an order of magnitude for a concentration in Cu^{2+} ions equal to 10^{-3} mol L^{-1}, there is the following:

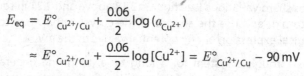

$$E_{eq} = E°_{Cu^{2+}/Cu} + \frac{0.06}{2} \log (a_{Cu^{2+}})$$

$$\approx E°_{Cu^{2+}/Cu} + \frac{0.06}{2} \log [Cu^{2+}] = E°_{Cu^{2+}/Cu} - 90 \, mV$$

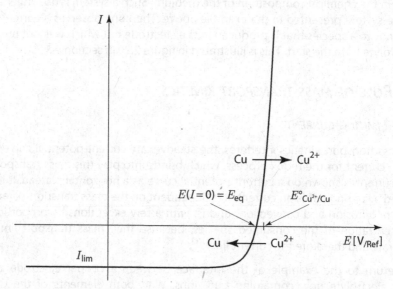

Figure 2.19 - Current-potential curve of a copper electrode
in an acidic aqueous solution containing Cu^{2+} ions

In another kind of experiment, starting with no Cu^{2+} ions in an aqueous solution, the limiting current can be used to follow the reaction progress during an electrolysis with a copper anode. This example is detailed in the next page and illustrated in figure 2.20.

There is another particular case concerning limiting currents where certain electroactive species are found in very large quantities compared to the other electroactive species studied (for example, the solvent in systems with electrolyte solutions). Since the limiting currents are proportional to the corresponding concentrations in solution, the plateau resulting from the mass transport limitation for these couples is much larger (in absolute value) than that of the other redox couples in the system. These large limiting currents are hardly visible and the current increases very sharply (in absolute value) on the scale of the current-potential curves. The definition of the electrochemical window of a system is based on these characteristics. The term solvent window is also used, even if this constitutes a misnomer when the corresponding reaction involves either the supporting electrolyte or the electrode itself (see section 2.3.6).

▼ Take the example of electrolysis being carried out in an electrochemical system, with copper oxidation as the anodic reaction in a non-renewed solution, which at the outset contained no Cu^{2+} ions. If we draw a quick steady-state current-potential curve at different times throughout the electrolysis process, then we can trace the electrolysis progress by measuring the limiting reduction current, which indicates the concentration of Cu^{2+} ions generated. The time required to draw each current-potential curve is very small compared to the electrolysis duration and the system can be considered in a quasi-steady state during this short time period. The system is no longer steady on the time scale of the electrolysis process as shown by the difference between the two current-potential curves in figure 2.20.

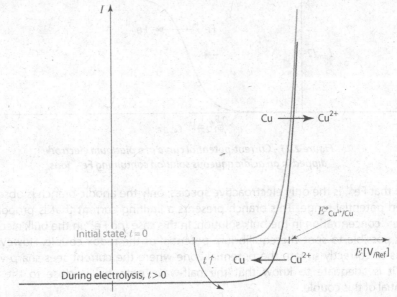

Figure 2.20 - Example of a copper electrode dipped in an aqueous solution containing initially no Cu^{2+} ions: changes during oxidation ◢

2.3.2.2 - HALF-WAVE POTENTIAL

Even though the aim of this presentation is not to outline quantitative relationships, it is nonetheless interesting to know a property which characteristically emerges when these curves are described in quantitative terms[67]. It is indeed possible to show that for simple redox systems with very close transport parameters (in the example below, the diffusion coefficients of the two ions of the Fe^{3+}/Fe^{2+} couples are taken as equal) the value of the current for the standard potential[68] is equal to half the sum of the limiting currents (see figure 4.26 in section 4.3.3.2). In other words, the half-wave potential, $E_{1/2}$, and the standard potential are identical. When systems contain both elements of a redox couple at the outset, then this particular detail has little impact on how the current-potential curves are plotted in qualitative terms, since the NERNST law allows one

---

[67] Quantitative analytic expressions for these curves are given for simple redox systems in section 4.3.3.

[68] More generally speaking, the value to consider here is the apparent standard potential, when it applies.

to spot the open-circuit potential. Yet in the opposite case, when only one species of a redox couple is present, then on the contrary this becomes a highly relevant detail in helping to correctly draw the shape of the current-potential curves. Take the example of the current-potential curve of a platinum electrode dipped in an aqueous solution containing Fe^{2+} ions and no Fe^{3+} ions, as shown in figure 2.21.

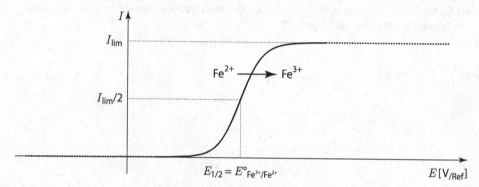

Figure 2.21 - Current-potential curve of a platinum electrode dipped in an acidic aqueous solution containing Fe^{2+} ions

Given that Fe^{2+} is the only electroactive species, only the anodic branch is observed in a limited potential range. This branch presents a limiting current that is proportional to the Fe^{2+} concentration in the bulk solution. In this case (no Fe^{3+} in the bulk electrolyte) it is not possible to spot the equilibrium potential of the redox couple. However, if one wishes to correctly situate the potential zone where the current sees sharp variations, then it is adequate to know that the half-wave potential is close to the standard potential of this couple.

The value of the couple's standard potential (or apparent potential when it applies) is what enables one to position the potential range where the current varies greatly, assuming that one is only interested in the shape of the current-potential curves and provided that the kinetics of the redox reactions does not interfere (see section 2.3.3). One must not forget that when the concept of apparent standard potential becomes relevant, then it has to be taken into account to locate the portions of current-potential in relation to each other. For example, in a very basic medium, the characteristics of the various couples may or may not depend on the pH of the system, and consequently the order of the apparent standard potentials can be reversed in relation to the order in highly acidic medium.

This reasoning, based on the value of half-wave potentials, leads to a qualitative but rigorous drawing of current-potential curves. This line of reasoning is in contrast with common habits found in scientific literature, especially in the field of corrosion. In this area, when drawing the current-potential curve shape in an example where only one redox species is present, one takes into consideration an arbitrary and low concentration of the missing species, most often 10^{-6} mol L^{-1} (but why not 10^{-7} mol L^{-1} or 10^{-9} mol L^{-1}?) as present, meaning that the equilibrium potential can be calculated using the NERNST law. It is not valid to take this value for the open-circuit potential.

REGULATING OF FUEL ENGINES

*Document written with the kind collaboration of J. FOULETIER,
professor at Joseph Fourier University of Grenoble, in France*

The 'lambda' sensor is used to monitor internal combustion engines. It is shaped like a candle and placed in the vehicle exhaust pipe at the outlet of the cylinder head. It is a potentiometric type of sensor using a closed-bottom ceramic tube. Several million examples of this type of sensor are produced every year.

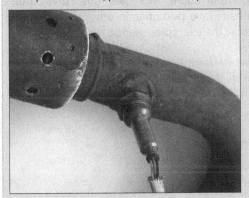 

Photograph of a lambda sensor extracted from a car after use

The type of solid electrolyte used is yttria-stabilised zirconia ($ZrO_2 - Y_2O_3$, 8% molar), and conducts with O^{2-} oxide ions. Both the interior and exterior surfaces of the tube are coated with platinum paint, and constitute the electrodes. The reference electrode of the cell is the oxygen in the air in contact with the inside wall of the tube. The external surface of the tube is in contact with the exhaust gases. The measuring electrode is protected by a metal cover and a porous oxide layer.

The electrochemical chain is: O_2(air),Pt | O^{2-} solid electrolyte | Pt,O_2(gas).

If the electrode were in equilibrium, the NERNST law would give the accurate partial pressure of oxygen in the analysed gases which is fixed by the mixture of combustion products. In fact, the exhaust gases are not in equilibrium, and therefore the device merely serves to indicate any oxygen excess or lack in the fuel mixture at the carburettor outlet. To prevent the measuring electrode from being contaminated, it is imperative that the fuel used is lead-free.

The term 'lambda' indicates that the sensor is designed to operate at the stoichiometric point, which has an air/fuel ratio equal to 14.5. The transition from a 'rich' mixture in fuel (therefore meaning poor in oxygen) to a 'poor' mixture, or indeed the opposite transition, entails a shift in the signal delivered by the sensor which is over 500 mV. Such a variation is sufficient for the sensor (which itself is placed in a feedback loop) to provide adequate control of the carburettor operation.

To finish, similar other sensors using stabilised zirconia are also developed to measure the partial pressure of oxygen to be found in gases (air, and waste gas) or in molten metals (in metallurgy). However, these sensors have a reference electrode made of a metal which is mixed with its oxide inside the zirconia tube.

2.3.3 - ROLE OF REDOX REACTION KINETICS

The redox reactions kinetics also plays a part in shaping the current-potential curves. Assume[69] that the main impact is the change seen in the potential zones which relate to the rising parts of the curve, themselves corresponding to the redox couple. Therefore, for the two limiting cases concerning the kinetics of redox reactions, most often referred to as fast and slow couples[70], we find the following:

▸ for fast couples, the kinetics of the redox reaction, in other words the electron transfer step, has no impact on the current-potential curve. Moreover this curve shows significant current variations around the half-wave potential, in other words around the standard potential. This case applies to all the examples seen so far;

▧ Consider the interface between an organic solvent containing Li^+ ions and the solid phase Li_xMnO_2 (manganese oxide which is a lithium insertion material). The system is in thermodynamic equilibrium when no current flows through it. The equilibrium potential is generally close to the standard apparent potential (see insertion isotherm in section 2.1.2.2):

$$E_{eq} = E^{\circ}{}_{app_{MnO_2/LiMnO_2}} + \frac{RT}{\mathscr{F}} \ln \frac{1-y}{y}$$

The shape of the current-potential curve is presented in figure 2.22.

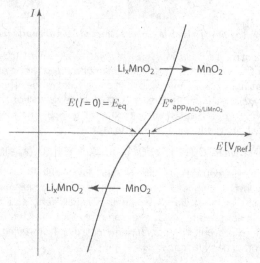

Figure 2.22 - Current-potential curve of a manganese oxide electrode
in an organic solution containing Li^+ ions

▸ for slow couples, however, the kinetics of the redox reaction has a strong influence on the characteristics of the current-potential curve. This curve shows a significant potential difference between the anodic and cathodic branches when both species of

---

[69] Section 4.3.3 gives quantitative analytic expressions of these curves for simple redox systems as a function of various kinetic and thermodynamic parameters.

[70] Section 4.3.2.6 gives a precise definition of the terms fast and slow, which characterise the charge transfer kinetics of simple redox systems.

the couple are present in the bulk of the electrolyte or electrode. In the anodic branch, large current variations are found around a potential called the anodic half-wave potential, $E_{1/2_{an}}$, which is higher than the thermodynamic standard potential[71]. The difference between the anodic half-wave potential and the standard potential of the couple in question increases as the kinetics of the redox reaction slows down. The cathodic branch shows similar behaviour, with a cathodic half-wave potential that is algebraically lower than the standard potential. All things being equal, if you take the example of a redox couple involving a simple mechanism and two species, it gives a change in current-potential curves as appears in figure 2.23, when the kinetics of the redox reaction varies. The cathodic and anodic half-wave potentials are indicated for the very slow system of curve **d**[72].

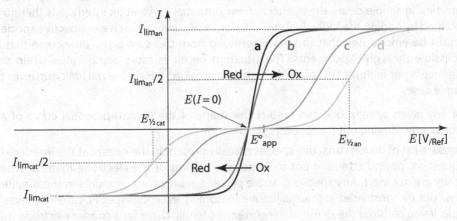

Figure 2.23 - Current-potential curve of an electrode containing both elements of a redox couple present for systems with different redox kinetics
a - very fast; b - fast; c - slow; d - very slow

Certain documents refer to the difference between the half-wave potential and the equilibrium potential as an activation overpotential[73]. This terminology is not used here because it is rarely possible to distinguish the various contributions to a system's overpotential in unambiguous terms[74].

---

[71] *Generally speaking, the value to consider is that of the apparent standard potential when it applies. Section 4.3.3.3 gives a precise definition of the anodic and cathodic half-wave potentials for slow systems and the relationship with the system's kinetic parameters.*

[72] *When establishing a method for determining redox kinetics, one can consider using the value of the slope of the current-potential curve for the open-circuit potential. However, it is impossible to assign a specific slope value to a fast couple because this slope is highly dependent on the limiting anodic and cathodic currents. To approach this issue from a quantitative point of view, one can start from the expression of the polarisation resistance around the equilibrium potential (see section 4.3.3.4).*

[73] *This derives from the notions of activation energy and activated complex (see section 4.3.2.3). It therefore refers to the influence of the redox half-reaction kinetics. The term activation polarisation can also be found, yet it is even less suitable if the system at open circuit is not in equilibrium.*

[74] *This question is approached from a more precise quantitative angle for simple systems in section 4.3.3.4.*

The kinetics of redox reactions naturally depends on the couple in question but may also depend on the electrode where the reaction is occurring, even if this electrode does not play an explicit role in the overall redox reaction (it then intervenes in the mechanism). This is the case for instance in the reduction of protons in acidic medium. Here, the characteristics vary depending on the nature of the electrode where the reaction is occurring (e.g., platinum or mercury, see figure 2.31 in section 2.3.6).

2.3.4 - ADDITIVITY OF FARADIC CURRENTS OR CURRENT DENSITIES

Remember that faradic currents are additive: if several electroactive species react at a given interface, the overall current density is the sum of the corresponding current densities. In simple cases, the overall current-potential curve at an interface is therefore obtained by adding the individual current-potential curves of each electroactive species. It must be emphasised that the curve resulting from this sum is the only curve that is accessible through experiments. The situation becomes more complicated when the interface is not uniform as is often the case in corrosion[75]. Here we will limit ourselves to simple cases.

For any given system, one can predict the shape of the current-potential curve of an interface by carrying out the following steps:

▸ make a list of the electroactive species actually present in the system at the interface in question, paying attention not to forget the solvent or the electrode material when they are not inert. Any species that are present in only trace amounts in the medium will not be considered in this qualitative reasoning, except in specific cases. For example, if you consider the shape of the current-potential curve for a copper electrode (see figure 2.18), even though the copper electrode used contains traces of another electroactive element, it is represented by taking into account the material as if it were pure;

▸ list all the possible redox couples relating to each species identified. In particular one must remember that for aqueous solution, water is involved in two redox couples[76]: $O_2/H_2O,OH^-$ and $H^+,H_2O/H_2$;

▸ position the standard potentials (apparent standard potentials when it applies[77]) for each of these couples;

▸ represent the current-potential curves of each electroactive system by positioning them using the standard potential (to see how this positioning is done refer to sections 2.3.2 et 2.3.3);

▸ finally, for each potential value, sum up the currents of all the curves previously drawn. The resulting curve is the only one accessible in an experiment.

---

[75] In corrosion phenomena, the same interface can frequently present both anodic and cathodic zones.

[76] The terms used to denote both the water redox couples presented in section 2.1.2.3 correspond with the terms used to write these two same couples in acidic or basic media.

[77] The concept of apparent standard potential, defined in section 2.1.2.2 (and 2.1.2.3 for water couples), can be useful when one or several chemical species written in the redox half-reaction are found in very large quantities compared to the other. This is often the case for instance for redox couples in very acidic or very basic solutions.

For a given potential, the construction of the overall current-potential curve from the different contributions of each electroactive species allows to view the concept of faradic yield as shown below.

▶ Consider the interface between an inert platinum electrode and a deaerated aqueous electrolyte (that is, without any dissolved oxygen left) containing H$^+$ ions ($pH = 2$) and Ni^{2+} ions. This example picks up again on the system previously chosen to illustrate the concept of faradic yield (see section 2.2.2.3). The shape of the overall current-potential curve is shown in figure 2.24 with a dashed and dotted line [78].

If we are only interested in the cathodic branch of the curve, then the relevant electroactive species are protons and Ni^{2+} ions. The grey lines represent the current-potential curves that would be seen if only one reducible species were present. Each branch is situated in the zone which corresponds to its apparent standard potential:

$$E°_{Ni^{2+}/Ni} = -0.23\,V_{/SHE} \quad \text{and} \quad E°_{app_{H^+,H_2O/H_2}} = -0.12\,V_{/SHE}$$

The overall current-potential curve (represented by a dashed and dotted line) is the sum of the two grey curves. As for the working point shown in figure 2.24, the current required for the reduction of Ni^{2+} ions, I_1 represents about 25% of the overall current, $I = I_1 + I_2$. This therefore means that the faradic yield for hydrogen production by reduction is equal to about 75%.

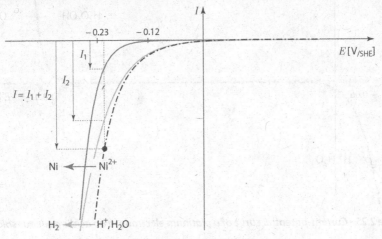

Figure 2.24 - Current-potential curve of a platinum electrode
in an acidic solution containing Ni^{2+} ions

2.3.5 - WATER REDOX COUPLES

As already emphasised, the water redox couples play a particularly important role in all electrochemical systems involving an aqueous electrolyte. The corresponding current-potential curves are rather complex, notably because water autoprotolysis occurs[79]. However some qualitative characteristics can be retained, especially for very acidic or

[78] Remember that figure 2.24 does not represent experimental data. It is merely a simplified shape illustrating the various principles involved.

[79] Appendix A.2.1 focuses in detail on the shape of current-potential curves of the H$^+$,H$_2$O/H$_2$ couple, which is supposed to be fast, thus justifying the results stated here.

very basic media in the case of fast redox reactions [80]. In these conditions, one can say that the apparent standard potential is what defines the position of the current-potential curves in qualitative terms.

For example, let us consider the interface between a very acidic aqueous electrolyte ($pH = 1$), containing neither dioxygen nor dihydrogen dissolved species, and a platinum electrode which is inert in this electrolyte. The only electroactive species present in the bulk electrolyte are protons and water. The corresponding redox couples are $O_2/H_2O,OH^-$ and $H^+,H_2O/H_2$. The shape of the current-potential curve (represented in figure 2.25 with a dashed and dotted line) is the sum of the contributions of each couple (grey curves). These couples are supposed to be fast, and the values of the apparent standard potentials are:

$$E^\circ_{app_{H^+, H_2O/H_2}} = -0.06 \, V_{/SHE} \quad \text{and} \quad E^\circ_{app_{O_2/H_2O,OH^-}} = +1.17 \, V_{/SHE}$$

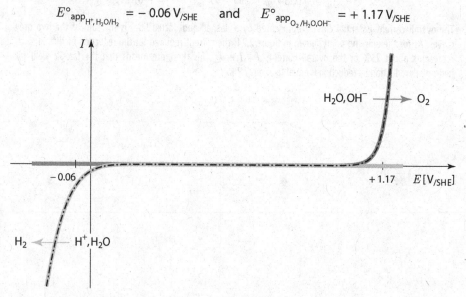

Figure 2.25 - *Current-potential curve of a platinum electrode in an acidic aqueous solution*

Since there are large quantities of both water and protons, no limiting current plateau can be seen (in the usual current ranges) which would be explained by a limited supply rate of water or protons at the interface. This current-potential curve shows the position of the electrochemical window in this system (see section 2.3.6).

Exactly the same type of current-potential curve would be found for a very basic solution. The only difference in this case would be the values of the apparent standard potential. Here again no limiting current is observed. However, when the pH values become less extreme, the shape of the curve changes and can become quite complex (see appendix A.2.1) as illustrated by the two examples discussed below (figures 2.26 and 2.27).

---

[80] *Here one can refer to figure 2.28 that gives an example where the redox reactions kinetics plays a role. It is worth recalling that the current-potential curves in the figures presented in this document are not experimental results but simplified representations to explain the various phenomena involved.*

▸ Consider the example of an inert platinum electrode and a deaerated electrolyte (with no dissolved dioxygen) containing H^+ ions ($pH = 3$) and involving redox couples that are considered to be fast. In these conditions, with a limited amount of protons, a limiting proton reduction current appears in the cathodic domain. However, given the specific characteristics of the water redox couples, H_2O can also be reduced in a second potential zone. Consequently, there is no limiting current plateau in this second zone, as shown in figure 2.26.

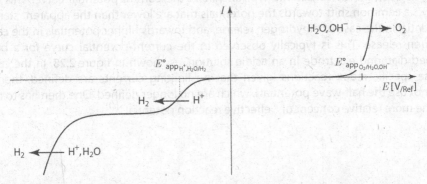

Figure 2.26 - Current-potential curve of a platinum electrode in a deaerated aqueous solution at $pH = 3$

▸ The behaviour is entirely symmetrical in moderately basic solutions ($pH = 11$ for example). The corresponding current-potential curve displays two anodic zones relating to the single $O_2/H_2O,OH^-$ couple, as shown in figure 2.27.

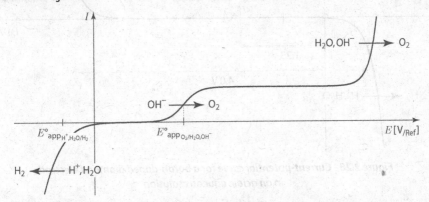

Figure 2.27 - Current-potential curve of a platinum electrode in a deaerated aqueous solution at $pH = 11$

In addition, reactions that produce gaseous dihydrogen (or dioxygen) *via* the reduction (or oxidation) of an aqueous solution are quite complex. The characteristics and even the mechanisms involved differ greatly depending on the nature of the electrode. These reactions are studied extensively, especially in the development of fuel cells whose performance is lowered by any kinetic limitation due to these couples. One of the main limitations faced by fuel cells at moderate temperatures, such as the PEMFC, is the dioxygen reduction reaction [81].

---

[81] PEMFC: *Proton Exchange Membrane Fuel Cell.*

Let us return to the more conventional examples of dioxygen and dihydrogen production, which most of the time form the electrochemical window bounds for
electrochemical systems with aqueous electrolytes. Depending on the mechanisms
involved, these couples can lead the current-potential curves to take on very different
shapes. The simplest case of kinetic limitation can be found when both reactants have
slow kinetics, which causes the typical shift in the current-potential curve. This curve
shows a common shift towards the potentials that are lower than the apparent standard
potential in the case of dihydrogen release, and towards higher potentials in the case of
oxygen release. This is typically observed in the current-potential curve for a boron-
doped diamond electrode in an acidic solution, as shown in figure 2.28. In the case of
these specific redox reactions, given that no limiting currents are defined, then one
cannot use the half-wave potentials, which are no longer defined. One then has to resort
to the more relative concept of ' effective reaction potential '[82].

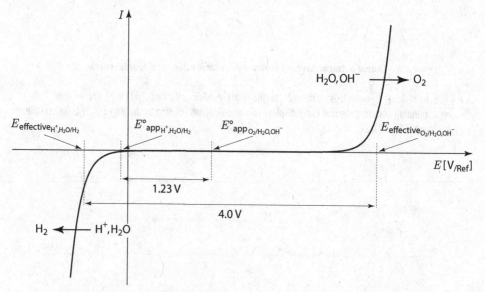

**Figure 2.28 - Current-potential curve for a boron-doped diamond electrode
in an acidic aqueous solution**

We can finally note that, even when no limiting current can be observed as in the case
of dihydrogen or dioxygen production, one can still frequently observe a limiting
reduction diffusion current (see figure 2.29). This relates to the reduction of dissolved
oxygen in the case where the electrolyte has not been correctly deaerated. Indeed, air is
made up of about 20% dioxygen and 80% dinitrogen. The solution can only be properly
de-oxygenated by bubbling an inert gas through it (most often dinitrogen or argon) if
the bubbling phase lasts at least ten minutes or so and if an inert gas atmosphere is
maintained above the solution.

---

[82] This 'effective reaction potential' is defined as the potential corresponding to a given current
value or current density value taken as reference, and with a typical value of 1 mA cm^{-2}.

ENERGY STORAGE:
THE Li-METAL-POLYMER (LMP) BATTERIES

Document written with the kind collaboration of K. GIRARD,
R&D engineer for the company Batscap based in Quimper, France

A secondary battery works on the basis of having two redox couples with different redox potentials. The two electrodes are separated by an electrolyte which is an electronic insulator as well as an ionic conductor. Lithium is a particularly interesting material because it is the lightest of the metals ($0.534 \, g \, cm^{-3}$) and the best at reducing making it an ideal negative electrode material. In the case of lithium-ion batteries, it is only used in its ionic form. The positive and negative electrodes are made up of insertion materials, where lithium ions are inserted in both discharge and charge modes respectively.

In LMP batteries, the negative electrode is a lithium-metal foil, meaning that it enjoys the full range of benefits offered by this material. Another peculiarity of the LMP batteries is the fact that the electrolyte used is a polymer-based solid film. Its main advantages include increased safety (no risk of explosion) and improved processing ability. Its main drawback is low conductivity which means that operations have to be carried out at a temperature between 60 and 80 °C.

Current collector

Positive electrode: vanadium oxide, carbon and polymer-containing compound

Electrolyte: polyethylene oxide (PEO) with lithium salts

Negative electrode: metallic lithium

Presentation of the film components in a LMP battery (© Batscap)

On discharge:
$$z \, Li \rightarrow z \, Li^+ + z \, e^-$$
$$Li_{x-z}VO_y + z \, Li^+ + z \, e^- \rightarrow Li_xVO_y$$

This means that a particularly high energy density can be reached, measuring about 100 Wh/kg for a 30 kWh pack. The various electrochemical phenomena at play in the batteries being operated, such as electron transfer or diffusion, limit the power density to around 200 W/kg in discharge for the same battery pack. The LMP's main application is the fully-electric vehicle, for which it works as an energy reservoir. When linked to a supercapacitor, which is a kind of power reservoir, it can provide the ideal performance for use in electric vehicles.

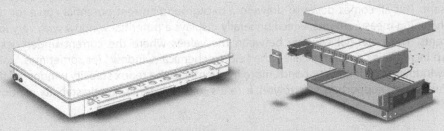

Presentation of a LMP Battery pack of 30 kWh (photo Batscap)

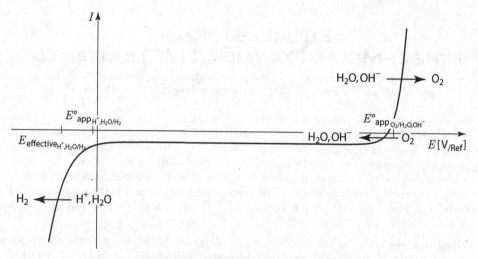

*Figure 2.29 - Current-potential curve of a platinum electrode
in a non deaerated acidic aqueous solution*

The degree to which dioxygen contributes to the current-potential curve often remains negligible, since the concentration level of dissolved dioxygen is quite low with a typical value of 3×10^{-4} mol L^{-1} at usual air pressures. The value for the diffusion limiting current in the case of dissolved dioxygen reduction is therefore around 1 mA cm^{-2} for systems with usual forced convection [83].

2.3.6 - ELECTROCHEMICAL WINDOW

In any electrochemical system the currents created by the redox reactions of the solvent or supporting electrolyte join to combine with the other phenomena, as illustrated in the example below. Figure 2.30 shows the shape of the current-potential curve of a platinum electrode in an acidic solution ($pH = 1$) containing Fe^{2+} ions with a concentration of 10^{-3} mol L^{-1}. The current-potential curve (indicated by a black dashed and dotted line) is the sum of the grey curves corresponding to the Fe^{2+} ions (see figure 2.21) and to the water couples (see figure 2.25).

In practice, when studying a system, it is recommended to first examine the characteristics without yet introducing the electroactive species that is studied. As a result one obtains the current-potential curve of the solvent, possibly with a supporting electrolyte present. This all comes down to drawing a background current-potential curve for the forthcoming measurements, which generally displays a potential window with very low currents. The potential window between the values where the current undergoes a sharp increase is commonly called the ' electrochemical window ' (or sometimes ' solvent window ' even if it corresponds more frequently to the redox stability window of the solvent with a supporting electrolyte). It refers to the system's kinetic stability window,

---

[83] *This value corresponds to an RDE experiment, for example, with a rotating speed of about 500 rpm and therefore a diffusion layer thickness of about 30 μm.*

and can be wider than the thermodynamic stability window if the solvent decomposition reactions (or possibly those of the supporting electrolyte) are slow. It provides the potential window that opens up the opportunity to making various experimental studies possible. In fact, in the medium in question, one cannot study any electroactive species with a current-potential curve that is beyond this range. In that instance one would then merely measure the overall current, the majority of which is derived from the solvent, which in itself would completely hide the contribution of the compound added. Figure 2.31 collects together the electrochemical windows of a few systems, giving orders of magnitude. Obviously these values depend on what choices are made for the maximal current/current densities authorized.

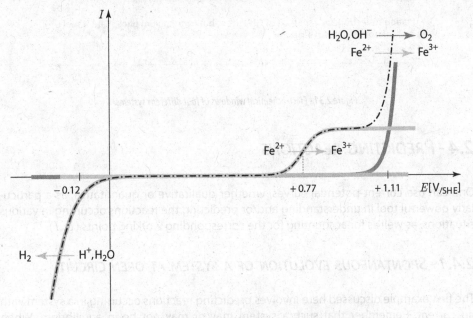

Figure 2.30 - Current-potential curve of a platinum electrode
in an acidic aqueous solution containing Fe^{2+} ions

The electrochemical window of a system is highly dependent on the composition of the electrolyte and on the nature of the electrode. This is illustrated below through four different examples: aqueous or organic electrolyte and platinum or mercury electrode. In the latter case, the limiting oxidation reaction is that of mercury oxidation, therefore the term solvent window would be misused. The limits for the potential ranges shown in figure 2.31 are estimated through experimentation for current density values of 10 µA cm^{-2}.

This figure highlights the following in particular:

▸ organic electrolytes have a redox stability window that is often wider than that of the aqueous electrolytes;

▸ mercury has the particular effect of considerably slowing down the rate of proton reduction in aqueous solutions. This property is used to advantage in several analytical techniques found in electrochemistry which are particularly effective for analysing trace metal ions, although obviously less used due to the toxicity of mercury. All analytical techniques using mercury are grouped under the term polarography.

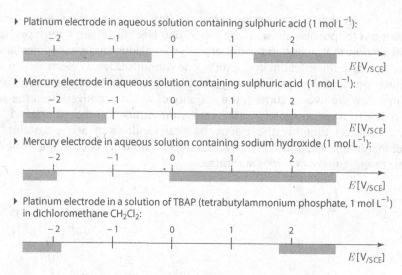

▶ Platinum electrode in aqueous solution containing sulphuric acid (1 mol L^{-1}):

▶ Mercury electrode in aqueous solution containing sulphuric acid (1 mol L^{-1}):

▶ Mercury electrode in aqueous solution containing sodium hydroxide (1 mol L^{-1}):

▶ Platinum electrode in a solution of TBAP (tetrabutylammonium phosphate, 1 mol L^{-1}) in dichloromethane CH_2Cl_2:

Figure 2.31 - Electrochemical windows of four different systems

2.4 - PREDICTING REACTIONS

One can use current-potential curves, whether qualitative or quantitative, as a particularly powerful tool in understanding and/or predicting the reactions occurring in various situations, as well as for accounting for the corresponding working points (U, I).

2.4.1 - SPONTANEOUS EVOLUTION OF A SYSTEM AT OPEN CIRCUIT

The first example discussed here involves predicting reactions occurring in a system with no current. Remember that such a system may or may not be in equilibrium. When describing an entire electrochemical system at open circuit, one usually considers what is occurring at each of the two interfaces as two separate entities. Here we will therefore focus on only one current-potential curve, in other words, one single interface.

Drawing a current-potential curve enables one to identify the processes occurring at open circuit. Indeed, the open-circuit working point (which is located at the intersection between the current-potential curve and the potentials axis) always corresponds to the sum of two opposite currents for oxidation and reduction. Three main scenarios can be distinguished.

▶ The interface in question is in thermodynamic equilibrium. For this to occur requires the two species of the couple to be present. In these conditions, the open-circuit potential is equal to the potential determined by the NERNST law applied to this redox couple. Figure 2.32 shows how the overall current-potential curve can theoretically be divided down into two contributions, one anodic and the other cathodic. This figure picks up on the example given of the interface between an organic electrolyte containing Li$^+$ ions and solid Li$_x$MnO$_2$ (refer back to figure 2.22). At open circuit, the overall current, which is zero, is the sum of two exactly opposite currents as indicated in the figure by the symbol //. The common absolute value of these currents indicates the rate of the redox kinetics, since it characterises the number of events occurring per unit time in opposite directions, which lead to the equilibrium state without involving any overall chemical transformation. If the couple in question is very slow, then the theoretical value of the open-circuit potential is still defined by the NERNST law. However, because the slope of the current-potential curve is very low, this potential can be significantly affected by small perturbations, with only a slow return to equilibrium. For instance, very low current flow or traces of another species can both make a significant change in the open-circuit potential of such a system.

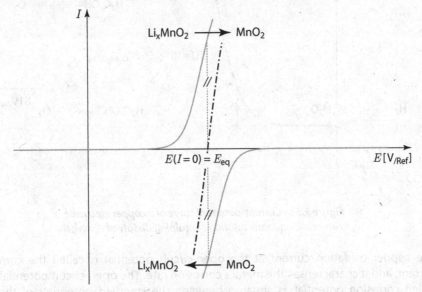

Figure 2.32 - Current-potential curve of a manganese oxide electrode in an organic solution containing Li^+ ions

▸ The interface in question is not in thermodynamic equilibrium. For this to occur at least two different redox couples must be responsible for the value of the open-circuit potential. The overall zero current then corresponds to a chemical balance which is not null. This is the case for instance in corrosion. The open-circuit condition is not in equilibrium and the system changes with time. The open-circuit potential, which is not given by the NERNST law, depends on the properties of at least two redox couples: in this instance one often refers to the term ' mixed potential '.

For example, in the case of a copper foil immersed in an acidic solution (with no Cu^{2+} ions at the outset) which has not been deaerated, dioxygen can be reduced. The resulting current-potential curve, which is shown by a dashed and dotted line in figure 2.33, is the sum of the three grey curves which relate to the three couples (supposedly fast) present in the system: Cu^{2+}/Cu, $H^+,H_2O/H_2$ and $O_2/H_2O,OH^-$. It is important to emphasise again that the resulting curve shown by a dashed and dotted line is the only one accessible through experiments. As far as the open-circuit potential is concerned, the overall current-potential curve for such a simple system is the sum of the copper oxidation current and the dioxygen reduction current. This accounts for the corrosion of the copper foil, as shown in the following:

$$4\,H^+ + 4\,e^- + O_2 \longrightarrow 2\,H_2O$$
$$Cu \longrightarrow Cu^{2+} + 2\,e^-$$

$$O_2 + 4\,H^+ + 2\,Cu \longrightarrow 2\,H_2O + 2\,Cu^{2+}$$

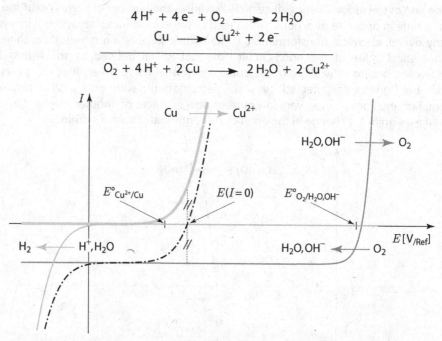

Figure 2.33 - Current-potential curve of a copper electrode
in an acidic aqueous solution containing dissolved dioxygen

The copper oxidation current at the open-circuit potential is called the corrosion current, and it characterises the metal's corrosion rate. The open-circuit potential, also called corrosion potential, is situated between the standard potentials of the two redox couples involved. There is no simple equation to link these values: more particularly there is no reason why the corrosion potential should be the average of both standard potentials, as illustrated in figure 2.33.

▶ The interface is not in thermodynamic equilibrium and the properties of the redox couples involved are such that the overall chemical reaction which expresses this situation progresses extremely slowly. If the time scale is not too large then the system undergoes practically no change. However rigorously speaking, this system is not in thermodynamic equilibrium. In particular, the open-circuit potential is a mixed potential which does not depend exclusively on thermodynamic data. Consequently traces of another species may significantly change its value.

▶ Let us look back at the example of an acidic aqueous electrolyte containing neither dioxygen nor dihydrogen in solution and a platinum electrode which is inert in this medium. The only electroactive species here are water and protons, and the redox couples involved are $O_2/H_2O,OH^-$ and $H^+,H_2O/H_2$. The shape of the current-potential curve is represented in figure 2.34 (the same as figure 2.25): the dashed and dotted line corresponds to the overall curve.

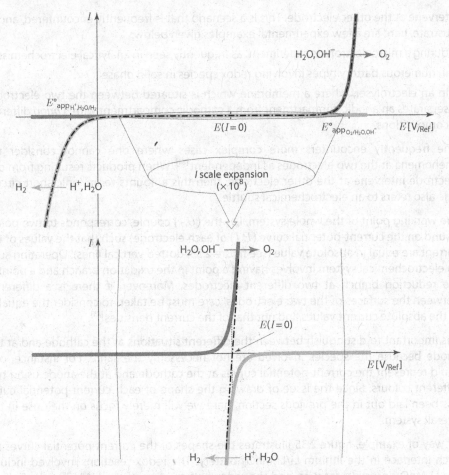

Figure 2.34 - *Current-potential curve of a platinum electrode in a deaerated acidic aqueous solution*

At open circuit the system is not in thermodynamic equilibrium, however the reaction occurring, namely the following:

$$2\,H_2O \longrightarrow 2\,H_2 + O_2$$

progresses extremely slowly.

The mixed potential measured at open circuit depends on the thermodynamic and kinetic parameters of the couples $O_2/H_2O,OH^-$ and $H^+,H_2O/H_2$ at the interface studied (platinum in this example). It is situated between the two apparent standard potentials of the two water couples.

In these experimental conditions the open-circuit potential is generally unstable. This can be explained by the slope of the current-potential curve which is extremely low in this potential zone. Traces of an electroactive species in this potential zone would significantly modify the value of the open-circuit potential. ◀

2.4.2 - WORKING POINTS OF A WHOLE ELECTROCHEMICAL SYSTEM

The following description is confined to simple systems in which both electrodes can be considered chemically independent. The products generated at a given electrode do not intervene at the other electrode. This is a scenario that is frequently encountered, and to illustrate, here are a few experimental examples given below:

▸ during a microelectrolysis experiment, as frequently seen in analytical electrochemistry,

▸ in numerous battery types involving redox species in solid phase,

▸ in an electrolyser, where a membrane which is situated between the two electrodes separates an anodic compartment from a cathodic compartment with *a priori* different compositions.

One frequently encounters more complex cases where one cannot consider the phenomena at the two electrodes as independent[84]. When products resulting from one electrode intervene at the other electrode, then this amounts to an ionic short circuit. One also refers to an electrochemical shuttle.

The working point of the whole system, i.e., the (U,I) couple, corresponds to two points (found on the current-potential curve (E,I) of each electrode) such that the values of the current are equal in absolute value (see figure 2.35, dotted vertical lines). Operating such an electrochemical system involves having a point in the oxidation branch and a point in the reduction branch at two different electrodes. Moreover, if there is a difference between the surfaces of the two electrodes, care must be taken to consider the equality of the absolute current values and not that of the current densities[85].

It is important to distinguish between the different situations at the cathode and at the anode because the species involved are not necessarily the same. For instance, one could represent the current-potential curves at the cathode and at the anode using two different colours. Since the issue of drawing the shape of each current-potential curve has been laid out in the previous section, here we will merely focus on their use in the overall system.

By way of example, figure 2.35 illustrates the shapes of the current-potential curves for each interface in the lithium Li/Li_xMnO_2 battery. The redox reactions involved include the Li_xMnO_2 insertion reaction and the redox half-reaction of the Li^+/Li couple. Each of these two curves has been represented in its entirety (anodic and cathodic branches), as both of them would be obtained independently of one another. At open circuit, the whole system is in thermodynamic equilibrium. The working point of the whole system represented in figure 2.35 corresponds to the electrolyser mode or to a charging battery[86]. Bear in mind that in a recharging setup the anode is the positive electrode (Li_xMnO_2) and the cathode is the negative electrode (lithium metal). This diagram shows again the polarisation signs previously introduced in section 2.3.1. The working points indicated here correspond to an anodic overpotential which is positive and to a negative cathodic overpotential.

---

[84] *A few examples of this type are studied in section 4.4.*

[85] *Examples illustrating the influence of the difference between anodic and cathodic surface areas are given in section 1.4.1.3.*

[86] *The example of an electrochemical system in power source mode or in the case of a discharging battery is illustrated later in figure 2.38 in section 2.4.4.*

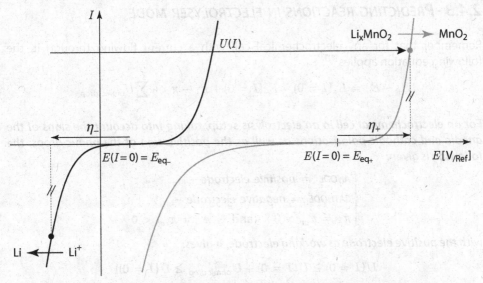

**Figure 2.35 - Current-potential curves of the two electrodes with a working point
in electrolyser mode (i.e., while the battery is charging)**

There is a particular case encountered in systems where two identical electrodes are
immersed in the same electrolyte. Both current-potential curves are identical and
both electrodes play a symmetrical role. However, unlike when studying open-circuit
conditions, as previously, here we must consider both curves, as opposed to just one. In
such a system, a particular detail is that the open-circuit condition always corresponds to
an overall system voltage equal to zero ($U = 0$ or short circuit).

▶ For instance, a system with two platinum electrodes dipped in an acidic deaerated solution gives the current-
potential curves as shown in figure 2.36.

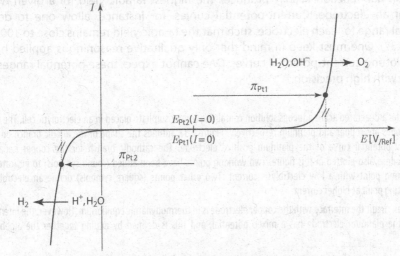

**Figure 2.36 - Working point in the electrolyser mode of a cell with two identical platinum electrodes
in an acidic deaerated aqueous solution**

2.4.3 - PREDICTING REACTIONS IN ELECTROLYSER MODE

Remember that for any electrochemical cell with a current flowing through it, the following equation applies[87]:

$$E_+ - E_- = E_+(I = 0) - E_-(I = 0) + \pi_+ - \pi_- + \sum U_{\text{ohmic drop}}$$

For an electrochemical cell in an electrolysis setup, taking into account the signs of the anodic and cathodic polarisations as well as the positive sign of the ohmic drops, the following is given:

$$\begin{cases} ANODE = positive\ electrode\ + \\ CATHODE = negative\ electrode\ - \\ \pi_+ = \pi_{an} > 0 \quad \text{and} \quad \pi_- = \pi_{cat} < 0 \end{cases}$$

with the positive electrode as working electrode, it gives:

$$U(I \neq 0) \geq U(I = 0) + U_{\text{ohmic drop}} \geq U(I = 0))$$

Starting at open circuit, when the electrolysis current is gradually increased, the voltage imposed across the system also increases in turn and several simultaneous half-reactions may possibly be observed at one electrode (or both of them). Remember that the currents for each half-reaction at the same interface must be added together. This results in faradic yields that are lower than 100%. When it is possible for several reactions to occur at one of the electrodes, the main half-reaction is the one that would lead to the lowest polarisation in absolute value, for the same individual current. When several reactions can be envisaged at both interfaces in the whole system, then the overall reaction which results from the two main half-reactions is the one that requires the lowest imposed voltage.

The main half-reaction usually produces the highest faradic yield for a given working point of an electrode. Current-potential curves, for instance, allow one to define a potential range for each electrode, such that the faradic yield remains close to 100% (see figure 2.37). One must keep in mind that only qualitative reasoning is applied here, as when plotting current-potential curves. One cannot expect these potential ranges to be defined with high precision.

▼ Consider a deaerated acidic aqueous solution containing copper sulphate placed in an electrolytic cell. The two electrodes are copper and platinum respectively. Figure 2.37 outlines the shape of the anodic branch on the current-potential curve of the platinum positive electrode. The cathodic branch for the copper negative electrode is also plotted in this figure. Two working points have been added (round symbols) to indicate an operating point with a low electrolysis current. Two other points (square symbols) define an electrolysis operating point at higher current.

At open circuit the interface with the copper electrode is in thermodynamic equilibrium. However, the interface with the platinum electrode has a mixed potential, and this is defined by adding together the algebraic

---

[87] *Refer to sections 1.5.2, 2.2.3, 2.2.4 and 2.3.1 to find the introductions to these notions with their respective signs corresponding to electrolysis.*

currents of both water oxidation and proton reduction. The polarity of each electrode is defined by the experimentally measurable potentials of each electrode *vs* a saturated calomel reference electrode.

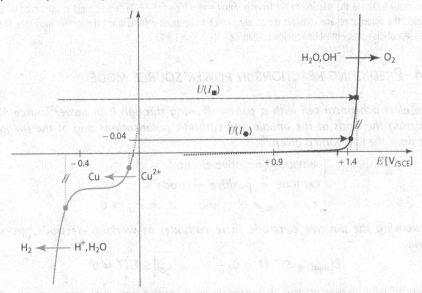

Figure 2.37 - Current-potential curve of a copper electrode (cathodic branch)
and of a platinum electrode (anodic branch) in a deaerated acidic aqueous solution containing Cu²⁺ ions

The curves in figure 2.37 show the following values:

$$E_{Cu}(I=0) = -0.04 \, V_{/SCE}$$
$$E_{Pt}(I=0) = +0.9 \, V_{/SCE}$$

At open circuit, the copper electrode is the negative electrode in the system whereas the platinum electrode is the positive electrode.

The anodic branch at the positive electrode corresponds to a single redox half-reaction, i.e., water oxidation: whatever the working point is, the anodic faradic yield along this branch is 100%. For the current to flow and therefore for electrolysis to occur, the following condition must be fulfilled: $E_{Pt}(I \neq 0) > +1.4 \, V_{/SCE}$. This value should be considered as an order of magnitude. In fact, for potentials between 0.9 and 1.4 $V_{/SCE}$, electrolysis occurs, but the current is very low.

The cathodic branch at the negative electrode corresponds to one or two redox half-reactions, depending on the potential zone: the reduction of Cu²⁺ ions and the reduction of protons (or of water). Here for the current to flow and therefore for electrolysis to occur, one must have the following: $E_{Cu}(I \neq 0) < -0.04 \, V_{/SCE}$. If the absolute value of the current increases, then the copper electrode potential decreases. When the potential becomes lower than $-0.4 \, V_{/SCE}$ (the working point is indicated by squares in figure 2.37) two half-reactions occur simultaneously and the faradic yield of each half-reaction drops lower than 100%.

To give a brief summary, the main overall reaction corresponding, for example, to the operating points indicated by round symbols is the following:

$$2\,H_2O \longrightarrow 4\,H^+ + 4\,e^- + O_2$$
$$Cu^{2+} + 2\,e^- \longrightarrow Cu$$

$$\overline{2\,H_2O + 2\,Cu^{2+} \longrightarrow O_2 + 4\,H^+ + 2\,Cu}$$

Such an overall reaction only makes sense if the anodic and cathodic faradic yields are identical and equal to 100%.

While the open-circuit voltage of this electrolysis cell is $0.9 - (-0.04) = +0.94$ V (the platinum electrode is chosen as the working electrode), the minimum electrolysis voltage is about $1.4 - (-0.04) = 1.44$ V. Here we can finally evaluate the maximum electrolysis voltage at which the 100% faradic yield is preserved. If we neglect the polarisation and consider the anodic branch to be quasi-vertical when the current increases, then the electrolysis voltage must be kept lower than $1.4 - (-0.4) = 1.8$ V. ◢

2.4.4 - PREDICTING REACTIONS IN POWER SOURCE MODE

For an electrochemical cell with a current flowing through it in power source mode, considering the signs of the anodic and cathodic polarisations and of the (negative) ohmic drop, the following is given[88]:

$$\begin{cases} ANODE \; = \; negative \; electrode \; - \\ CATHODE \; = \; positive \; electrode \; + \\ \pi_+ = \pi_{cat} < 0 \quad and \quad \pi_- = \pi_{an} > 0 \end{cases}$$

and choosing the positive electrode (here cathode) as working electrode, gives the following:

$$U_{supplied} \leq U(I = 0) - |U_{ohmic\,drop}| \leq U(I = 0)$$

When gradually increasing the current delivered by the power source (though all the while keeping it lower than the short-circuit current, see section 2.4.5), one may observe several half-reactions occurring simultaneously at one or both of the electrodes. As in the previous case, when it is possible for several half-reactions to occur at an interface, then the main half-reaction is the one with the lowest polarisation in absolute value for the same individual current. In power source mode, when several half-reactions can be envisaged at the interfaces, the main overall reaction is the one delivering the highest voltage to the external circuit.

As in electrolysis mode, the main half-reaction usually offers the highest faradic yield. The current-potential curves define a potential range for each electrode such that the faradic yield remains close to 100%, as illustrated in figure 2.38.

▶ Let us consider an electrochemical system with two compartments separated by a membrane or a salt bridge, and whose impact in terms of ohmic drop is disregarded. A zinc electrode is dipped in a dearerated acidic solution in the first compartment. The second compartment contains a silver electrode dipped in a deaerated acidic solution containing Ag^+ and Cu^{2+} ions. Figure 2.38 shows the shape of the cathodic branch of the current-potential curve of the positive silver electrode together with that of the anodic branch of the negative zinc electrode. Two points on the curves (round symbols) define a working point with a low delivered current. Two other points (square symbols) define a working point with a high delivered current.

At open circuit the interface with the silver electrode is in thermodynamic equilibrium. The interface with the zinc electrode has a mixed potential, which is defined by adding the zinc oxidation and proton reduction currents, with the latter reaction characterised as being very slow at the zinc electrode. The polarities of the electrodes in open-circuit conditions are defined by the experimentally measured potentials of each electrode *vs* a saturated calomel reference electrode.

---

[88] Refer to sections 2.2.3, 2.2.4 and 2.3.1 to find the introductions to these notions with their respective signs corresponding to power source mode.

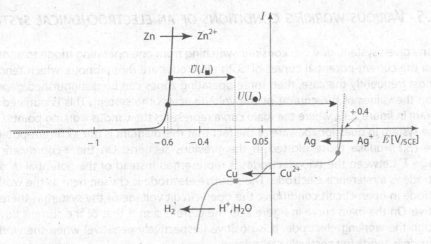

Figure 2.38 - Current-potential curve of a silver electrode (cathodic branch) in a deaerated acidic solution containing Ag⁺ and Cu²⁺ ions and of a zinc electrode (anodic branch) in a deaerated acidic solution

The curves in figure 2.38 show the following values:

$$E_{Zn}(I=0) = -1.0\ V_{/SCE}$$
$$E_{Ag}(I=0) = +0.4\ V_{/SCE}$$

At open circuit, zinc is the negative electrode in the system while silver is the positive electrode.

In the potential range plotted, the anodic branch on the negative electrode side corresponds to a single redox half-reaction, i.e., zinc oxidation. The simultaneous oxidation of water would equate to a working point with an extremely large current. One can therefore consider that, regardless of the working point in question, the anodic faradic yield is equal to 100%. For an anodic current to flow, the condition $E_{Zn}(I\neq0) > -0.6\ V_{/SCE}$ must be fulfilled, and this value should be considered as an order of magnitude. In fact, for potentials between -1.0 and $-0.6\ V_{/SCE}$, a spontaneous reaction occurs, but the current is very low. In the following part we will assume that this anodic branch is quasi-vertical.

The cathodic branch on the positive electrode side corresponds to one, two or even three redox half-reactions depending on the potential range considered: the reduction of Ag⁺, Cu²⁺ ions, or the reduction of protons (or of water). For a cathodic current to flow, one must have the following: $E_{Ag}(I\neq0) < +0.4\ V_{/SCE}$. If the current delivered by the electrochemical system increases in absolute value, then the silver electrode potential decreases. When this potential reaches values lower than about $-0.05\ V_{/SCE}$, two simultaneous half-reactions occur: the reduction of Ag⁺ ions and of Cu²⁺ ions, which are represented by squares in figure 2.38. The faradic yield of the main half-reaction becomes lower than 100%. When the potential gets lower than about $-0.4\ V_{/SCE}$ three half-reactions occur simultaneously.

The overall main reaction, corresponding to the working point indicated by round symbols, is expressed as:

$$Zn \longrightarrow Zn^{2+} + 2\,e^-$$
$$Ag^+ + e^- \longrightarrow Ag$$
$$\overline{Zn + 2\,Ag^+ \longrightarrow 2\,Ag + Zn^{2+}}$$

Writing such an overall reaction only makes sense if anodic and cathodic faradic yields are identical and equal to 100%.

While the open-circuit voltage of the cell is $0.4 - (-1.0) = 1.4\ V$, with the silver electrode chosen as working electrode, the maximum voltage that can be delivered by this system working as power source is about $0.4 - (-0.6) = 1.0\ V$. The conditions required for preserving faradic yields of 100% equate to a voltage value between $-0.05 - (-0.6) = 0.55\ V$ and $1.0\ V$.

2.4.5 - VARIOUS WORKING CONDITIONS OF AN ELECTROCHEMICAL SYSTEM

For any given system, one can consider switching from one operating mode to another. When the current-potential curves of both electrodes are monotonous, which tends to be most frequently the case, then three operating zones can be distinguished depending on the values of the current or the voltage across the system. This is outlined in a diagram in figure 2.39, where the main curve represents the various working points (U,I) of the system in question. Beware of the fact that this diagram is not a current-potential curve (E,I), unlike those plotted in the previous sections. On the x-coordinate the voltage U between the two electrodes is represented instead of the potential E of an electrode *vs* a reference electrode. The positive electrode is chosen here as the working electrode in open-circuit conditions: the open-circuit voltage of the system is therefore positive. On the main curve in figure 2.39 the current sign is that of the current flowing through the working electrode. It is positive (respectively negative) when the working electrode is anode (respectively cathode).

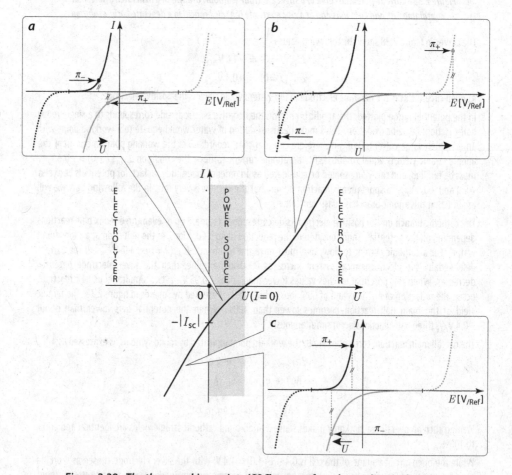

Figure 2.39 - *The three working points* (U,I) *zones of an electrochemical system*

In this figure the current-potential curves of each electrode are represented by three medallions, with a working point that is typical of each of the three zones. Medallion *b*, for instance, refers to figure 2.35 in electrolyser mode. The three possible (U,I) types of operating points are delimited by two particular working points corresponding respectively to open circuit ($U(I=0)$, $I=0$) and to short circuit ($U=0$, $I=-|I_{sc}|$).

Two of the three zones correspond to the usual power source and electrolyser modes. When the system works as a power source, the positive electrode is the cathode [case *a*, $-|I_{sc}| < I < 0$ and $0 < U < U(I=0)$]. Yet when the system operates as an electrolyser, the positive electrode is the anode [case *b*, $I>0$ and $U > U(I=0)$]. The third zone [case *c*, $I < -|I_{sc}|$ and $U < 0$] provides a more singular case whereby the electrode polarisations are reversed in relation to the open-circuit setup. As long as the working electrode (which switched from positive to negative polarity) is the cathode, then this particular operating mode is possible only on condition that there is a sufficient external energy source used. In this respect, the system works as an electric load. This type of case applies to a battery with several cells in series. If there is a defective cell then it may cause a reversal in polarity, yet all the while the overall battery still functions as a power source. The overall reaction occurring in the defective cell is a forced reaction which is only made possible by the spontaneous reactions occurring in the other cells.

The diagram in figure 2.40 shows the two specific situations which involve open-circuit and short-circuit conditions. Here it can be underlined that the point for $U=0$ corresponds to an ideal short circuit. Therefore, in other words it is achieved using a conductor, meaning that the ohmic drop is insignificant in this setup, which is no easy result to obtain in experimental conditions. It is worth recalling that the curves presented here correspond to systems whose ohmic drop in the electrolyte can be disregarded.

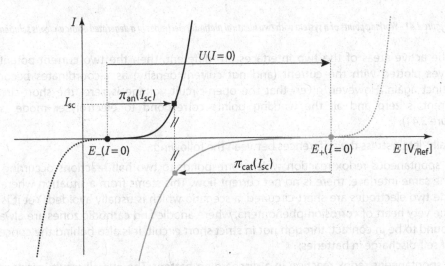

**Figure 2.40 - *Particular working points (U, I) of an electrochemical system:*
*open circuit (●) short circuit (■)***

In systems where two electrodes which are strictly identical are placed in the same electrolyte, the working points in open circuit and short circuit are identical (in other words, the short-circuit current is zero). All the working points here therefore correspond to electrolysis. Both electrodes play perfectly symmetrical roles, and equally symmetrical is the (U,I) curve of the working points of the overall system. Only the role of anode or cathode is swapped between the two electrodes, but the system never works as a power source.

▶ Let us consider the example of an electrolyte and two strictly identical electrodes: the open-circuit voltage is zero, then the corresponding open-circuit and short-circuit points are identical. The two electrodes are indiscernible, and the curve representing the working points is symmetrical, as shown in figure 2.41.

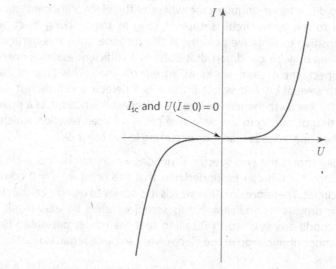

Figure 2.41 - *Working points of a system with two identical platinum electrodes in a deaerated acidic aqueous solution* ◢

If the active areas of the two interfaces are different, then the two current-potential curves plotted with the current (and not current density) as y-coordinates become distinct again. However, given that the open-circuit voltage is zero, the short-circuit current is zero and all the working points correspond to electrolyser mode (see figure 2.41).

Finally, let us stress the differences between the following:

▶ a spontaneous redox reaction which corresponds to two half-reactions occurring at the same interface: there is no net current flow. This stems from a situation whereby the two electrodes are short-circuited, a scenario which is usually avoided. Yet it is at the very heart of corrosion phenomena, where anodic and cathodic zones are always found to be in contact, through not in strict short circuit. It is also behind the concept of self-discharge in batteries;

▶ a spontaneous redox reaction in a discharging battery. The overall result is indeed a spontaneous reaction, but it represents the sum of two half-reactions occurring at two different interfaces (one working as cathode and the other as anode). For the overall reaction to actually occur, the two electrodes must be connected *via* an external circuit which can retrieve electric energy. The idea of electric energy retrieving was impossible in the previous case.

Questions on Chapter 2

1 - The three mass transport processes are:

▸ ....

▸ ....

▸ ....

2 - One studies an interface between cobalt (metal) and an Na^+Cl^- solution in acetonitrile (an organic solvent) where the following reaction occurs:

$$Co + 4Cl^- \longrightarrow CoCl_4^{2-} + 2e^-$$

▸ The interface is reactive	true	false
▸ By convention the current sign is positive	true	false

3 - FARADAY's law expresses, for a redox reaction, the amount of substance transformed as a function of the amount of electric charge which crosses the interface in question. The coefficient of proportionality at the numerator involves:

▸ the temperature	true	false
▸ the FARADAY constant	true	false
▸ the number of electrons	true	false
▸ the stoichiometric number of the species in question	true	false

4 - In an industrial aluminium production plant, the main cathodic reaction involves the Al(III)/Al couple with a faradic yield of 90%. The amount of aluminium produced per hour in an electrolysis cell working with a current of 300 000 A is:

10^3 mol 3.4×10^3 mol 3.6×10^3 mol 10^4 mol 3.4×10^6 mol

5 - The overall polarisation of an electrochemical chain can be split into different terms. In a system with no ionic junction, what do you call the term which adds itself to the two interfacial polarisations so as to gain the final overall polarisation value in the electrochemical chain? ....

6 - The concentration of a solution containing a species with a concentration of 0.1 mol L^{-1} is also equal to

100 mol m^{-3} 10^{-4} mol m^{-3} 100 mol cm^{-3} 10^{-4} mol cm^{-3}

7 - Assuming that the molar conductivity of Cu^{2+} ions in aqueous solution is a constant equal to 10 mS m^2 mol^{-1}, then the conductivity value of these same ions in a solution with a concentration of 0.1 mol L^{-1} is:

1 S cm^{-1} 10^{-2} S cm^{-1} 1 S m^{-1} 10 S m^{-1}

8 - Adding a supporting electrolyte to an electrochemical system causes, for the electroactive ions, the decrease in:

▸ their transport number	true	false
▸ their ionic conductivity	true	false

9 - In the following diagram:
- ▸ hatch the half-plane corresponding to an anodic operating mode
- ▸ indicate the half-reactions occurring in each half-plane with the usual writing conventions, taking the example of the Fe^{3+}/Fe^{2+} couple

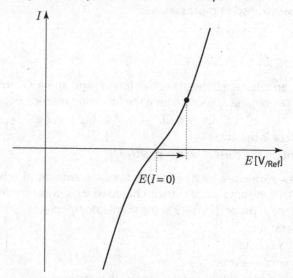

- ▸ what does the black arrow represent for the working point which is identified by a black dot?

....

10 - Except in very specific cases, one can predict the signs for each of the two interfacial polarisations in a given system. Therefore, in most cases, one can say that:
- ▸ the polarisation of the positive electrode is positive   true   false
- ▸ the polarisation of the anode is positive   true   false

11 - On the following diagram, draw the shape of the steady-state current-potential curves of three systems with the same open-circuit potential and the same diffusion limiting currents: a fast system (**a**), a slow system (**b**) and a very slow system (**c**).

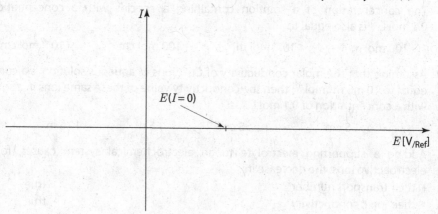

12 - On the following diagram, plot the steady-state current-potential curve of a system containing an inert working electrode dipped in a deaerated acidic aqueous solution ($pH = 0$), with no Fe^{3+} ions and an amount of Fe^{2+} ions befitting the existence of a limiting current. The reference electrode is a saturated calomel electrode. It will be assumed that the electrochemical window is determined by the fast half-reactions of water.

Indicate in the diagram the relevant numerical values of the potentials as well as the half-reactions involved.

$$E\,(\text{SCE}) = +\,0.24\ \text{V}_{/\text{SHE}} \qquad E^{\circ}_{Fe^{3+}/Fe^{2+}} = +\,0.77\ \text{V}_{/\text{SHE}}$$

$$E^{\circ}_{H^{+}/H_2} = 0\ \text{V}_{/\text{SHE}}\ (\text{at } pH = 0) \qquad E^{\circ}_{O_2/H_2O} = +\,1.23\ \text{V}_{/\text{SHE}}\ (\text{at } pH = 0)$$

13 - Using arrows, complete the following diagram (which represents the steady-state current-potential curves of the two electrodes in a given electrochemical cell) to indicate the following:

▸ the system's open-circuit voltage, $U(I=0)$

▸ the polarisation π_- of the negative electrode at the working point indicated by a black dot

▸ the polarisation π_+ of the positive electrode at the corresponding working point of this electrode (indicate this second dot in the diagram)

▸ the corresponding working voltage, $U(I \neq 0)$

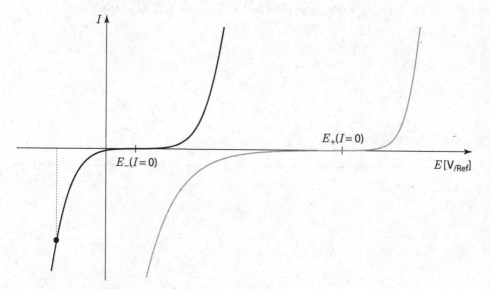

▸ the operating mode represented in the diagram corresponds to electrolysis   true   false

3 - THERMODYNAMIC FEATURES

A state of equilibrium can be defined in any system where the intensive variables (pressure, temperature, concentration, etc.) remain constant over time. In this chapter we will describe the state of equilibrium in an electrochemical system and give examples of conditions in which equilibrium can be observed [1]. In fact, the state of equilibrium is not always necessarily observed, since it cannot always be reached on the time scale of the experiment.

When a system is in equilibrium there is no movement of species on the macroscopic scale, resulting in a zero current. However, when the current is zero this does not necessarily mean that the system is in equilibrium, since a compensation of charge fluxes of different carriers can occur. This is the case when the interfacial open-circuit potential is a mixed potential [2] or for a liquid junction, a salt bridge, in an electrochemical system at open circuit [3].

3.1 - CONCEPTS OF POTENTIAL

The term potential is used frequently in different fields of physics and chemistry. It always refers to the energy per unit quantity related to the matter in question (number, mass, charge, etc.). Here we are particularly interested in electric and chemical energies. Therefore we will speak on the one hand in terms of electric potential: by using the quantity of charge (in C) as a quantity parameter, the electric potential is expressed in volts ($V = J\,C^{-1}$). On the other hand, by using the amount of substance (in mol) as a quantity parameter, the chemical potential is expressed in $J\,mol^{-1}$.

In a material these potentials are macroscopic, and represent an average across a large volume compared to atomic sizes (higher than a few tens of nm^3).

---

[1] As presented in section 2.1, an electrochemical system is defined as being isolated without any possibility of exchanging energy or matter with the surroundings. To make a different comparison to that in section 2.1, imagine an apple on a table. One can say it is in equilibrium, yet if the table is removed, then its equilibrium state is its new position on the ground. In the same way, all isolated electrochemical systems have an equilibrium state, yet this state is often very different after a current has been allowed to flow through an electric load, in other words once the system has functioned as a power source.

[2] Section 2.4.1 outlines this type of behaviour, which can be seen at an interface that is at open circuit however not in equilibrium.

[3] As described in qualitative terms in section 2.1.1, when there is a salt bridge or a liquid junction through a porous membrane, it takes an extremely long time to reach a state of equilibrium where a perfect mix of solutions is achieved. Generally a quasi-steady state is observed at open circuit (i.e., at zero current) where the molar fluxes are low but not zero (see appendix A.1.1).

3.1.1 - ELECTRIC POTENTIAL

In equilibrium, the electric charges found in a medium or in vacuum generate an electric field which itself derives from a scalar called the electric potential. From a mathematical point of view, this electric potential is defined up to a constant. In physics this constant is generally defined by taking vacuum at infinite distance as the potential reference. Thereupon, the term potential is used for any value using this reference and the term voltage is used for the difference between two potentials. A voltage is therefore independent of the reference used (vacuum, earth or any other).

More generally speaking, when equilibrium is disrupted (in physics this constitutes the change from electrostatics to electrokinetics), a magnetic field is associated *a priori* with the electric field. These two fields are then described using the MAXWELL equations. In electrochemistry, other than in exceptional cases[4], one can disregard the effects of magnetic field, and so in this scenario the electric field is derived from a potential, and therefore the MAXWELL equations are reduced to the POISSON law:

$$\nabla^2 V = -\frac{\rho_{ch}}{\varepsilon}$$

with: V the electric potential, in the general sense   [V]
         ρ_{ch} the charge density   [C m^{-3}]
         ε the dielectric permittivity of the medium   [A^2 s^4 kg^{-1} m^{-3}]

3.1.1.1 - ELECTRIC POTENTIAL AND ELECTRONEUTRALITY

In equilibrium all conductors are equipotential and uncharged in volume. In other words, electroneutrality applies at any macroscopic point in the volume. However, once on the atomic scale, this rule can no longer be upheld since here there are negatively charged electrons and positively charged nuclei. If there is a macroscopic perturbation in the charge distribution, then electroneutrality is restored, though only outside the interfacial zones, with a very fast transient state (lasting about a femtosecond[5] in a metal, see below).

▶ Take the example of an electric conductor with a fixed initial volume charge distribution (denoted by $\rho_{ch} = \rho_0 \neq 0$), such that there is no electroneutrality. It is assumed that the OHM law can be written here using a constant conductivity and relative dielectric permittivity of the medium in question. By applying simple calculations one can estimate the characteristic time constant for the spontaneous process of returning to equilibrium, where the conducting volume is equipotential and electroneutrality applies throughout the volume. The charge is then spread over the surface, i.e., at the interface with the external medium.

By combining the OHM law ($\mathbf{j} = \sigma \mathbf{E} = -\sigma \, \mathbf{grad} \, V$), the POISSON law ($\nabla^2 V = -\frac{\rho_{ch}}{\varepsilon}$) and the equation for

charge preservation [6] ($\frac{\partial \rho_{ch}}{\partial t} = -\operatorname{div} \mathbf{j}$), the following differential equation is obtained:

---

[4]   *The aluminium industry is an example where magnetic fields in an electrochemical system cannot be disregarded. During the process of industrial aluminium electrolysis the extremely high current flow builds up a strong magnetic field. This in turn triggers waves on the liquid surface which can subsequently create short circuits between the anode and cathode.*

[5]   *That is to say 10^{-15} s.*

$$\frac{\partial \rho_{ch}}{\partial t} = \sigma \ \mathrm{div}\,(\mathbf{grad}\,V) = -\frac{\sigma}{\varepsilon}\,\rho_{ch}$$

Electroneutrality, $\rho_{ch} = 0$, is therefore reached *via* an exponential law of the following type:

$$\rho_{ch} = \rho_0\,e^{-\frac{t}{\tau}}$$

with a time constant of $\tau = \dfrac{\varepsilon}{\sigma} = \dfrac{\varepsilon_0\,\varepsilon_r}{\sigma}$.

For an electrolytic aqueous solution, the order of magnitude of σ is $1\,\mathrm{S\,m^{-1}}$ [7] and ε_r is around 80. Therefore this results in a time constant of around 1 ns [8].

For a metal such as copper, the order of magnitude of σ is $10^7\,\mathrm{S\,m^{-1}}$ and $\varepsilon_r \approx 1$. The time constant is therefore around 10^{-18} s.

The current flow through an electrochemical cell is capable of sustaining a slight deviation from electroneutrality. However, it must be remembered that although this deviation is very low in numerical terms, it can nonetheless trigger a significant change in the potential profile, due to the very low value of ε [9]. Consequently, in a usual electrochemical system one can apply the electroneutrality equation to describe the concentration profiles in the electrolyte, outside the interface zones. However, it should not be inferred that it can be used to define the electric potential through the LAPLACE law ($\nabla^2 V = 0$) which is applicable only when electroneutrality is strictly respected (see appendix A.4.1).

3.1.1.2 - VOLTA AND GALVANI POTENTIALS

The laws of electrostatics show that a conductor that is in vacuum is equipotential in volume: it can therefore only be charged on the surface. The electrostatic potential is constant throughout the conductor volume, but then it suddenly varies at the conductor surface, before finally seeing slower changes in vacuum further away from the conductor. The potential difference between the conductor volume and vacuum at infinite distance, φ, can be divided into two terms, as depicted in figure 3.1 with $\varphi = \psi + \chi$. These terms are respectively called:

▸ GALVANI potential or internal potential, φ, which is the electric potential difference between the conductor inside and vacuum at infinite distance;

---

[6] *Charge preservation can be expressed by combining the mass balances in volume for different charged species, as described in section 4.1.2.*

[7] *Table 4.1 in section 4.2.2.3 gives conductivity values for various charge carriers in different conductors.*

[8] *With $\dfrac{1}{4\pi\,\varepsilon_0} = 9 \times 10^9$ SI .*

[9] *Appendix A.4.1 outlines an example of electrolysis with a diluted electrolyte containing only Ag+ and NO₃⁻ ions. It gives several orders of magnitude for the impact on the potential profiles caused by deviation from electroneutrality (see figure A.17 in appendix A.4.1). In particular, a deviation from electroneutrality of around 10^{-14} mol L⁻¹ in aqueous solution does not cause a shift from the LAPLACE law (linear potential profile), while a deviation of around 10^{-11} mol L⁻¹ cannot be overlooked when describing the potential profile.*

▸ VOLTA potential or external potential, ψ, which is the electric potential difference between the outlet of the interfacial zone and vacuum at infinite distance;

▸ surface electric voltage, χ, which is the voltage across the interfacial zone[10]. The surface voltage, χ, is notably linked to the surface charge[11]. However, even without surface charge, surface dipoles can lead to a significant surface electric voltage.

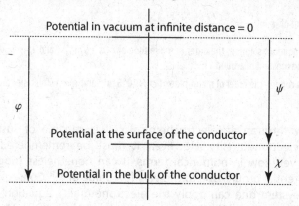

Figure 3.1 - Potential levels for a conducting volume in vacuum

3.1.2 - CHEMICAL AND ELECTROCHEMICAL POTENTIALS

3.1.2.1 - CHEMICAL POTENTIAL

In thermochemistry, the chemical potential of an uncharged species i is defined by the following equation:

$$\mu_i = \left(\frac{\partial G}{\partial n_i} \right)_{n_{j \neq i}, T, P}$$

with: G the system's GIBBS energy   [J]
             n_i the amount of substance of the species i [mol]
             n_j the amount of substance of each other species   [mol]

μ_i is the partial molar GIBBS energy. Using EULER's identity, valid for all homogeneous functions of degree 1, one can show that the GIBBS energy can be expressed by the following sum:

$$G = \sum_i n_i \, \mu_i$$

Therefore μ_i represents the contribution of species i (relative to one mole) to the system's GIBBS energy, i.e., in interaction with other species in the medium in question.

---

[10] *The term surface potential is sometimes used, although the correct term is surface voltage.*

[11] *Appendix A.3.1 outlines the calculations for the electrostatic potential profile, as well as the values of the three voltages described here, relating to the case of a conducting surface-charged sphere that is in equilibrium, in vacuum.*

This is also the work required for the reversible transport of one mole of species i to switch from infinite distance in vacuum to the bulk of the system, at T and P constant.

The activity of species i, a_i (a dimensionless number), is defined in terms of its chemical potential by the following equation:

$$\mu_i = \mu_i^\circ + RT \ln a_i$$

where μ_i° is the standard chemical potential of species i at temperature T.

A standard state corresponding to $a_i = 1$ should be defined for each species. Except in the case of solutions (see below), most of the time the standard state of species i at temperature T corresponds to the pure compound in its physical state (gas, solid, liquid) at the same temperature and under the standard pressure of $P^\circ = 1$ bar. A standard state of reference is also defined for each pure compound, and this corresponds to the standard state of the pure compound in its most stable physical state at the temperature in question.

Furthermore, one can also define ideal behaviours. In order to describe a compound in a real system, the degree of deviation from the ideal situation is characterised by the activity coefficient, which is a dimensionless number as the activity itself.

▶▶ For gases

The standard state is the ideal gas corresponding to the standard pressure P° [12]. The activity (also called the fugacity) of a compound i in a gas mixture under total pressure P is therefore defined by:

$$a_i = \gamma_i \, x_i \, \frac{P}{P^\circ} = \gamma_i \, \frac{P_i}{P^\circ}$$

with: γ_i the activity coefficient, or the fugacity coefficient of the gas i in the mixture, taking the interactions in the gas mixture into account

x_i the molar fraction of the compound i in the gas mixture

P_i the partial pressure of the gas i in the mixture [bar]

▶▶ For solutions

In a liquid mixture, when there is a fairly balanced composition in the compounds then the pure compounds can be kept as standard states using the molar ratios to define the ideal activities. Consequently, the activities are expressed as $a_{i,x} = \gamma_{i,x} \, x_i$. However, the solutions most commonly seen in electrochemistry involve a compound, called a solvent, which is found in much greater quantity than other compounds, called solutes. Here one often distinguishes between these two types of compounds by choosing a different standard state for the solvent and the solutes:

▶ for the solvent, the standard state is the pure compound at standard pressure P°. For a moderately concentrated solution, $x_{solvent} \approx 1$, therefore $\gamma_{solvent} \approx 1$ and it is consequently often recognised that $a_{solvent} = \gamma_{solvent} \, x_{solvent} \approx 1$;

[12] At very low pressure all gases behave like an ideal gas. This remains true up to 1 bar $(= P^\circ)$ for most usual gases, but deviations occur at high pressure.

▸ for the solutes, the standard state is a hypothetical solution without any interaction between the solutes' ions and molecules (this solution is called infinitely diluted), under $P°$, extrapolated at $C° = 1$ mol L^{-1} (or $m° = 1$ mol kg^{-1}):

$$a_i = \gamma_i \frac{C_i}{C°} \qquad \text{with} \qquad \gamma_i \xrightarrow[\sum_k C_k \to 0]{} 1 \qquad (\text{or} \quad a_{i,m} = \gamma_{i,m} \frac{m_i}{m°})$$

▸▸ **For solids**

The standard state corresponds to a solid alone in its own phase at standard pressure $P°$. Solids are frequently found in their standard state. However in electrochemistry there are cases where the solid phase is a solid solution, for example with metal alloys or insertion materials used in many batteries. Here therefore the activities are defined using composition parameters (for example with the atomic fraction for alloys, or the concentration for solid electrolytes, etc.).

3.1.2.2 - ELECTROCHEMICAL POTENTIAL

The state of an electrochemical system depends on an additional parameter, namely the GALVANI potential, φ. Therefore, when referring to a species in electrochemical systems, one frequently sees the term 'electrochemical potential' used instead of 'chemical potential', even though they represent exactly the same concept. Namely, both represent the contribution of species i, charged or not, to the GIBBS energy in the medium in question. The tilde mark is then placed over the symbol to point out that the intensive parameter φ is taken into account:

$$\widetilde{G} = \sum_i n_i \widetilde{\mu}_i$$

For a metal, the electrochemical potential of electrons corresponds to a measurable quantity[13]. However it is more difficult to envisage the experimental determination of an ion's electrochemical potential in an electrolyte. In fact, due to the media's electroneutrality, it is impossible for an ion intake to occur in an experiment without simultaneously there being a counter-ion, at the very least. Therefore, one can only access the electrolyte's electrochemical potential, this term being identical to its corresponding chemical potential. There are different consequences entailed here, which will be outlined later in this chapter: the definition of a solute's mean activity (section 3.2.1.1), and the convention for the thermodynamic data of ions in a solution (section 3.1.2.3).

The interactions between species i and other surrounding species involve electric forces which themselves are not simply the result of the influence of the local macroscopic electric potential.

However the electrochemical potential of ions can be divided into two terms, one called 'purely electric' and the other called 'chemical potential'. This is because the latter term shares the same form as that of the equations seen above:

$$\widetilde{\mu}_i = \mu°_i + RT \ln a_i + z_i \mathcal{F} \varphi = \mu_i + z_i \mathcal{F} \varphi$$

---

[13] *This quantity is related to the electron work function as described in section 3.2.2.3.*

One must remember that this common language can be confusing. In particular, the activity of an ion can depend *a priori* on the value of the electric potential. By way of simplifying, the so-called ' chemical potential ' is often said not to depend on the electric potential but rather depend exclusively on the medium's chemical composition (as in the DEBYE-HÜCKEL law, see section 3.2.1.3).

Once electroneutrality applies, it is no longer necessary to describe the system in terms of electrochemical potential, and the only quantities experimentally available then become the electrolytes' chemical potentials. On the other hand, if one wishes to describe media that demonstrate a deviation from electroneutrality, for instance inter-facial zones in electrochemistry, then the concept of electrochemical potential becomes a very useful tool [14].

3.1.2.3 - CONVENTION FOR THERMODYNAMIC DATA TABLES

In thermodynamic data tables (standard enthalpies or GIBBS energies of formation and standard molar entropies) which relate to compounds other than ions in a solution, the common convention that is applied involves setting the values of standard enthalpy and standard GIBBS energy of formation (or chemical potential) equal to 0 J mol^{-1} for all simple pure elements in their stable physical state at the temperature in question. The data therefore refer to the formation of substances from simple elements.

The thermodynamic quantities relating to the anion and cation cannot be measured separately for ions in a solution due to the electroneutrality (experimental thermody-namic data only concern electrolytes). Therefore, an additional convention is introduced here, known as LATIMER's convention. This involves setting the proton's chemical potential equal to zero:

$$\forall T \quad \mu^{\circ}_{H^+} = 0 \: J\,mol^{-1}$$

where H^+ represents the solvated proton in the solvent in question (aqueous or not) [15].

It is possible to deduce the chemical potential values for other ions in a solution by using experimental measurements of standard GIBBS energy of reaction. This convention can be stretched to include values for enthalpy, entropy and the heat capacity of protons in a solution, all of which are set equal to zero.

---

[14] *Such is the case when describing interfaces in equilibrium (see section 3.3) or when splitting a current into migration and diffusion currents (see section 4.2.1).*

[15] *This convention, which is similar to the one used for choosing the H^+/H_2 couple for the SHE, enables one to give a clear definition to thermodynamic data, even if these data may at times be virtual. However, one must be aware that such choices do not enable one to make easy comparisons between thermodynamic data in different solvents. Indeed, the energy of the solvated proton strongly depends on the medium in question, because the ion is very small in size and therefore very sensitive to electrostatic effects. Other reference compounds are used for making comparisons across different media, for instance in electrochemistry in linking different standard potentials scales to each other. In this instance, it is common for the standard GIBBS energy of solvation of BPh_4^- $AsPh_4^+$ (two large size ions of opposite charges) to be divided into two equal parts for each ion, regardless of the solvent. This extra-thermodynamic assumption then enables one to compare thermodynamic data in different solvents.*

FUELCELLS

Document written with the kind collaboration of E. ROSSINOT, head of the fuel cells team and A. RENARD, communications manager for Axane, Sassenage, France

A fuel cell is an electrochemical system which supplies electrical energy generated by spontaneous electrochemical reactions that take place between oxygen found in air and hydrogen. This is the case for the simplest fuel cell systems, though other fuels are also currently being studied. Insofar as only water is produced in this main reaction, it is the system of choice for a non-polluting power supply.

As in many batteries, several individual cells are connected in series or in parallel to obtain the electric characteristics (energy, power) necessary for applications. It should be noted that fuel cells are particular in the field of batteries given their use of gas reagents: they are open systems with a continuous gas flow feeding in. They are therefore more complicated to manage than a closed battery. However, the fact of decoupling the system's power and energy functions does provide an advantage. The power supplied depends on the active surface of the electrodes used, while the energy depends on the size of the hydrogen tank that is connected to the core of the cell inside the system.

In fuel cell systems where the internal working temperature is around 70 °C, the electrolyte is a polymer material which allows the conduction process to take effect via protons. A cell is then made up of this electrolyte placed between two electrodes containing a catalyst.

At the anode, hydrogen is oxidized:
$$2\,H_2 \longrightarrow 4\,H^+ + 4\,e^-$$

At the cathode, oxygen is reduced:
$$O_2 + 4\,H^+ + 4\,e^- \longrightarrow 2\,H_2O$$

The balance is:
$$2\,H_2 + O_2 \longrightarrow 2\,H_2O$$

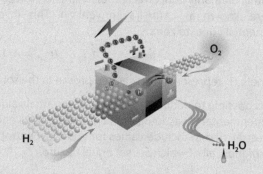

An energy solution with many advantages...

▸ High energy efficiency, excellent current quality

▸ Autonomy, mobility and reliability

▸ No harmful emissions (in the case of hydrogen fuel)

▸ Modularity (elementary cells either in series or in parallel)

▸ Simple and noiseless when in operation

▸ Indoor or outdoor use (high resistance to weather conditions)

... and great opportunities.

Considering that it remains a priority to comply with environmental constraints the fuel cell can be seen as a very promising power source in the fields of transport, small portable objects and energy distribution.

▶ **Mobile applications**

By way of example one can cite applications in telecommunications, propelling small electric vehicles or boats, uses in scientific or rescue missions (firefighters, rescue teams, medical emergencies, etc.) as well as in the cinema and building industry or indeed for any type of work, at any time and in any place.

Examples of nomadic applications (picture provided by Axane)

Axane is a wholly-owned subsidiary of the Air Liquide group. It aims to develop fuel cells systems powered by hydrogen, which are the core of the power supply offer proposed by Air Liquide Hydrogen Energy. These turnkey energy solutions are for new emerging markets (telecom antennas, one-off outdoor events, off-grid sensors, etc.)

▶ **Stationary or sedentary applications**

These cover continuous power supply in isolated areas; for example telecommunication antennas or relays, lighting critical sites or even providing power supply to measuring equipment, etc.

Moreover they are used for backup power supply (i.e., capable of taking over a main energy supply system in the event of power failure): for instance for telecommunications, health or the banking sector, etc.

Examples of sedentary applications (picture provided by Axane)

A system installed in difficult climatic conditions (–5°C to 33°C and with a humidity of 26% to 100%) performed over 10 000 h of continuous operation (about one year and 10 MWh of cumulative energy supplied) without maintenance work.

Table 3.1 shows an extract from a thermodynamic data table at 25 °C.

Table 3.1 - Thermodynamic data

Compound		$\Delta_f H°$ [kJ mol^{-1}]	$\Delta_f G° = \mu°$ [kJ mol^{-1}]	$S°$ [J K^{-1} mol^{-1}]	$C_P°$ [J K^{-1} mol^{-1}]
F_2	gas	0	0	202.7	31.3
F^-	gas	− 270.7			
F^-	aq. sol.	− 332.6	− 278.8	− 13.8	− 106.7
H^+	gas	1536.2	1523	109	20.8
H^+	aq. sol.	0	0	0	0
H^+	non-aq. sol.	0	0	0	0
H_2	gas	0	0	130.6	28.8
H_2O	liquid	− 285.2	− 237.2	69.9	75.3

Note that for any given ion, different values can be found depending on its physical state, i.e., whether it is in a solution or in gaseous plasma. In the latter case, a particular convention is needed to produce thermodynamic data which is similar to the LATIMER convention. Therefore, the gaseous H^+ data given above are obtained by setting the electrochemical potential of electrons in the plasma at zero at any temperature.

3.2 - THERMODYNAMIC EQUILIBRIUM IN A MONOPHASIC SYSTEM

As previously stated, any conducting phase in equilibrium is equipotential in its volume and electroneutrality applies at each point [16].

Through an analogy with thermochemistry, the thermodynamic equilibrium is expressed in these conducting systems by the following:

▸ in the absence of any possible chemical reaction,

$$\forall i \quad \widetilde{\mu}_i = \text{Cst} \quad \text{in the volume}$$

▸ in the presence of one or several possible chemical reactions (index r), the thermodynamic equilibrium is expressed by the following equation:

$$\forall r \quad \Delta_r \widetilde{G} = \sum_i v_{i,r} \, \widetilde{\mu}_i = 0 \qquad \text{in the volume}$$

whereby $v_{i,r}$ is the algebraic stoichiometric number of species i in the reaction r in question [17].

Because a chemical reaction always preserves the charge (identical overall charge on both sides of the balanced equation), then the electrochemical GIBBS energy of reaction, $\Delta_r \widetilde{G}$, and the chemical GIBBS energy of reaction, $\Delta_r G$, are identical (see the example below).

---

[16] *Describing the interfacial thermodynamic equilibrium is covered in section 3.3.*

[17] *Some authors prefer the concept of affinity to that of GIBBS energy of reaction. In fact, using affinity helps avoid possible confusion between variations in GIBBS energy, ΔG (J) and GIBBS energy of reaction, $\Delta_r G$ (J mol^{-1}).*

The previous equation is then expressed in the following form:

$$\Delta_r \widetilde{G} = \Delta_r G = \Delta_r G° + RT \ln \prod_i a_i^{\nu_{i,r}} = 0$$

or:

$$K_{eq}(T) = \prod_i a_i^{\nu_{i,r}} \qquad \text{mass action law}$$

with

$$0 = \Delta_r G°(T) + RT \ln K_{eq}(T)$$

Chemical reactions may be redox reactions as shown below.

▶ For example, in an aqueous solution containing fully dissociated potassium iodide, KI, and dissolved diiodine, I_2, the redox equilibrium of the couples I_2/I_3^- and I_3^-/I^- is rapidly observed in the homogeneous phase:

$$I^- + I_2 \rightleftharpoons I_3^-$$

In equilibrium, one can write:

$$0 = \Delta_r \widetilde{G} = \widetilde{\mu}_{I_3^-} - \widetilde{\mu}_{I^-} - \widetilde{\mu}_{I_2}$$

$$= \mu_{I_3^-} - \mathcal{F}\phi - \mu_{I^-} + \mathcal{F}\phi - \mu_{I_2} = \Delta_r G$$ ◢

3.2.1 - ELECTROLYTIC SOLUTION

3.2.1.1 - MEAN ACTIVITY AND MEAN ACTIVITY COEFFICIENT

Remember that in experimental conditions it is impossible to measure the electro-chemical potential of an ion in an electrolytic solution. An ion's amount cannot be modified without a simultaneous change in the amount of at least one other ions. Here therefore one defines mean values, which can be determined based on experiments [18].

For example, one mole of a fully dissociated solute of the type $A_{p_+} B_{p_-}$, with the respective algebraic ion charge numbers z_+ and z_-, corresponds to p_+ mole(s) of A^{z+} and p_- mole(s) of B^{z-}.

We then have the equation: $\qquad z_+ p_+ + z_- p_- = 0$

The only experimentally accessible quantity is the solute's electrochemical potential, which is identical to its chemical potential:

$$\widetilde{\mu}_{A_{p_+} B_{p_-}} = p_+ \widetilde{\mu}_{A^{z+}} + p_- \widetilde{\mu}_{B^{z-}}$$

$$= p_+ \mu_{A^{z+}} + p_+ z_+ \varphi + p_- \mu_{B^{z-}} + p_- z_- \varphi$$

$$= \mu_{A_{p_+} B_{p_-}}$$

[18] Appendix A.3.2 gives a few examples to show how common equations written using individual ion activities can wrongly suggest that these quantities are measurable. Indeed, the most rigorous equations always include quantities that involve mean activities. Two typical examples of these include potentiometric measurements either using selective electrodes or as in cells with a liquid junction (or salt bridge).

The mean electrochemical potential of the solute is defined by the following equation, with $p = p_+ + p_-$ representing the number of ions in the compound $A_{p_+} B_{p_-}$:

$$\widetilde{\mu}_\pm = \mu_\pm = \frac{1}{p_+ + p_-} \mu_{A_{p_+} B_{p_-}} = \frac{1}{p} \mu_{A_{p_+} B_{p_-}}$$

then,

$$\widetilde{\mu}_\pm = \mu_\pm = \frac{p_+ \mu^\circ{}_{A^{z+}} + p_- \mu^\circ{}_{B^{z-}}}{p} + \frac{RT}{p} \ln a_+{}^{p_+} + \frac{RT}{p} \ln a_-{}^{p_-}$$

$$= \mu^\circ{}_\pm + \frac{RT}{p} \ln a_+{}^{p_+} \ln a_-{}^{p_-}$$

The definition for the solute's mean activity derives from the previous equation:

$$\widetilde{\mu}_\pm = \mu_\pm = \mu^\circ{}_\pm + RT \ln a_\pm$$

hence: $a_\pm = \sqrt[p]{a_-{}^{p_-} a_+{}^{p_+}}$ or $a_{A_{p_+} B_{p_-}} = a_\pm{}^p = a_+{}^{p_+} a_-{}^{p_-}$

The mean concentrations and activity coefficients can be defined based on the concentrations of each ion, which are measurable in an experiment:

$$C_\pm = \sqrt[p]{C_-{}^{p_-} C_+{}^{p_+}}   \text{ and }   \gamma_\pm = \sqrt[p]{\gamma_-{}^{p_-} \gamma_+{}^{p_+}}$$

with the usual equation linking activity, concentration and activity coefficient as being:

$$a_\pm = \gamma_\pm C_\pm$$

▶ For example, for a 1–1 electrolyte such as KCl, the mean quantities correspond to geometric means:

$$p_+ = p_- = 1   \text{ hence }   \begin{cases} a_\pm = \sqrt{a_+ a_-} \\ \gamma_\pm = \sqrt{\gamma_+ \gamma_-} \\ C_\pm = \sqrt{C_+ C_-} \end{cases}$$

These equations can also be applied to electrolyte mixtures. In these cases, even if anions and cations can have different concentrations, yet their mean values are always expressed in a similar way (see KCl in the following example in section 3.2.1.2).   ◢

3.2.1.2 - IONIC STRENGTH

LEWIS, RANDALL and BRÖNSTED showed through experiments in diluted solutions, that each mean activity coefficient depends on all the different kinds of ions in the electrolyte, through the ionic strength, I_s[19]:

$$\log \gamma_\pm = A \, z_+ z_- \sqrt{I_s}$$

with

$$I_s = \frac{1}{2} \sum_i C_i \, z_i{}^2$$

---

[19] In many documents, the symbol typically used to indicate ionic strength is I. We prefer I_s to avoid confusion between this quantity and the current I.

Now let us turn to the case of aqueous solutions at 25 °C: $A \approx 0.5$ $L^{1/2}$ $mol^{-1/2}$, with the ionic strength I_s and the concentrations C_i expressed in mol L^{-1}. This relationship, which has been verified *via* experiment, is quite accurate for media with an ionic strength such as $I_s \leq 10^{-3}$ mol L^{-1} (or exceptionally up to 10^{-2} mol L^{-1}). Remember that the charge numbers are algebraic, therefore the product $z_+ z_-$ is always negative. Consequently, when this equation applies, the mean activity coefficient is a number below 1.

▶ We can estimate the mean activities and activity coefficients for an aqueous solution containing a mixture of zinc chloride, $ZnCl_2$, with a concentration of 0.001 mol L^{-1} and potassium nitrate, KNO_3, with a concentration of 0.003 mol L^{-1}.

The ionic strength of the solution is equal to:

$$I_s = \frac{1}{2}\left[0.001\,(1\times 2^2 + 2\times 1^2) + 0.003\,(1\times 1^2 + 1\times 1^2) \right] = 0.006 \text{ mol } L^{-1}$$

The mean activities and activity coefficients can be calculated for the electrolytes $ZnCl_2$ and KNO_3, but also for KCl and $Zn(NO_3)_2$.[20]

For example for KCl, $\log \gamma_\pm = -A\sqrt{0.006}$ hence $\gamma_\pm = 0.91$

$a_\pm = \gamma_\pm C_\pm$ and $C_\pm = \sqrt{0.003 \times 0.002} = 0.0024$ mol L^{-1} $a_\pm = 0.0022$

For example for $ZnCl_2$, $\log \gamma_\pm = -2A\sqrt{0.006}$ hence $\gamma_\pm = 0.84$

$a_\pm = \gamma_\pm C_\pm$ and $C_\pm = \sqrt[3]{0.001\,(2\times 0.001)^2} = 0.0016$ mol L^{-1} $a_\pm = 0.0013$ ◀

The ions' charges play a very important role in the electric interactions and therefore in the values of the activity coefficients. Table 3.2 illustrates this property by giving values for ionic strength and mean activity coefficients (calculated from the previous equations) for different types of electrolytes in a concentration equal to 10^{-3} mol L^{-1}.

Table 3.2 - Examples of ionic strength and mean activity coefficient values

Type of electrolyte : $z_+ - z_-$	Examples	I_s (10^{-3} mol L^{-1})	$\gamma_\pm$
1–1	NaCl	1	0.96
1–2 or 2–1	Na_2SO_4 or $ZnCl_2$	3	0.88
1–3 or 3–1	Na_3PO_4 or $CeCl_3$	6	0.76
2–2	$CuSO_4$	4	0.74
2–3 or 3–2	$Ca_3(PO_4)_2$ or $Ce_2(SO_4)_3$	15	0.42

[20] The concepts of mean activity or mean activity coefficient of a given electrolyte in a solution containing more than two types of ions are relevant and moreover correspond to quantities which are accessible via experiment. For example in the case of an aqueous solution obtained by dissolving silver nitrate, potassium chloride and sodium perchlorate, the solubility of silver chloride is directly related to the mean activity of AgCl in this medium. Another example, which is very important in the aluminium industry, is that of molten salt used in industrial electrolysis. At about 1000°C, the salt contains Na^+, F^-, AlF_5^{2-}, AlF_6^{3-} and AlF_4^- ions. The gas atmosphere found above such an electrolyte is mainly composed of $NaAlF_4$. The corresponding pressure measurement enables one to determine the mean activity of sodium tetrafluoro-aluminate.

The ionic strength of a given solution is unique. Therefore, when respecting the validity limits of the previous equation, one can state that electrolytes of the same type (i.e., with the same value for the product $z_+ z_-$), all have the same mean activity coefficient for a given ionic strength, regardless of their concentration.

When a supporting electrolyte is used in electrolytic solutions, its advantages go beyond the direct impact upon the properties of mass transport[21]. In fact, the supporting electrolyte fixes the ionic strength of the solution. In other words, a solution with a supporting electrolyte is a solution with a buffered ionic strength.

3.2.1.3 - DEBYE-HÜCKEL's MODEL

The theoretical model put forward by DEBYE and HÜCKEL helps one to understand the experimental law outlined above. In order to determine any deviation from the ideal situation, the approach consists firstly in calculating the interactions exerted by all other ions on a central ion[22]. These interactions then appear as a corrective term in the electrostatic potential around the central ion, and are subsequently expressed as an activity coefficient for the ion.

Let us specify the different hypotheses laid out by the DEBYE-HÜCKEL model and the subsequent approximations which help one develop a means of expressing the activity coefficient of various ions in an electrolyte. Nearly all the assumptions, expressed below, are based on the fact that this theory is developed for dilute solutions:

▸ all solutes are strong electrolytes, and therefore fully dissociated. In particular, it is assumed that there are no neutral molecules, except for the solvent, and that there are no ion pairs;

▸ the electric permittivity of the medium is constant and equal to that of the pure solvent, regardless of its position in relation to the central ion;

▸ the interactions between the different ions in the solution are exclusively electrostatic (COULOMB's law);

▸ the system's layout is spherically symmetric around each ion, which is considered as a point in relation to the distances in question;

▸ the COULOMB energy is, in absolute value, much lower than the thermal energy.

The assumptions above lead to a law called the limiting DEBYE-HÜCKEL law:

$$\log \gamma_k = -A \, z_k^2 \sqrt{I_s}$$

For aqueous solutions at 25 °C, the theoretical expression of A leads to the same value, $A \approx 0.5 \, \mathrm{L}^{1/2} \, \mathrm{mol}^{-1/2}$, if the ionic strength I_s and the concentrations C_i are expressed in $\mathrm{mol \, L}^{-1}$. This law is applied generally when the ionic strength is lower than $10^{-3} \, \mathrm{mol \, L}^{-1}$.

---

[21] See section 2.2.4.4.

[22] Appendix A.3.3 lays out the full reasoning and calculations behind DEBYE-HÜCKEL's limiting law, covering all the various steps, based on the assumptions listed here.

When the ionic strength value is higher than 10^{-3} mol L^{-1}, then one can use the following equation, which is known as the extended DEBYE-HÜCKEL law[23]:

$$\log \gamma_k \approx -A\, z_k^{\,2}\, \frac{\sqrt{I_s}}{1+\sqrt{I_s}}$$

This equation results from a calculation similar to the previous one, although the charges are not considered as point charges. At room temperature, when the ionic strength is higher than 10^{-3} mol L^{-1} one needs to use the extended law (which corresponds to an adjustment of about 10% on the logarithm, for an ionic strength of 10^{-3} mol L^{-1}). This extended law is generally applied to cases showing ionic strengths of up to 10^{-2} mol L^{-1} (and exceptionally 10^{-1} mol L^{-1}).

It is easy to switch from the theoretical expression of an ion's activity coefficient to the theoretical expression of an electrolyte's mean activity coefficient, which is the only expression that can be compared to experimental data. For example, the limiting law gives the following:

$$\log \gamma_{\pm} = \frac{1}{p}\left(p_- \log \gamma_- + p_+ \log \gamma_+\right) = -A\sqrt{I_s}\,\frac{p_-\, z_-^{\,2} + p_+\, z_+^{\,2}}{p_+ + p_-}$$

and because $p_-\, z_- + p_+\, z_+ = 0$, with z_+ and z_- algebraic, then:

$$\log \gamma_{\pm} = -A\sqrt{I_s}\,\frac{-p_+\, z_+\, z_- - p_-\, z_-\, z_+}{p_+ + p_-}$$

hence:
$$\log \gamma_{\pm} = A\, z_-\, z_+ \sqrt{I_s}$$

Scientific literature shows how the experimental curves that represent the variation in the mean activity coefficients as a function of the molality (i.e., the amount of substance in mol per kilogram of solvent[24]), often have a minimum. Moreover, the value of the mean activity coefficient may exceed 1 (see figure 3.2). Therefore, when dealing with concentrated electrolytes, one can see how the experimental results rapidly seem to deviate from the variation laws as presented above (they all show a monotonic variation in the mean activity coefficient as a function of the molality or the concentration).

In reality, these experimental results are not wholly inconsistent with the DEBYE-HÜCKEL model, whereby the mean activity coefficient decreases as the molality increases. This comes down to taking into account the fact that a usual solute/solvent description is no longer satisfactory for concentrated electrolytes. Indeed, when dealing with concentrated electrolytes, the number of solvent molecules involved in the solvation sphere, close to the ions, cannot be ignored when compared to the total number of solvent molecules. Once one takes this phenomenon into account, then three types of adjustment emerge, each of which are laid out in detail below.

---

[23] The extended DEBYE-HÜCKEL law is described in more detail in appendix A.3.3.

[24] In an experiment, it is much easier to obtain an accurate value for the molality of a concentrated solution (that is, the amount of solute introduced in 1 kg of solvent) than it is to know its concentration, and this due to the volume variations caused by the mixing. Frequently as a result, experimental thermodynamic data relating to concentrated solutions are presented using the molality scale. Then, in order to return to a concentration scale, one firstly needs to know the data relating to the solutions' density changes, as a function of their composition.

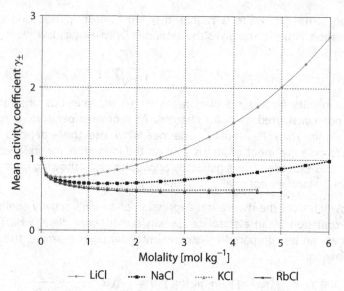

**Figure 3.2 - Evolution of the mean activity coefficient
as a function of the molality for some solutes in aqueous solutions**

▼ When dealing with concentrated solutions, it is necessary to:

▸ replace the apparent molality (the amount of ions per unit mass of solvent) by the actual molality of solvated ions, i.e., the ratio of the amount of solvated ions to the mass of free solvent molecules. The actual molality is not datum that can be directly obtained from experiments, because one must firstly assume a value for the hydration number, which is *a priori* different for each solute. The actual molality is higher than the apparent molality.

$$m_{actual} = \frac{m_{apparent}}{1 - \frac{18}{1000}\, n_{hydratation}\, m_{apparent}}$$

▸ calculate the actual activity coefficient corresponding to the actual molality, assuming that the energies (chemical potentials) are equal in both the real system and the non-hydrated apparent system:

$$A^+_{hydrated} + Cl^-_{hydrated} \equiv A^+ + Cl^- + n_{hydratation}\, H_2O$$

$$2RT \ln a_{\pm_{actual}} = 2RT \ln a_{\pm_{apparent}} + n_{hydratation} RT \ln a_{water}$$

or

$$\left(a_{\pm_{actual}}\right)^2 = \left(a_{\pm_{apparent}}\right)^2 \left(a_{water}\right)^{n_{hydratation}}$$

hence

$$\gamma_{\pm_{actual}} = \gamma_{\pm_{apparent}} \frac{m_{\pm_{apparent}}}{m_{\pm_{actual}}} \left(a_{water}\right)^{n_{hydratation}/2}$$

▸ and finally take into account that the water activity cannot be considered as equal to 1:

$$a_{water} \approx x_{free\,water} = \frac{n_{free\,water}}{n_{free\,water} + n_+ + n_-} = \frac{1 - \frac{18}{1000}\, n_{hydratation}\, m_{apparent}}{1 - \frac{18}{1000}\,(n_{hydratation} - 2)\, m_{apparent}}$$

These last two adjustments give the following equation for the mean actual activity coefficient of the electrolyte:

$$\gamma_{\pm_{actual}} = \gamma_{\pm_{apparent}} \frac{\left(1 - \frac{18}{1000} n_{\text{hydratation}} \, m_{\text{apparent}}\right)^{\left(1 + n_{\text{hydratation}}/2\right)}}{\left(1 - \frac{18}{1000} (n_{\text{hydratation}} - 2) m_{\text{apparent}}\right)^{\left(n_{\text{hydratation}}/2\right)}}$$

Figure 3.3 shows how the ions hydration phenomenon works on experimental data relating to aqueous electrolytes containing lithium chloride in variable amounts, and assuming that the hydration number is equal to 7.

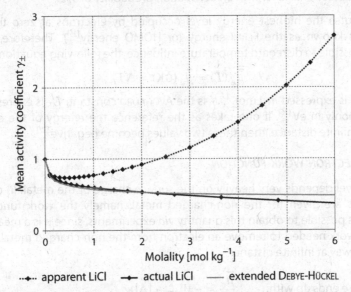

Figure 3.3 - Variations of the apparent and actual mean activity coefficients
as a function of the apparent molality for aqueous solutions of LiCl

Also shown here is the curve calculated from the extended DEBYE-HÜCKEL law, having taken into account the experimental density variations in the concentrated solutions (i.e., the relationship between molality and concentration). This figure shows how, if one applies the correct terms to define concentration in highly concentrated media, then the experimental mean activity coefficients are not too far off those prescribed by the extended DEBYE-HÜCKEL law. There is no minimum on the curve with actual values, and the mean activity coefficient never exceeds 1.

3.2.2 - METALLIC ELECTRODE

3.2.2.1 - ELECTROCHEMICAL POTENTIAL

Electron activity is considered to be constant in the bulk of a metal. This is a realistic assumption, given that there is a large quantity of charge carriers present, i.e., of mobile electrons.

Therefore, by setting the activity of the electrons as equal to 1, their standard state is defined as their actual state in the metal:

$$\mu_e = \mu^\circ_e$$

thus:

$$\widetilde{\mu}_e = \mu^\circ_e - \mathscr{F}\varphi$$

3.2.2.2 - FERMI'S ENERGY

The FERMI-DIRAC statistics dictate that the electrochemical potential of electrons corresponds to the electronic level with an occupation probability of 1/2.

This constitutes the highest energy level occupied by electrons at zero temperature (0 K), and is known as the FERMI energy (or HOMO energy[25]). Therefore, since it is generally possible to disregard temperature influence, the following equation emerges:

$$\widetilde{\mu}_e(T) \approx \widetilde{\mu}_e(0K) = \mathcal{N}E_F$$

whereby $\widetilde{\mu}_e$ is expressed in J mol^{-1}, $\mathcal{N}$ is the AVOGADRO constant, E_F is expressed in J, or more commonly in eV[26]. If one takes as the reference the energy of the electrons in vacuum at infinite distance, then these two values become negative[27].

3.2.2.3 - ELECTRON WORK FUNCTION

The FERMI level depends very heavily on the surface charge of the metal. In data tables, these values are given for the non-charged metal: namely the work function of an electron. It is possible to obtain this quantity *via* experiments, since it is a measure of the amount of work needed to remove an electron from the non-charged metal in vacuum and take it away at infinite distance.

One therefore ends up with: $$\frac{\widetilde{\mu}_e}{\mathcal{N}} = -W_{extr} - |e|\psi$$

with ψ as the VOLTA potential or external potential,

or: $$W_{extr} = -\frac{\mu^\circ_e}{\mathcal{N}} + |e|\chi$$

with χ as the surface electric voltage, representing the difference between the GALVANI potential and the VOLTA potential ($\varphi = \psi + \chi$).

W_{extr} is a positive quantity, expressed in J or more commonly in eV.

In addition, one must remember that:

▸ E_F, W_{extr}, ψ, $\widetilde{\mu}_e$ are experimentally accessible quantities,

▸ φ, μ°_e, χ are not experimentally accessible quantities.

---

[25] HOMO is the acronym for Highest Occupied Molecular Orbital.

[26] 1 eV = 1.6×10^{-19} J.

[27] The typical reference used in solid state physics is the energy level of the fundamental state, i.e., a quantum state with the lowest energy. So, the FERMI energy values of different metals are usually positive in data tables.

As for the metals most often used in electrochemistry (namely silver, copper or zinc) the electron work function values are 4 to 5 eV. For single-crystals this value is influenced by the particular type of crystalline face. This is hardly surprising given that the metal's surface properties affect the electric surface voltage χ. The differences that can be observed between different types of crystalline faces vary by about 0.1 eV.

3.3 - THERMODYNAMIC EQUILIBRIUM AT AN INTERFACE

In equilibrium, an interface never provides a seat for the macroscopic displacement of species. In particular, it does not have a current flowing through[28]. These macroscopic properties can relate to a very different set of microscopic situations. In the case of reactive interfaces (where one or more reactions can occur), here the equilibrium state is a dynamic phenomenon whereby, on a microscopic scale, there are more or less species reacting at the surface. In this instance therefore, the equilibrium corresponds to a situation whereby at any given moment, there is an equal number of species reacting in one direction as in the other. Let us take the example of the transfer reaction of a species from one phase to another[29]. Here the equilibrium is reached whenever there are exactly as many species crossing the interface in one direction as in the opposite direction, at any given moment. On the other hand, when an interface is non-reactive, there is a total absence of microscopic movement across the interface at all times, and in particular in thermodynamic equilibrium. In this instance the thermodynamic equilibrium is not written in the same manner, as will be explained in detail in the next section.

3.3.1 - THERMODYNAMIC EQUILIBRIUM AT A NON-REACTIVE INTERFACE

In the case of a non-reactive interface, it is not possible to reach thermodynamic equilibrium *via* mass exchange. Species accumulate on both sides of the interface, subsequently creating a surface composition that minimizes the energy level. However this phenomenon is necessarily transient. The same arguments apply to the thermodynamics of surfaces, which are for example used to define oil/water emulsions, soap bubbles, etc. Here, a surface energy term is brought into play, which is based on the surface tension and surface excess concentrations.

When an electrochemical interface is non-reactive it is called an ideally polarisable interface or blocking electrode. This situation is generally only found in a limited potential range. Moreover, while the thermodynamic equilibrium is being established, charged species can be transiently seen to accumulate on both sides of the interface, while the overall interface remains neutral. There is therefore a transient current reflecting the variation in the surface charge excesses on both sides, depending on the voltage across

---

[28] Remember that the opposite is not true: an interface with no current flowing through it is not necessarily in equilibrium (see introduction to chapter 3).

[29] Let us recall (see section 2.2.1.2) that we have chosen to look at interfacial exchange phenomena as reactive processes, and therefore not use the terms that differentiate between permeable and impermeable interfaces: all interfaces here are considered as impermeable, with or without reactive processes.

the interface. The interfacial zone behaves like a capacitor and is called the electro-chemical double layer.

This phenomenon is identical to the one found in n/p, p/n or SCHOTTKY's diodes, where it is called the space charge phenomenon. The GOUY-CHAPMAN-STERN model is the name given to the detailed description of the double layer structure, when dealing with a simple, ideally polarisable interface[30]. This theory is based on principles that are relatively close to principles behind the DEBYE-HÜCKEL model. It will be not described in detail here. It essentially boils down to writing that the electrochemical potential of each species is constant in the phase where it is present, which in turn creates a link between the potential and concentration profiles of charged species in the interfacial zone. This is a satisfactory way of accounting for the experimental values of double layer capacitances in a typical setup. Based on this theory, the electrochemical double layer consists of two zones, hence its name. The first one, located next to the electrode, is called the compact or HELMHOLTZ layer. Its thickness L_H is about a few Å, which constitutes the negligible distance at which solvated ions can approach the surface. The potential profile is linear inside this layer. The second one is called the diffuse layer[31], and the order of magnitude of its thickness is the DEBYE length L_D [32]. Inside this second layer, the potential varies more slowly.

Without actually studying the whole mathematical equation of the potential profile in the double layer, one can remember that when the potential difference between the terminals of the diffuse double layer is moderate, then the variations inside the latter are close to an exponential function[33]:

$$\varphi - \varphi_{solution} = \text{Cst } e^{-\frac{x-L_H}{L_D}}$$

Figure 3.4 shows the shape of the potential variations, as a function of the distance from the interface on the double layer thickness scale.

For concentrated solutions (for instance those involving a supporting electrolyte with a concentration higher than 0.1 mol L^{-1}), the DEBYE length is less than one nanometer and one can consider the double layer as being reduced to the compact layer. For very dilute solutions (for instance with an ionic strength of about 10^{-5} mol L^{-1}), the DEBYE length is about 0.1 μm and the diffuse layer represents most of the volume in the double layer. It should be remembered that the electrochemical double layer is typically about a few nanometers in thickness.

---

[30] It is more complicated to describe the interfacial zone if specific adsorption phenomena are involved. The description often follows a model using several capacitors in series or in parallel.

[31] This terminology is rather awkward because it may cause confusion with the diffusion layer. However it is commonly used to designate this part of the double layer.

[32] The DEBYE length is defined in mathematical terms in appendix A.3.3 which describes the DEBYE-HÜCKEL theory. Some orders of magnitude are given later in the text.

[33] Typically, potential profiles of the exponential type are found in the diffuse layer, when the overall interfacial voltage has values of about 50 mV for a 1–1 electrolyte. This applies to numerous experimental situations.

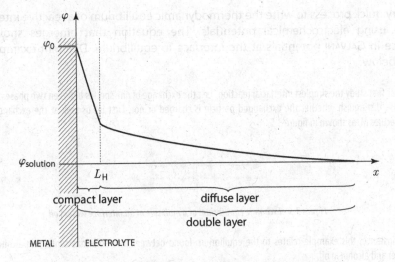

Figure 3.4 - Potential profile in the electrochemical double layer

This book will make little reference to phenomena relating to this double layer. Yet a number of applications in electrochemistry, such as electrophoresis, electroosmosis, supercapacitors, etc. are rooted in these same phenomena.

In the following section we will focus on reactive interfaces. Here again, thermodynamic equilibrium corresponds to particular potential profiles in the interfacial zone. Although double layer phenomena do indeed also come into play for these reactive interfaces, here our analysis will be carried out on a much broader scale than merely the double layer alone. Therefore the potential profile appears simply as flat in both extreme phases, showing discontinuity when crossing the interface. This discontinuity, this potential difference between the two phases, is what thermodynamics connects up to the system's composition, as explained below.

3.3.2 - THERMODYNAMIC EQUILIBRIUM AT A REACTIVE INTERFACE

By analogy with the case of chemical reactions in volumes, the state of thermodynamic electrochemical equilibrium corresponds to a zero value for the electrochemical GIBBS energy of reaction:

$$\forall r \quad \Delta_r \widetilde{G} = \sum_i \nu_{i,r} \widetilde{\mu}_i = 0$$

When dealing with the particular case of reactive phenomena at an interface, the species in question (reactants or products) are not all in the same phase. This has a significant impact in electrochemistry because the species involved in the reaction are not subjected to the same electric potential. Therefore, although the charge balance in the chemical reaction is zero, as soon as charged species are involved then the chemical GIBBS energy of reaction is not zero:

$$\Delta_r \widetilde{G} = 0 \quad \text{but} \quad \Delta_r G \neq 0$$

It is a very quick process to write the thermodynamic equilibrium of a reactive interface if you are using electrochemical potentials. The equation that emerges shows the difference in GALVANI potentials at the interface in equilibrium. Different examples are studied below.

▶ Let us first study the simplest interfacial reaction, i.e., the exchange of one species between two phases. One has to distinguish whether the exchanged particle is charged or not. First let us look at the exchange of molecules M, as shown in figure 3.5.

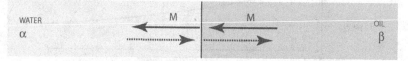

Figure 3.5 - Exchange equilibrium of a molecule at an interface water | oil

For instance, this example relates to the equilibrium found between two solutions containing alcohol in water and alcohol in oil.

The surface reaction is the following:
$$M \rightleftharpoons M$$
$$\alpha \qquad \beta$$

The thermodynamic equilibrium results in the following equation:
$$\Delta_r \widetilde{G} = \widetilde{\mu}_{M_\beta} - \widetilde{\mu}_{M_\alpha} = 0$$

or
$$\widetilde{\mu}_{M_\beta} = \widetilde{\mu}_{M_\alpha}$$

and since it is a molecule (neutral species):
$$\mu_{M_\beta} = \mu_{M_\alpha}$$

▶ Secondly, let us now consider the exchange of one particular charged species, assuming that it is the only reactive phenomenon at the interface. This is the case, for example, of an AgCl solid crystal in equilibrium with an aqueous solution containing Ag^+ ions, as shown in figure 3.6.

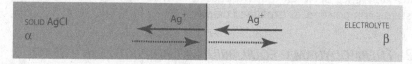

Figure 3.6 - Exchange equilibrium of Ag^+ ions between a crystal of AgCl and an aqueous solution

The surface reaction is the following:
$$Ag^+ \rightleftharpoons Ag^+$$
$$\alpha \qquad \beta$$

The thermodynamic equilibrium results in the following equation:
$$\Delta_r \widetilde{G} = \widetilde{\mu}_{Ag^+_\beta} - \widetilde{\mu}_{Ag^+_\alpha} = 0$$

or, since Ag^+ is a cation:
$$\mu_{Ag^+_\beta} + \mathscr{F} \varphi_\beta = \mu_{Ag^+_\alpha} + \mathscr{F} \varphi_\alpha$$

The difference in GALVANI potentials between the two phases is therefore:
$$\varphi_\beta - \varphi_\alpha = \frac{1}{\mathscr{F}} \left(\mu_{Ag^+_\alpha} - \mu_{Ag^+_\beta} \right)$$

▶ Now let us take the example of the phase transfer reaction between an aqueous solution containing potassium chloride and a solution containing crown ether L (a complexing agent of K^+ ions) in an organic solvent, as shown in figure 3.7[34].

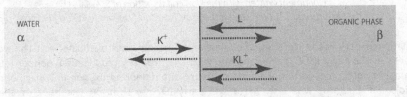

Figure 3.7 - Complexation equilibrium at a water | organic phase interface

The surface reaction is the following: $K^+ + L \rightleftharpoons KL^+$
                                              $\alpha \quad \beta \qquad \beta$

The thermodynamic equilibrium results in the following equation:

$$\widetilde{\mu}_{L_\beta} + \widetilde{\mu}_{K^+_\alpha} = \widetilde{\mu}_{KL^+_\beta}$$

or:
$$\mu_{L_\beta} + \mu_{K^+_\alpha} + \mathscr{F}\varphi_\alpha = \mu_{KL^+_\beta} + \mathscr{F}\varphi_\beta$$

The difference in internal potentials (or GALVANI) between the two phases is therefore:

$$\varphi_\alpha - \varphi_\beta = \frac{1}{\mathscr{F}}\left(\mu_{KL^+_\beta} - \mu_{K^+_\alpha} - \mu_{L_\beta}\right)$$

▶ Finally, let us consider the example of the redox equilibrium between ferric and ferrous ions at a platinum electrode, as shown in figure 3.8.

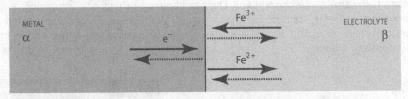

Figure 3.8 - Redox equilibrium at a metal | electrolyte interface

The surface reaction is the following: $Fe^{2+} \rightleftharpoons e^- + Fe^{3+}$
                                             $\beta \qquad \alpha \quad \beta$

The thermodynamic equilibrium results in the following equation:

$$\widetilde{\mu}_{e_\alpha} + \widetilde{\mu}_{Fe^{3+}_\beta} = \widetilde{\mu}_{Fe^{2+}_\beta}$$

or :
$$\mu_{e_\alpha} - \mathscr{F}\varphi_\alpha + \mu_{Fe^{3+}_\beta} + 3\mathscr{F}\varphi_\beta = \mu_{Fe^{2+}_\beta} + 2\mathscr{F}\varphi_\beta$$

The difference in internal potentials (or GALVANI) between the two phases is expressed in the following equation:

$$\varphi_\alpha - \varphi_\beta = \frac{1}{\mathscr{F}}\left(\mu_{Fe^{3+}_\beta} + \mu_{e_\alpha} - \mu_{Fe^{2+}_\beta}\right)$$

---

[34] *This kind of interface is found in some specific electrodes with a liquid (or polymer) membrane containing complexes. Valinomycin is a complexing compound which is selective to K+ ions. It is used in membranes of selective electrodes based on PVC.*

ELECTROCHEMISTRY AND NEUROBIOLOGY

Document written with the kind collaboration of S. ARBAULT,
of the Department of Chemistry, at the Ecole Normale Supérieure in Paris, France

Electrochemistry made its first contribution to biology and medicine with the works of L. CLARK in the early 1950's. He came up with the first electrochemical sensor based on an amperometric method in order to measure out dissolved oxygen in water, especially in biological fluids. It was only after the early 1970's that on the back of this type of device, other electrochemical biosensors were developed (specifically for detecting glucose, or chemical neurotransmitters). In the case of neurobiology, these developments reflected the need to understand how mechanisms worked for the chemical transmission of neuronal information, a field of study which has been in evidence since the late 1960's.

Moreover, the 1970's marked the development of new electrochemical tools, including microelectrodes. These tools quickly found a new field of applications: *in vivo* measurements. Indeed, if you take the case of carbon fibers measuring from 5 to 30 µm in diameter, you can see how microelectrodes share the same dimensions as living cells. They therefore have a slightly invasive effect when inserted into cellular tissue. Fundamental studies have also demonstrated how during analysis the time constants offered by microelectrodes are significantly lower than those offered by conventional millimetre electrodes. When a very high voltammetry scanning rate is applied (several hundreds of volts per second) it is possible to study transient phenomena in the range of milliseconds. Several research groups have developed this type of analysis using carbon microelectrodes to understand the mechanisms behind how dopamine, a major neurotransmitter in the central nervous system, is released. The studies which were pioneered on slices of rodent brain were then able to be carried out *in vivo* on the brains of anaesthetized animals.

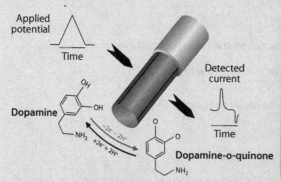 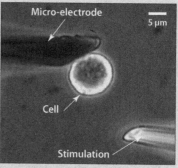

The principle of detecting dopamine at the surface of a carbon fiber microelectrode and the view under a microscope of a detection experiment carried out on a single cell

*On the left, dopamine detection: reprinted from Trends Anal. Chem., **22**, PHILLIPS P.E.M. and WIGHTMAN R.M., Critical guidelines for validation of the selectivity of in-vivo chemical microsensors, 509-514, © 2003, with permission from Elsevier; On the right, microelectrode: AMATORE C., ARBAULT S., BONIFAS I., BOURET Y., ERARD M. and GUILLE M.: Dynamics of full fusion during vesicula exocytotic events: release of adrenaline by chromaffin cells. ChemPhysChem. 2003. **4**. 147-154. © Wiley-VCH Verlag GmbH & Co. KGaA. Reproduced with permission)*

During the 1980-1990's science highlighted and demonstrated the essential neuro-biological mechanisms behind dopamine systems. The research developed by R.M. WIGHTMAM and his team contributed greatly to this domain, in particular in showing the process of reuptake by neurons of dopamine that is released in excess quantities by the synapse. This phenomenon has been quantified, and studied kinetically and pharma-cologically. It also helped to understand the cocaine-induced delay effects on the release of dopamine in the zones of brain related to learning activities, memory and sensations of pleasure.

Thank to recent technical developments, particularly in the field of miniaturizing meas-urement tools and data transfer via wireless communication, tests can now even be carried out on animals which are awake and mobile. Today it is possible to make a link between measurements in brain activity and an animal behaviour when it is stimulated by the presence of a fellow creature of the opposite sex or when it enters a process of drug addiction.

However, these in vivo studies are based on the principle of detection experiments carried out in the brain's extracellular fluid. They therefore do not allow for the mecha-nism of the chemical transmission of nervous stimuli to be analysed. In the early 1990's, several teams were able to study this phenomenon on a living cell by placing the surface of an electrochemical sensor, such as a carbon microelectrode, a few micrometers away from a cell membrane. Currently, when the results of this detection process are analysed in quantitative terms, fluxes of about a thousand molecules per millisecond are shown. Therefore, the various kinetic steps involved in the release process, which is called exocytosis, can be then differentiated and analysed in terms of the biological and physicochemical parameters of the cell and its environment.

Due to its advantages and synergy with other research techniques in membrane processes, this technique has been incorporated in neurobiology laboratories and has been used to study exocytosis of several types of neurotransmitters as well as the release processes of many other molecules of biological interest. These molecules can be neuromodulators (nitrogen monoxide, ascorbic acid, etc.), metabolites of cellular energy (oxygen, ATP, glucose, etc.), or even highly reactive molecules derived from immune defence processes (hydrogen peroxide, peroxynitrite, etc.).

Finally, over and above its significant contribution to the discovery and understanding of fundamental neurobiological mechanisms, this domain of electrochemistry has pos-sibly achieved its greatest success in its use in clinical practice and operating theatres. Today electrochemical sensors are routinely used to investigate cerebral pathology in patients following a stroke or those with other suspected brain abnormalities.

These equations only make links between quantities in equilibrium. For charged species, they link the difference in GALVANI potentials to the chemical compositions of the two phases in equilibrium. Most of the time, these equations are insufficient because they don't make the connection between the system's initial conditions and the equilibrium state that is finally reached. Taken from this point of view, it becomes all the more important to make the distinction between charged species and neutral species, as shown in the following paragraphs.

3.3.3 - THERMODYNAMIC EQUILIBRIUM AT A REACTIVE INTERFACE INVOLVING A SINGLE REACTION BETWEEN NEUTRAL SPECIES

Here we will present the simplest case of molecule exchange between two immiscible phases (see previous figure 3.5). Thermodynamic equilibrium is established from any initial state by means of exchanging molecules through the junction, until the chemical potentials are equal, i.e., by transient molecule fluxes. Take the example of a membrane which only allows molecules M to be exchanged, and which separates two solutions, α and β, with the following initial condition[35]:

$$\mu_{0_\alpha} > \mu_{0_\beta}$$

When the two parts of the system are placed in contact, then the molecules M move from the solution with the highest chemical potential (α) towards the solution with the lowest chemical potential (β), as shown in figure 3.9.

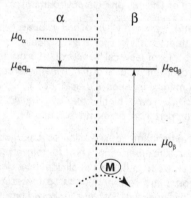

Figure 3.9 – Junction with single exchange of molecule M

In this case, the thermodynamic equilibrium is reached once the molecule being exchanged has equal chemical potential on both sides of the membrane. This generally requires a large number of molecules to be displaced, as well as a large change in the chemical potentials of the species M in both phases between the initial state and the equilibrium state. It may therefore take a long time to reach this equilibrium state. In

---

[35] *Be careful with the different symbols used: μ_{0_α} (initial value) and μ°_α (standard value) should not be confused.*

equilibrium, the following equation indicates the activities ratio between the two phases for the molecule being exchanged [35]:

$$\ln\left(\frac{a_{eq_\alpha}}{a_{eq_\beta}}\right) = \frac{\mu^\circ_\alpha - \mu^\circ_\beta}{RT}$$

In the particular case where both solutions in contact are in the same solvent [36], then the standard chemical potentials are identical in both phases. Here, the thermodynamic equilibrium is then achieved once the solutions are identical on both sides of the membrane, since in equilibrium the activities of the molecule being exchanged must be equal in both phases.

In more general cases the activities ratio in the two media in equilibrium is called the partition coefficient. This can hold very different values (varying especially from 1) depending on the solvents used, as shown in examples in table 3.3.

Table 3.3 - Examples of partition coefficients

Molecule	Solvent	$a_{solvent}/a_{water}$
diiode	CCl_4	10^2
salicylic acid	$CHCl_3$	2
decanoïc acid	$CHCl_3$	2×10^4
oxine	$CHCl_3$	5×10^2

3.3.4 - THERMODYNAMIC EQUILIBRIUM AT A REACTIVE INTERFACE INVOLVING A SINGLE REACTION BETWEEN CHARGED SPECIES

Firstly we will study the basic case of only one charged species being exchanged at the interface between two phases. Then we will go on to examine the most typical case found in electrochemistry of a redox reaction at a metal | electrolyte interface.

3.3.4.1 - JUNCTION WITH THE EXCHANGE OF A SINGLE CHARGED SPECIES

The thermodynamic equilibrium is reached by exchanging charges through the junction, up until the point where the electrochemical potentials of the ion being exchanged are equal on both sides of the junction. This process generates a transient current. For example, imagine a membrane which only allows M^+ cations to be exchanged. As shown in figure 3.10, the membrane separates two solutions, α and β, where M^+ starts out initially with different electrochemical potentials, with the following condition:

$$\widetilde{\mu}_{0_\alpha} > \widetilde{\mu}_{0_\beta}$$

---

[36] Here one can imagine a system with a membrane of minimal thickness, which would prevent the two solutions from being rapidly mixed via convection.

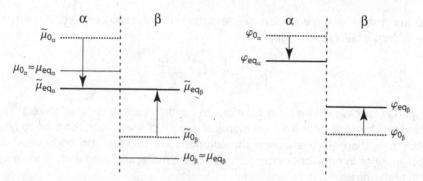

Figure 3.10 - Junction showing the single exchange of a M+ cation from phase α to phase β
Changes in chemical and electrochemical potentials (diagram on the left)
and in GALVANI potentials (diagram on the right) between initial and equilibrium states

When the system is set in contact, M^+ cations can be seen to move from the solution with the highest electrochemical potential (α) towards the solution with the lowest electrochemical potential (β). Given that conducting media can only be charged at the surface, then the charge movement through the junction results in a shift in the surface charge excess of both phases, therefore resulting in a significant variation in electric potentials (internal and external). In thermodynamic equilibrium one can see a difference *a priori* between the GALVANI potentials in each of the two phases. This is called the junction voltage.

It is reasonable to assume [37] that during the mass exchange process between two media, where at the outset there is a difference in the electrochemical potential of the species being exchanged, the process of reaching equilibrium leads to:
▸ *a significant concentration variation between the initial state and the equilibrium state for a molecule exchange (neutral species);*
▸ *a negligible concentration variation for the exchange of a single charged species, yet a significant variation in the junction voltage.*

Above all, one must keep in mind the following significant consequence for thermodynamic laws at such interfaces: for exchanges involving a single charged species, the concentrations, and therefore the chemical potentials, remain unchanged between the system's initial and equilibrium states. Therefore, the junction voltage in equilibrium is expressed as a function of the chemical potentials in the two phases, through the following equation:

$$\varphi_\alpha - \varphi_\beta = \frac{\mu_{i_\beta} - \mu_{i_\alpha}}{z_i \, \mathscr{F}}$$

with the chemical potentials corresponding to the initial compositions of the phases α and β. This junction voltage is generally non-zero. Moreover, from a practical point of view, it is generally reached quite rapidly because to establish this thermodynamic equilibrium very few charges are required to move.

---

[37] *Appendix A.3.4 focuses on examples that give the calculations in the case of fictive systems, though also highlighting this difference in behaviour between molecules and charged species.*

Two particular examples of such single-exchange junctions are given below:

- the metallic junction: here only electrons are exchanged between two different metals at room temperature. An equilibrium state is rapidly reached (about 10^{-18} s) when two different metals are set in contact[38]. In equilibrium state, the FERMI levels are identical, as are the electrons' electrochemical potentials. This results in the following electric junction voltage:

$$\varphi_\alpha - \varphi_\beta = \frac{1}{\mathcal{F}}(\mu_{e_\alpha} - \mu_{e_\beta}) = \text{Cst}$$

  It is not possible to obtain such a junction voltage *via* experiment. The voltage measurement taken with a voltmeter always corresponds to a measurement between two identical metals (the internal terminal connections of the voltmeter). Therefore, for a system with a metallic junction, the voltage measurement always includes at least two electronic junctions, and only the algebraic sum of the two junction voltages is measurable. On a chain of electronic conductors at the same temperature, the voltage measured is therefore always zero. Nevertheless, if the junctions involved in the chain are not at the same temperature, then a non-zero voltage can be measured (generally about 10 µV for a temperature difference of 1 °C): this is the operating principle of a thermocouple.

- the single-exchange ionic junction: an example of this type can be found in the contact made between an electrolyte solution and a barely soluble ionic conducting solid, with a common ion. Equally, another example could even be two electrolyte solutions with different concentrations separated by a membrane that only lets one of the ions through.

  Equilibrium is rapidly reached and the ionic junction voltage is given by:

$$\varphi_\alpha - \varphi_\beta = \frac{1}{z_i \mathcal{F}}(\mu_{i\beta} - \mu_{i\alpha}) = \underbrace{\frac{1}{z_i \mathcal{F}}(\mu^\circ_{i\beta} - \mu^\circ_{i\alpha})}_{\text{Cst}} + \frac{RT}{z_i \mathcal{F}} \ln \frac{a_{i\beta}}{a_{i\alpha}}$$

  The order of magnitude of this type of junction voltage can be relatively high (mV to V). However, one should keep in mind that it cannot be measured directly, just as in the case of electronic junctions.

Now imagine a junction where there is a selective membrane separating two solutions in the same solvent with different concentrations. Given that the standard chemical potentials are equal, then the constant is zero in the equation for the junction voltage, and the voltage is therefore directly linked to the activities ratio of the ion being exchanged between the two solutions.

An example of a single-exchange ionic junction can be found in the contact between a silver chloride ionic solid which is poorly soluble in water (phase α, including two sub-lattices of Ag^+ and Cl^- ions) and an aqueous solution of silver chloride (phase β, containing solvated Ag^+ and Cl^- ions). In fact, even if there are two types of ions common to both phases, only Ag^+ ions are mobile in the solid phase. At thermodynamic equilibrium, the following equation can be written (relating to the Ag^+ exchange reaction):

[38] *The same phenomenon can be seen involving two metals of the same chemical nature but with different structures, such as two silver single-crystal faces: Ag (100) and Ag (110).*

$$\varphi_\alpha - \varphi_\beta = \frac{1}{\mathcal{F}}(\mu^\circ_{Ag^+_\beta} - \mu^\circ_{Ag^+_\alpha}) + \frac{RT}{\mathcal{F}}\ln\frac{a_{Ag^+_\beta}}{a_{Ag^+_\alpha}}$$

Following the definition of standard states in solid AgCl, $a_{Ag^+_\alpha} = a_{Cl^-_\alpha} = a_{AgCl_\alpha} = 1$. The voltage is therefore:

$$\varphi_\alpha - \varphi_\beta = \frac{1}{\mathcal{F}}(\mu^\circ_{Ag^+_\beta} - \mu^\circ_{Ag^+_\alpha}) + \frac{RT}{\mathcal{F}}\ln a_{Ag^+_\beta}$$

The junction voltage can also be presented in an equation as a function of the activity of Cl^- ions in the solution. For these purposes, one uses the solubility equilibrium of AgCl:

$$AgCl_\alpha \rightleftharpoons Ag^+_\beta + Cl^-_\beta$$

$$\ln(a_{Ag^+_\beta} a_{Cl^-_\beta}) = \ln K_{eq} = \frac{\mu^\circ_{AgCl_\alpha} - \mu^\circ_{Ag^+_\beta} - \mu^\circ_{Cl^-_\beta}}{RT} = \frac{\mu^\circ_{Ag^+_\alpha} + \mu^\circ_{Cl^-_\alpha} - \mu^\circ_{Ag^+_\beta} - \mu^\circ_{Cl^-_\beta}}{RT}$$

Combining the previous two equations gives the following:

$$\varphi_\alpha - \varphi_\beta = \frac{1}{\mathcal{F}}(\mu^\circ_{Cl^-_\alpha} - \mu^\circ_{Cl^-_\beta}) - \frac{RT}{\mathcal{F}}\ln(a_{Ag^+_\beta} a_{Cl^-_\beta}) + \frac{RT}{\mathcal{F}}\ln a_{Ag^+_\beta}$$

and therefore in thermodynamic equilibrium, the following is obtained:

$$\varphi_\alpha - \varphi_\beta = \frac{1}{\mathcal{F}}(\mu^\circ_{Cl^-_\alpha} - \mu^\circ_{Cl^-_\beta}) - \frac{RT}{\mathcal{F}}\ln a_{Cl^-_\beta}$$

This last equation is identical to the one that would be obtained if Cl^- ions were exchanged instead of Ag^+ ions [39]. Thermodynamic measurements cannot distinguish these different mechanisms, therefore they do not enable one to determine the nature of the species being exchanged.

The junction voltage of this type of interface (single exchange) is non-zero in equilibrium. Assuming that the activity coefficients of the anions and cations are equal in same phase [40] and taking into account the electroneutrality of the media α and β, then this junction voltage can be written in the following simple equation:

$$\varphi_\alpha - \varphi_\beta = \frac{1}{2\mathcal{F}}(\mu^\circ_{Ag^+_\beta} - \mu^\circ_{Cl^-_\beta} - \mu^\circ_{Ag^+_\alpha} + \mu^\circ_{Cl^-_\alpha})$$

3.3.4.2 - REACTIVE ELECTROCHEMICAL INTERFACE WITH A SINGLE REACTION

The same reasoning can be applied to the case of an electrochemical interface, i.e., for a metallic phase in contact with an electrolyte containing electroactive species that are able to perform an electrochemical reaction at the interface. Furthermore, just as in the case of single-exchange junctions described above, here we will assume that the process of reaching equilibrium at a reactive electrochemical interface, whatever its initial state, only leads to negligible concentration variations [41]. In other words a very small quantity of species is transformed, yet a significant variation can be seen in the interfacial voltage.

---

[39] Section 3.4.2.2 outlines other examples of thermodynamic equilibrium equations for this system. Moreover, keep in mind that the GALVANI voltage between the two sides of an interface cannot be obtained in an experiment. The activity of a single ion appears in the equation, and is also a non-measurable quantity (see appendix A.3.2).

[40] Assuming that the cation and anion activity coefficients are equal, which is in the case particular in the context of the DEBYE-HÜCKEL theory, simplifies the calculation yet without making the reasoning any less general.

[41] Appendix A.3.4 highlights this property using the example of calculations on fictive systems.

Let us remember the following main consequence at these interfaces according to thermodynamic laws: the concentrations and therefore the chemical potentials are not modified by the process of reaching equilibrium.

Therefore, for example, if an inert electrode is immersed in an aqueous solution containing Fe^{2+} and Fe^{3+} ions, the potential difference between the metal and the solution spontaneously adopts the value given by thermodynamics. Moreover, in practical terms, equilibrium is generally reached quite rapidly, since very few charges are required to move.

In the case of any redox couple, the equilibrium interfacial voltage is given by the following equation, which is close to the usual expression of the NERNST law:

$$\varphi_{metal} - \varphi_{electrolyte} = \frac{1}{\mathscr{F}}\left(\mu_{e_{metal}} + \frac{1}{\nu_e}\sum_i \nu_i\,\mu_i\right) = Cst + \frac{RT}{\nu_e\,\mathscr{F}}\,ln\prod_i a_i{}^{\nu_i}$$

where ν_i is the algebraic stoichiometric number of species i in the redox half-reaction and the activities are those of the compounds in the initial state. Given that the coefficient ν_e is algebraic in this equation, the sign for the interfacial voltage is not dependent on the direction chosen to write the reaction. The chemical potential of electrons, which is not zero, is included in the constant. As in the case of the previous junctions, it is not possible to directly measure this voltage during an experiment nor is it possible to have tables giving values for the constant for each different type of interface [42]. Seen from this point of view, the equation does not demonstrate the NERNST law (see section 3.4.1.2), which indicates a measurable quantity for a complete electrochemical chain, and not just for a single interface.

3.3.5 - MULTI-REACTIVE JUNCTION OR INTERFACE

When several reactive phenomena co-exist at the same interface, it is necessary to express the equilibrium state by writing that the electrochemical GIBBS energies of reaction are zero for each phenomenon. In particular, keep in mind that when writing the equation for a combination of different equilibria in the form of a single overall equilibrium, this does not provide all the information necessary to give a thorough description of the interface equilibrium, as explained below in the example of multiple junctions.

As for a junction where several species can be exchanged, the ionic junction voltage in equilibrium relates to the differences between the chemical potentials of each charged species, based on the following equations:

$$\forall i \quad \varphi_\alpha - \varphi_\beta = \frac{1}{z_i\,\mathscr{F}}\left(\mu_{i_\beta} - \mu_{i_\alpha}\right)$$

$$= \frac{1}{z_i\,\mathscr{F}}\left(\mu^\circ{}_{i_\beta} - \mu^\circ{}_{i_\alpha}\right) + \frac{RT}{z_i\,\mathscr{F}}\,ln\frac{a_{i_\beta}}{a_{i_\alpha}}$$

$$= Cst_i + \frac{RT}{z_i\,\mathscr{F}}\,ln\frac{a_{i_\beta}}{a_{i_\alpha}}$$

---

[42] *For example, remember that the electrochemical potential of electrons in a metal is a measurable quantity and that their chemical potential is not measurable (see section 3.2.2).*

Therefore, relationships are formed between the chemical potentials of two types of exchanged species, X^{z_i} and Y^{z_j},

$$z_i\,\mu_{j_\beta} - z_j\,\mu_{i_\beta} = z_i\,\mu_{j_\alpha} - z_j\,\mu_{i_\alpha}$$

and corresponding to the following chemical equilibrium:

$$z_i\,Y^{z_j}{}_\alpha + (-z_j)\,X^{z_i}{}_\alpha \;\;\rightleftharpoons\;\; z_i\,Y^{z_j}{}_\beta + (-z_j)\,X^{z_i}{}_\beta$$

Depending on the value of the equilibrium constant, the equilibrium state is more or less different from the initial state, and is reached after a transfer of variable amounts of species. Two extreme examples can be given:

▸ two solutions containing electrolytes with different concentrations in the same solvent, separated by a non-selective membrane;

⊿ Let us imagine the contact between two aqueous solutions containing sodium chloride, each with a different concentration (C_α and C_β). The activity coefficients for the anions and cations in the same medium are set as equal. This is a valid approximation if applying the DEBYE-HÜCKEL theory, as it is a way of simplifying the calculations without distorting the reasoning in any way. The ion activities in each of the two phases are denoted respectively by a_α and a_β.

In thermodynamic equilibrium,
$$\varphi_\alpha - \varphi_\beta = 0 + \frac{RT}{\mathscr{F}}\ln\frac{a_\beta}{a_\alpha} \qquad \text{for Na}^+$$

$$\varphi_\alpha - \varphi_\beta = 0 - \frac{RT}{\mathscr{F}}\ln\frac{a_\beta}{a_\alpha} \qquad \text{for Cl}^-$$

Therefore, when this system is in equilibrium it results in the following equation:

$$\varphi_\alpha = \varphi_\beta \quad \text{and} \quad a_\alpha = a_\beta \qquad\qquad ⊿$$

The equilibrium is only reached once the GALVANI potential and the chemical potentials of each type of species are both identical in the two phases. The junction voltage is therefore zero in equilibrium, which indeed can take a long time to reach.

▸ two solutions containing the same electrolyte in two different solvents.

⊿ For example, consider the interface between two types of polymer electrolytes which are both different in nature, yet contain the same type of ions (e.g. the cation Li$^+$ and the anion bis-(trifluoromethanesulfonyl)-imide written TFSI$^-$). As above, the activity coefficients of the anions and cations in the same medium are set as equal. The ion activities will be denoted by a_α and a_β in each of the two phases.

In thermodynamic equilibrium,

$$\varphi_\alpha - \varphi_\beta = \frac{1}{\mathscr{F}}(\mu^\circ{}_{\text{Li}^+{}_\beta} - \mu^\circ{}_{\text{Li}^+{}_\alpha}) + \frac{RT}{\mathscr{F}}\ln\frac{a_\beta}{a_\alpha} = \frac{1}{\mathscr{F}}(\mu^\circ{}_{\text{TFSI}^-{}_\alpha} - \mu^\circ{}_{\text{TFSI}^-{}_\beta}) - \frac{RT}{\mathscr{F}}\ln\frac{a_\beta}{a_\alpha}$$

so,
$$2RT\ln\frac{a_\beta}{a_\alpha} = \mu^\circ{}_{\text{Li}^+{}_\beta} + \mu^\circ{}_{\text{TFSI}^-{}_\beta} - \mu^\circ{}_{\text{Li}^+{}_\alpha} - \mu^\circ{}_{\text{TFSI}^-{}_\alpha} = \text{Cst}$$

This equation is nothing more than the law of mass action applied to the following equilibrium:

$$\text{Li}^+{}_\alpha + \text{TFSI}^-{}_\alpha \;\;\rightleftharpoons\;\; \text{Li}^+{}_\beta + \text{TFSI}^-{}_\beta$$

and there is a non-zero junction voltage at equilibrium:

$$\varphi_\alpha - \varphi_\beta = \frac{1}{2\mathscr{F}}(\mu^\circ{}_{\text{Li}^+{}_\beta} - \mu^\circ{}_{\text{TFSI}^-{}_\beta} - \mu^\circ{}_{\text{Li}^+{}_\alpha} + \mu^\circ{}_{\text{TFSI}^-{}_\alpha}) \qquad\qquad ⊿$$

In this case, when the multiple junction reaches equilibrium, the latter's junction voltage is not zero.

3.4 - ELECTROCHEMICAL SYSTEMS IN EQUILIBRIUM

The equilibrium state of electrochemical cells at open circuit is described below. This section will therefore be focusing on the equilibrium state of these systems when no current is allowed to be supplied to an external circuit. When dealing with an electrochemical cell that enables electric energy exchange with the environment (for instance when cell terminals are connected by a resistor) the system shifts towards a new equilibrium state, which will not be described here.

Remember that in an electrochemical chain in thermodynamic equilibrium, the conducting volumes are equipotential and the overall equilibrium voltage is the sum of all the potential differences at the interfaces [43]. In order to be rigorous when writing the thermodynamic equations for electrochemical cells in equilibrium, it is particularly vital in this section to remember to include the terminal electronic junctions which are inevitable during experimental measurements using a voltmeter. In fact, one must keep in mind that a voltmeter's internal connections are always made of the same material. The measurement corresponds to the VOLTA potential difference, which is equal to the GALVANI voltage for two identical materials. For example, in a DANIELL cell the voltage measured corresponds to the following:

$$Cu' \,|\, Zn \,|\, ZnSO_4 \text{ aqueous solution } |\,| CuSO_4 \text{ aqueous solution } |\, Cu$$

Numerous applications have been developed in the field of chemical analysis using potentiometric measurements as indicators, including the production of potentiometric sensors and titration devices. In this chapter, we will focus on the defining principles of these potentiometric methods at zero current when these systems are in thermodynamic equilibrium, which is not necessarily true for all potentiometric measurements. In particular, the following description is confined to electrochemical cells with no ionic junction. In practice, these results will also be applied to many experimental cases in which ionic junction voltages can be neglected [44].

3.4.1 - ELECTROCHEMICAL CELLS WITH NO IONIC JUNCTION

This section studies systems without any ionic junction. Depending on the particular redox couples in question, this category may correspond to an experimental system or to a system which cannot be built, yet whose thermodynamic characteristics can still be defined nonetheless.

---

[43] The distribution of potentials in an electrochemical system in equilibrium is shown in section 2.1.1.

[44] Appendix A.3.2 addresses these questions in more thorough detail: even if the ionic junction voltage is negligible in numerical terms, it may nonetheless be important to consider it from a fundamental point of view.

3.4.1.1 - THERMODYNAMIC REACTION QUANTITIES

For simple electrochemical cells, the electromotive force, emf, can be easily linked to the thermodynamic reaction quantities by using the same equations as those previously demonstrated for GALVANI voltages at an interface in equilibrium.

▼ For example for the following electrochemical cell:

$$Cu' \mid Pt, H_2 \mid \text{aqueous solution containing } H_2SO_4 \text{ and } CuSO_4 \mid Cu$$

the equilibrium at both electrochemical interfaces results in:

$$\varphi_{Cu} - \varphi_{solution} = \frac{1}{2\mathscr{F}}\left(2\mu_{e_{Cu}} + \mu_{Cu^{2+}} - \mu_{Cu}\right)$$

$$\varphi_{Pt} - \varphi_{solution} = \frac{1}{\mathscr{F}}\left(\mu_{e_{Pt}} + \mu_{H^+} - \frac{1}{2}\mu_{H_2}\right)$$

The equilibrium at the Cu' | Pt electronic junction equalises the electrochemical potentials of the electrons in both metals ($\widetilde{\mu}_{e_{Pt}} = \widetilde{\mu}_{e_{Cu'}}$) and therefore the previous equation can be written as a function of the parameters of Cu':

$$\varphi_{Cu'} - \varphi_{solution} = \frac{1}{\mathscr{F}}\left(\mu_{e_{Cu'}} + \mu_{H^+} - \frac{1}{2}\mu_{H_2}\right)$$

Here, when a rigorous demonstration is required, one can see quite how important it is to write the complete chain with the terminal electronic junctions. Indeed, since the electrons' chemical potential terms concern connexions that are made of the same metal, these terms, which cannot be obtained *via* experiment, end up cancelling each other out ($\mu°_{e_{Cu}} - \mu°_{e_{Cu'}}$). Consequently, all that remains in the voltage expression is the chemical potentials of chemical compounds.

By writing the emf with Cu as the working electrode and Cu' as the counter-electrode, $U = \varphi_{Cu} - \varphi_{Cu'}$, the following equation is obtained:

$$-2\mathscr{F}U = \mu_{Cu} + 2\mu_{H^+} - \mu_{Cu^{2+}} - \mu_{H_2} = \Delta_r G$$

with the GIBBS energy of reaction written for the following overall reaction:

$$Cu^{2+} + H_2 \;\rightleftharpoons\; Cu + 2H^+$$

Note that the writing convention above is used to clearly define the sign of the quantities in question.   ◢

To take a more general case, let us imagine an electrochemical cell, which may possibly be fictive, where the sum of the ionic junction voltages is zero. This would give the following equations:

$$\Delta_r G = \nu_e \,\mathscr{F}U$$

$$\Delta_r S = -\nu_e \,\mathscr{F}\left(\frac{\partial U}{\partial T}\right)_P$$

$$\Delta_r H = \nu_e \,\mathscr{F}\left[U - T\left(\frac{\partial U}{\partial T}\right)_P\right]$$

whereby ν_e is the algebraic stoichiometric number of electrons in the redox couple's half-reaction at the working electrode, and U is the difference in internal potentials between the working electrode and the counter-electrode, $U = \varphi_{WE} - \varphi_{CE}$.

Note that the previous equations imply that the reaction is written in the direction of the redox half-reaction at the working electrode. Yet this choice is nothing more than a convention to help ensure there are no ambiguities over the signs. This choice in direction does not by any means represent a real reaction. Keep in mind that these relationships are valid in thermodynamic equilibrium, therefore when no current flows. Therefore in particular the terms cathode and anode have no meaning in this case.

When the different species involved in the overall reaction are taken in their standard state, then a standard emf, $U°$, can also be defined:

$$U° = \frac{\Delta_r G°}{\nu_e \mathscr{F}}$$

A typical order of magnitude for standard GIBBS energy of reaction is 100 kJ mol^{-1}. At room temperature, this equates to a standard voltage of about one volt.

Note that for redox couples involving H$^+$ or OH$^-$ in their redox half-reaction, it is vital to specify whether the standard state is set for H$^+$ or OH$^-$. In other words, this means that the relevant pH has to be specified to define $U°$:

$pH=0$ standard conditions for H$^+$ ($a_{H^+} = 1$)

$pH=14$ standard conditions for OH$^-$ ($a_{OH^-} = 1$)

Remember that in thermodynamic data tables (standard GIBBS energies and enthalpies of formation) the following conventions are used:

▸ the standard GIBBS energy of formation (or chemical potential) of all simple elements is zero, at any temperature;

▸ the standard GIBBS energy of formation (or chemical potential) of protons in a solution is zero:

$$\forall T \quad \mu°_{H^+} = 0 \text{ J mol}^{-1}$$

▶ For example, the standard emf of the following cell, at $pH= 0$, can be given using thermodynamic data (assuming the ionic junction voltage is zero [44]):

Pt | Cu | aqueous solution containing Cu^{2+} | | aqueous solution containing Mn^{2+}, MnO$_4^-$ and H$^+$ | Pt′

The thermodynamic data give these values at 25 °C, following the convention $\mu°_{H^+} = 0$ J mol^{-1}:

$$\mu°_{H_2O} = -237.2 \text{ kJ mol}^{-1} \quad \mu°_{MnO_4^-} = -447.3 \text{ kJ mol}^{-1}$$

$$\mu°_{Mn^{2+}} = -228.1 \text{ kJ mol}^{-1} \quad \mu°_{Cu^{2+}} = +65.5 \text{ kJ mol}^{-1}$$

The two redox half-reactions are:

$$5\,e^- + MnO_4^- + 8\,H^+ \rightleftharpoons Mn^{2+} + 4\,H_2O$$
$$(\qquad\qquad Cu \rightleftharpoons Cu^{2+} + 2\,e^- \qquad\qquad) \times 5/2$$

$$\overline{\tfrac{5}{2}Cu + MnO_4^- + 8\,H^+ \rightleftharpoons Mn^{2+} + 4\,H_2O + \tfrac{5}{2}Cu^{2+}}$$

For this equilibrium, by choosing the Pt′ electrode as the working electrode, you get the following equation:

$$\Delta_r G° = -5\mathscr{F}U°$$

Consequently, when taking into account that copper is a simple element ($\mu°_{Cu}=0$ J mol^{-1}), the emf at 25 °C is:

$$U° = -\frac{1}{5\mathscr{F}}\left(\mu°_{Mn^{2+}} + 4\,\mu°_{H_2O} + \frac{5}{2}\mu°_{Cu^{2+}} - \mu°_{MnO_4^-} \right) = +1.17\,V$$

CORROSION OF REINFORCED CONCRETE

Document written with the kind cooperation of E. CHAUVEAU,
R&D engineer for the company Ugitech, based in Ugine, France

The HASSAN II mosque in Casablanca was built in the 1990's. The architect, inspired by a verse from the Koran that says 'God's throne was on water', designed a luxurious building with two thirds of its total surface above the waters of the Atlantic Ocean. The foundations, which were exposed to the effects of sea water and wind, required 26 000 m^3 of concrete and 59 000 m^3 of rip-rap. By 1998, signs of the building degradation began to show mainly related to corrosion of the structures submitted to wave movements and contact with sea water.

Degradation shown on a pillar due to the marine environment (© Christine RAYNAUD - BTPM)

The mechanisms which led to this highly rapid deterioration process are briefly laid out below. Concrete is made up of a mixture of cement, sand and aggregates (stones and gravel). The process which allows concrete to set is caused by the hydration of calcium orthosilicate, based on the following reaction:

$$4\ H_2O + 2\ Ca_2SiO_4 \rightleftharpoons Ca_3Si_2O_7,3\ H_2O + Ca(OH)_2$$

This reaction explains the basic character of concrete. Because concrete has poor mechanical tensile properties, the idea of introducing a mesh of carbon steel bars (called rebars) emerged in 1870. The concrete is separated from each bar by an interface layer of cement, whose porosity can vary mainly depending on the precise composition of the concrete as well as on the water content of the structure. Care is usually taken to leave a minimum thickness of 30 mm of concrete between each steel bar close to the surface and the external surface of the structure. In fact, when concrete is exposed to humid conditions, water penetrates by means of capillarity, and there is a risk that the hydrated area may eventually spread as far as the steel bars. When this occurs, a second mechanism usually comes into play which helps prolong the durability of the structures: due to the chemical properties particular to concrete, the pH in the damp zones is close to 12, meaning the steel bars become coated with a passivating and protecting layer of FeOOH.

The two major causes of the more or less accelerated corrosion of the bars are:

▶ if there are high atmospheric levels of carbon dioxide, the gas diffuses through the porous hydrated zone, thus causing the pH to drop locally. From a pH value of below about 9, the metal protection stops because FeOOH becomes hydrated to form non-protecting Fe(OH)$_3$;

▶ if the surrounding medium is charged with chloride ions (saline water), these ions can also diffuse through to the metal-concrete interface and depassivate the metal on account of forming chlorinated compounds.

These two phenomena can occur simultaneously, but the risk of carbonation is low in the case of submerged concrete. When the bars are depassivated, corrosion caused by dioxygen can occur, triggering reaction products. The swelling caused by these products as they gradually increase in volume, eventually lead to cracks forming, and even causing the concrete to break. Thereafter, the corrosion process may then spread along the reinforced elements, creeping through the interface layer between the metal and concrete, and damaging the building.

The measures most commonly taken to prevent corrosion include opting for compact concrete formulas with low levels of porosity, protecting the steel surfaces through galvanisation, and by introducing soluble anodes (zinc or magnesium). Owing to their lower redox potentials, these sacrificial anodes are more vulnerable to oxidation, thus protecting the steel from corroding. Other techniques are used such as concrete surface metal plating, epoxy resin coatings on steel bars or even choosing stainless steel for the bars. In some situations, the steel bars mesh can be treated using cathodic protection, namely by setting them at a reducing potential.

In the case of the mosque in Casablanca, the deterioration was so rapid that a prompt decision had to be taken regarding repair work to maintain the building. Since 2005, a temporary dyke has been built so as to provide a dry working environment in which to operate (using $200\,000$ m^3 of a mixture of various materials, 2800 concrete tetrapods each with a volume of 6.3 m^3), and entailing the demolition of 8000 m^3 of concrete, the injection of 1200 m^3 of grout and mortar, and above all the development of 10 000 m^3 of high-performance concrete reinforced with 1300 tons of ferrito and austenitic Duplex stainless steel bars (Ugitech). The estimated cost of this work is 50 million Euros, that is about 5% of the original construction costs.

Reconstruction, using high-performance stainless steel reinforced concrete, of the 100 external combs supporting the peripheral slabs. The columns are called 'combs' because they have a wave-breaking effect. In the background you can see the sea and temporary dyke that was built to keep the construction site dry, as well as the tetrapods maintaining the dyke. (© Christine RAYNAUD - BTPM).

3.4.1.2 - NERNST'S LAW

Remember that the potential reference chosen in electrochemistry is the virtual system called the standard hydrogen electrode (SHE):

$$H^+/H_2 \text{ with, } \forall T \begin{cases} a_{H^+} = a_{H_2} = 1 \\ \gamma_{H^+} = \gamma_{H_2} = 1 \end{cases}$$

The virtual nature of this system essentially arises from the fact that molar acidic solutions are non-ideal. Indeed, if one can obtain an acidic solution with a mean activity equal to 1, the mean activity coefficient will be different from 1 nonetheless.

By definition, the potential of a redox couple *vs* SHE is the emf of the fictive electro-chemical cell, whereby the working electrode is in the half-cell involving the redox couple in question. The counter-electrode is the standard hydrogen electrode at the same temperature. The terminals are made of the same metals and the sum of the possible ionic junction voltages are considered to be equal to zero. In this case we therefore have:

$$E_{/SHE} = \varphi_{metal} - \varphi_{metal\,SHE}$$

▼ For example, the potential of the Ag^+/Ag couple *vs* SHE is equal to the emf of the electrochemical cell:

$$Ag', H_2 \,|\, H^+ \,(std) \,|\,|\, Ag^+ \,|\, Ag$$

When writing the equilibrium: $\qquad Ag^+ + \tfrac{1}{2}H_2 \rightleftharpoons Ag + H^+$

we obtain: $\qquad E = -\dfrac{\Delta_r G}{\mathscr{F}} = \dfrac{1}{\mathscr{F}}\left(\mu_{Ag^+} + \dfrac{1}{2}\mu^\circ_{H_2} - \mu^\circ_{H^+} - \mu_{Ag} \right) = \dfrac{\mu_{Ag^+}}{\mathscr{F}}$

and: $\qquad E = E^\circ + \dfrac{RT}{\mathscr{F}} \ln a_{Ag^+}$

with: $\qquad E^\circ = \dfrac{\mu^\circ_{Ag^+}}{\mathscr{F}}$
◢

Therefore, the potential of a redox couple only depends on its own thermodynamic data. More generally, by applying the conventions previously outlined, we end up with the following NERNST law equation:

$$E_{/Ref} = E^\circ_{/Ref} + \dfrac{RT}{v_e \mathscr{F}} \ln \prod_i a_i^{v_i}$$

where i represents each species involved in the redox half-reaction, except for the electrons, v_i is the stoichiometric number of the species i, a_i is its activity in equilibrium and v_e is the electron's stoichiometric number.

Thermodynamic data enable one to predict the direction in which a system will sponta-neously evolve, based on the thermodynamic criteria. When keeping within the bounds of these criteria, it is possible to take such an approach based on a potential scale[45] (applying a similar line of reasoning to that used for acido-basic reactions analysed on a pK_a scale). Presenting the data in this way, using what we call the γ rule, is particularly

---

[45] Just as in the qualitative description of current-potential curves in section 2.3, it is the apparent standard potential, when relevant, that is positioned and used for this type of reasoning.

useful for redox reactions in a solution[46]. However, one can rapidly see the limits of its application when dealing with electrochemical systems because the kinetics of different reactions can upset these predictions. In this case it is therefore better to apply reasonings based on current-potential curves (see sections 2.3 and 2.4), which enables one to visualize the results obtained from a potential scale. However at the same time it also gives a broader view, which immediately incites one to take into account the thermodynamic and kinetic aspects.

▶ Let us take one of the examples previously outlined involving spontaneous evolution, using current-potential curves (see figure 2.33 in section 2.4.1). This example relates to the corrosion of a copper piece by dioxygen, when immersed in an acidic solution which is not deaerated. Three redox couples are involved in the system: Cu^{2+}/Cu, $H^+, H_2O/H_2$ and $O_2/H_2O,OH^-$. They are positioned on the scale of thermodynamic apparent potentials, as shown in figure 3.11. We will assume that the pH of the system is equal to 1: the apparent standard potential of the $H^+,H_2O/H_2$ couple is equal to -0.06 V_{SHE} and that of the $O_2/H_2O,OH^-$ couple is $+1.17$ V_{SHE}. As far as copper is concerned, the apparent standard potential is equal to the standard potential of the Cu^{2+}/Cu couple, i.e., $+0.34$ V_{SHE}. The species present in the system are circled in the diagram (by convention, oxidants are positioned on the left and reductants on the right).

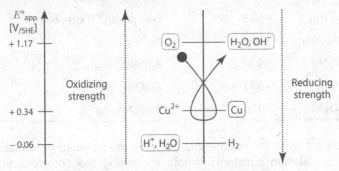

Figure 3.11 - Using a scale of apparent thermodynamic potential

In this type of diagram, one can immediately visualize whether a spontaneous reaction can occur or not: a reaction is possible if an oxidant (O_2) is present and if its couple has an apparent standard potential higher than that of the couple corresponding to the reductant (Cu) which must also be present.

Here therefore we can obtain the result for spontaneous evolution in the system at open circuit. This equates to the balance of both the oxidation and reduction half-reactions which corresponds to the corrosion phenomenon:

$$4\,H^+ + 4\,e^- + O_2 \longrightarrow 2\,H_2O$$
$$(\qquad\qquad Cu \longrightarrow Cu^{2+} + 2\,e^- \qquad\qquad) \times 2$$
$$\overline{O_2 + 4\,H^+ + 2\,Cu \longrightarrow 2\,H_2O + 2\,Cu^{2+}}$$

It must be remembered that this reasoning is based on purely thermodynamic criteria.

[46] The γrule enables one, at a glance, to identify which reactions could spontaneously occur between redox species belonging to different couples. Its name is so called because a minuscule gamma letter is written on the redox potential axis which starts from the Ox₁ species of the couple with the highest potential, and moves towards the Red₂ species of a couple with a lower potential, thus identifying both reactants. Then the two reaction products are found by finishing the γwriting via Ox₂ and ending with Red₁ in the upper right-hand part. The reaction $Ox_1 + Red_2 \rightarrow Red_1 + Ox_2$ obviously requires Ox_1 and Red_2 to actually be present in the system.

3.4.1.3 - CONSIDERING MULTIPLE CHEMICAL EQUILIBRIA

When several chemical equilibria are involved, the equilibrium concentrations depend on all of them, and the redox equilibria can be displaced by other possible reactions (for example, precipitation, complexation or acido-basic reactions). For example, when poorly soluble redox species are involved, they can cause a highly significant change in the corresponding couple's standard potential, as shown in table 3.4. When stable complexes in one (or two) species of a redox couple are involved, then they can also cause a highly significant change in the corresponding standard potential, as shown in table 3.5.

Examples of standard potentials

Table 3.4 - Influence of precipitation

Couple	$E°$ [$V_{/SHE}$]
Ag^+/Ag	+ 0.80
$AgCl/Ag$	+ 0.22
Hg_2^{2+}/Hg	+ 0.79
Hg_2SO_4/Hg	+ 0.61
Hg_2I_2/Hg	− 0.04
Pb^{2+}/Pb	− 0.13
$PbSO_4/Pb$	− 0.36

Table 3.5 - Influence of complexation

Couple	$E°$ [$V_{/SHE}$]
Fe^{3+}/Fe^{2+}	+ 0.77
$Fe(CN)_6^{3-}/Fe(CN)_6^{4-}$	+ 0.36
$Fe(o\text{-}phen)_3^{3+}/Fe(o\text{-}phen)_3^{2+}$	+ 1.14
Ag^+/Ag	+ 0.80
$Ag(NH_3)_2^+/Ag$	+ 0.37
$Au(CN)_2^-/Au$	− 0.60
$Au(SCN)_2^-/Au$	+ 0.70

One can rapidly establish the relationship between the standard potentials of different couples with equilibrium constants simply by writing the corresponding balanced reactions. The NERNST law can also be applied here provided that all the species involved in the solution are considered, however low their concentrations.

For example, the standard potential value of the $Fe(CN)_6^{3-}/Fe(CN)_6^{4-}$ couple can be calculated based on that of the Fe^{3+}/Fe^{2+} couple and on the complexation constants of $Fe(CN)_6^{3-}$ ($pK_{d_1} = 42$) and $Fe(CN)_6^{4-}$ ions ($pK_{d_2} = 35$).

▸ *First method* - When cyanide ions are involved, the major species are the complexes, yet with very low quantities of Fe^{3+} and Fe^{2+} ions in equilibrium. The NERNST law can equally be written for these two couples, giving the same electrode potential:

$$E = E°_{Fe(CN)_6^{3-}/Fe(CN)_6^{4-}} + \frac{RT}{\mathscr{F}} \ln \frac{a_{Fe(CN)_6^{3-}}}{a_{Fe(CN)_6^{4-}}} \quad \text{and} \quad E = E°_{Fe^{3+}/Fe^{2+}} + \frac{RT}{\mathscr{F}} \ln \frac{a_{Fe^{3+}}}{a_{Fe^{2+}}}$$

which leads to:
$$E°_{Fe(CN)_6^{3-}/Fe(CN)_6^{4-}} = E°_{Fe^{3+}/Fe^{2+}} + \frac{RT}{\mathscr{F}} \ln \frac{a_{Fe^{3+}}\,a_{Fe(CN)_6^{4-}}}{a_{Fe^{2+}}\,a_{Fe(CN)_6^{3-}}}$$

$$= E°_{Fe^{3+}/Fe^{2+}} + \frac{RT}{\mathscr{F}} \ln \frac{K_{d_1}(a_{CN^-})^6}{K_{d_2}(a_{CN^-})^6}$$

therefore, at 25 °C: $E°_{Fe(CN)_6^{3-}/Fe(CN)_6^{4-}} = E°_{Fe^{3+}/Fe^{2+}} - 0.059\ pK_{d_1} + 0.059\ pK_{d_2} = + 0.36\,V_{/SHE}$

▶ **Second method** - more elegant and faster (especially if there is a significant number of simultaneous equilibria):

$$Fe^{2+} \rightleftharpoons Fe^{3+} + e^- \qquad \Delta_r G^\circ = \mathscr{F} E^\circ_{Fe^{3+}/Fe^{2+}}$$

$$Fe^{3+} + 6\,CN^- \rightleftharpoons Fe(CN)_6^{3-} \qquad \Delta_r G^\circ = RT \ln K_{d_1}$$

$$Fe(CN)_6^{4-} \rightleftharpoons Fe^{2+} + 6\,CN^- \qquad \Delta_r G^\circ = -RT \ln K_{d_2}$$

$$\overline{Fe(CN)_6^{4-} \rightleftharpoons Fe(CN)_6^{3-} + e^- \qquad \Delta_r G^\circ = \mathscr{F} E^\circ_{Fe(CN)_6^{3-}/Fe(CN)_6^{4-}}}$$

therefore: $E^\circ_{Fe(CN)_6^{3-}/Fe(CN)_6^{4-}} = E^\circ_{Fe^{3+}/Fe^{2+}} - 0.059\,pK_{d_1} + 0.059\,pK_{d_2} = +0.36\,V_{/SHE}$ ◀

3.4.1.4 - PARTICULAR CASES INVOLVING ACIDO-BASIC EQUILIBRIA

It is very common for acido-basic equilibria, and therefore for the pH of electrolytes, to have an influence on potential values. In thermodynamic equilibrium, this influence can be depicted *via* various diagrams, usually called POURBAIX diagrams or E/pH diagrams.

These diagrams can be particularly useful in predicting the thermodynamic behaviour of systems involving several redox couples, each depending on pH in a different way. Obviously this study is focused on redox chemistry in a solution. However, this issue also has an important role in electrochemistry, for example in corrosion or when predicting reactions (main and parasitic) using apparent standard potentials (see section 2.4). Of course these diagrams only give information on a thermodynamic point of view. A system can only be fully described if its kinetic parameters are also included (see section 2.3). Here we will not go into detail about the plotting and use of such diagrams, as this can be found in many other books. Only one example is outlined briefly below.

◤ In the Cl(+I)/Cl(−I) system, you can find the chlorine species with the oxidation number (+I) in the form of HClO or ClO⁻, depending on the medium's pH. Therefore, depending on the predominance areas for the acidic and basic forms of Cl(+I), which themselves are governed by the pH, the following two couples emerge:

▶ HClO/Cl⁻ couple: $\qquad Cl^- + H_2O \rightleftharpoons HClO + H^+ + 2\,e^-$

Applying the NERNST law to this couple gives the following equation:

$$E = E^\circ_{HClO/Cl^-} + \frac{RT}{2\mathscr{F}} \ln \frac{[HClO][H^+]}{[Cl^-]}$$

which at room temperature, and when following the graph conventions of the POURBAIX diagrams (here [HClO] = [Cl⁻]), therefore gives:

$$E = E^\circ_{HClO/Cl^-} - 0.03\,pH = (1.5 - 0.03\,pH)\,V_{/SHE}$$

▶ ClO⁻/Cl⁻ couple: $\qquad Cl^- + H_2O \rightleftharpoons ClO^- + 2H^+ + 2\,e^-$

Applying the NERNST law to this new couple gives the following equation:

$$E = E^\circ_{ClO^-/Cl^-} + \frac{RT}{2\mathscr{F}} \ln \frac{[ClO^-][H^+]^2}{[Cl^-]}$$

when following the graph conventions of the POURBAIX diagrams (here [ClO⁻] = [Cl⁻]) and with the continuity of E at $pH = pK_a$, one obtains:

$$E = E^\circ_{HClO/Cl^-} + 0.03\,pK_a - 0.06\,pH$$

Figure 3.12 shows the shape of the POURBAIX diagram as applied to the Cl(+I)/Cl(−I) system. In the same pH range, the two straight lines symbolizing the water redox couples are also indicated using grey lines.

One example of when the POURBAIX diagram is applied involves studying the stability domains of aqueous solutions. To illustrate, the diagram below clearly shows that JAVEL water (equimolar mixture of Cl⁻ and ClO⁻ ions with a pH close to 10) is not a thermodynamically stable solution since its redox potential is higher than that of the $O_2/H_2O,OH^-$ couple. However, given the slow kinetics of water oxidation by ClO⁻ hypochlorite ions, the JAVEL water remains relatively stable on a usual time scale.

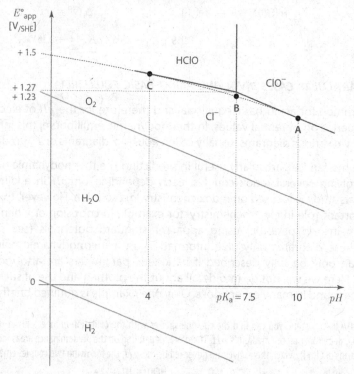

Figure 3.12 - The POURBAIX diagram of chlorine in aqueous solution, restricted to the Cl(+I)/Cl(−I) system

The POURBAIX diagrams are also used in electrochemistry to determine the curve depicting the change in open-circuit potential of an inert electrode (for example a platinum electrode) immersed in a dearated solution of JAVEL water when sulphuric acid is gradually added to this solution.

To give an idea of values, let us take a JAVEL water solution with an initial concentration of 0.5 mol L⁻¹ in ClO⁻ and Cl⁻ ions, at $pH = 10$. One can assume that it does not react with water, since the kinetics involved is extremely slow.

▸ This solution is symbolized on the POURBAIX diagram in figure 3.12 by point A ($pH = 10$; $E = 1.12$ V/SHE). ClO⁻ and Cl⁻ ions are the predominant chlorine species in the solution. From the pK_a value, the concentration in HClO (minority) can be calculated: 1.6×10^{-3} mol L⁻¹.

When the pH is lowered by adding sulfuric acid, then the following reaction is first seen:

$$ClO^- + H^+ \rightleftharpoons HClO$$

When the ClO⁻ concentration gradually decreases, then the corresponding point moves along a curve slightly below the straight line of the POURBAIX diagram, since this straight line represents the potential whenever Cl⁻ and ClO⁻ are equimolar (which is no longer the case).

▸ At point B, i.e., when the pH is equal to pK_a of the acido-basic HClO/ClO$^-$ couple ($pH = 7.5$), these two species share equal concentration. The potential is $E = 1.26$ V$_{/SHE}$. HClO, ClO$^-$ and Cl$^-$ are the major chlorine species in the solution (with a Cl$^-$ concentration of 0.5 mol L^{-1} and HClO and ClO$^-$ concentrations of 0.25 mol L^{-1}).

▸ While the pH is still decreasing (between B and C), again due to the previous main reaction, ClO$^-$ becomes the minor species compared to HClO. For example at point C at $pH = 4$ ($E = 1.38$ V$_{/SHE}$), HClO and Cl$^-$ are the major species of the chlorine element in the solution (concentration of 0.5 mol L^{-1}). The concentration of minor species can therefore be calculated: ClO$^-$ has a concentration of 2×10^{-4} mol L^{-1}.

If the pH decreases even further then other chlorinated species will appear, such as Cl$_2$ gas, a phenomenon which is not addressed here.

3.4.2 - EXPERIMENTAL ASPECTS

Remember that for electrochemical measurements, voltmeters with field effect transistors which have particularly high input impedance are used[47]. When taking a voltage measurement, one can identify two vital but still insufficient experimental criteria of equilibrium:

▸ no evolution over time in the measured voltage,

▸ the voltage measured is independent of the system's hydrodynamic conditions (mechanical stirring).

3.4.2.1 - IONIC JUNCTIONS

In general, the voltages of ionic junctions in equilibrium are not zero and must be taken into account in the overall voltage value. However, in many typical experimental cases (e.g., the junction between two solutions through a salt bridge), the junctions involved are multiple ionic junctions for which the sum of junction voltages plays an insignificant role in quasi-steady state conditions[48]. For such cases, the voltage measured at the system's terminals is very close to the algebraic sum of the equilibrium voltages of the electrochemical interfaces[49].

3.4.2.2 - REFERENCE ELECTRODES

In experimental conditions any redox couple can be used to create a reference electrode, as long as its composition is perfectly monitored and no redox half-reaction

---

[47] The input impedance is typically higher than 10^{12} Ω for instruments used in electrochemistry.

[48] In the case of solutions separated by a membrane or a porous material, the equilibrium state corresponds to a perfect mixture of the two solutions. However, it takes a very long time to reach this state, if the porosity and/or the geometric configuration of the interfacial zone are well chosen (see section 4.4.2). In such a case, the key phenomenon to consider therefore is the fact that the quasi-steady state is rapidly reached. This latter state has different solutions compositions and non-zero junction voltages. From an experimental point of view, the aim is to minimize this quasi-steady-state ionic junction voltages. Different examples are outlined in appendix A.1.1.

[49] Appendix A.3.2 addresses these points in more thorough detail: even if the ionic junction voltage is negligible in numerical terms, it may nonetheless be important to consider it from a fundamental point of view.

kinetics intervenes[50]. Indeed, the corresponding electrochemical interface should reach thermodynamic equilibrium rapidly on the measurement's time scale which is typically much less than one second.

Two other interesting criteria may come into play:

▸ the sum of the voltages of the ionic junctions with the other half-cell is negligible;

▸ the reference half-cell's potential remains unchanged if accidental current has flown through this electrode[51].

Here we will briefly summarise three typical examples which have already been laid out in section 1.5.1.2.

▸▸ Hydrogen electrode (HE)

The hydrogen electrode is made by bubbling dihydrogen into a solution with a known pH, on a platinum electrode (or platinised platinum[52]), corresponding to the following half-cell:

$$\text{Platinised Pt, } H_2 \text{ (1 bar) } | \, H^+ \text{ of concentration } C |$$

Therefore, the half-cell involves the following H^+/H_2 couple:

$$H_2 \rightleftharpoons 2\,H^+ + 2\,e^-$$

In equilibrium, the NERNST law can be used to express the half-cell's potential with the following equation:

$$E_{/\text{SHE}} = \frac{RT}{\mathscr{F}} \ln \frac{a_{H^+}}{\sqrt{a_{H_2}}}$$

Remember that even if a molar concentration of a strong mono-acid is chosen (the HE is called a NHE in this instance), then this reference electrode is not a SHE because the real compounds are not in their standard state. Figure 3.13 shows this difference on the mean activity coefficient of hydrogen chloride acidic solutions. In practice, if one wishes to use a hydrogen electrode whose potential is as close as possible to that of the SHE, then one would use an acidic solution whose concentration is slightly higher than 1 mol L^{-1}. For example, if you take the case of HCl, a concentration equal to 1.2 mol L^{-1} is chosen because of a mean activity equal to $1.2 \times 0.84 \approx 1$ (0.84 is the mean activity coefficient found in figure 3.13). These values all depend on the system's temperature.

Platinised platinum is used to make the reaction very fast. In practice, the gaseous dihydrogen, the dihydrogen dissolved in the solution and the hydrogen adsorbed at the electrode surface must also reach a state in which they are in equilibrium. In experimental conditions this is not an easy situation to manage, and therefore it is tricky to use such an electrode as reference.

---

[50] Here one often refers to a fast or reversible couple, which is defined in precise detail in sections 4.3.2.5 and 6.

[51] This can be achieved, for example, with a SCE. If a current flows through the electrode, then chloride ions are either consumed or produced at the electrode, but the solid potassium chloride keeps the chloride concentration constant in the saturated solution.

[52] Note [66] of section 1.5.1.2 gives a rapid definition of the characteristics and interest of using platinised platinum in electrochemistry.

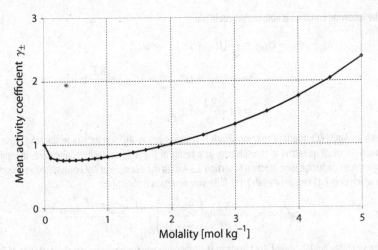

Figure 3.13 - Mean activity coefficient at 25 °C of hydrogen chloride acid in aqueous solutions
at different concentrations

▸▸ Silver chloride electrode

The silver chloride electrode is made up of a silver wire coated with silver chloride,
immersed in an aqueous solution of potassium chloride. Usually a solution is used with a
high concentration of chloride ions (from 1 to 3 mol L^{-1}), yet with levels below saturation
in KCl. This is done to prevent silver chlorocomplexes (AgCl$_2^-$ and AgCl$_3^{2-}$) from forming,
as this would consume AgCl and chloride ions, and in turn fix a different potential value
from the value originally expected.

The half-cell's electrochemical chain is therefore the following:

$$\text{Ag} \mid \text{AgCl} \mid \text{KCl aqueous solution of concentration } C \mid$$

AgCl is a pure separate phase (with an activity equal to 1). It is an ionic solid where only
Ag$^+$ ions are mobile.

The following equation gives the overall equilibrium developing in this half-cell,
involving the AgCl/Ag couple:

$$\text{Ag} + \text{Cl}^- \rightleftharpoons \text{AgCl} + e^-$$

At equilibrium, the NERNST law can be used to express the half-cell's potential with the
following equation:

$$E = E°_{\text{AgCl/Ag}} - \frac{RT}{\mathscr{F}} \ln a_{\text{Cl}^-}$$

▰ It may be interesting to focus in greater detail on the structure of the half-cell, which involves several interfaces
all in equilibrium. The potential expression as indicated above should be applicable to all cases.

▸ The first way of examining the half-cell's equilibria is to develop it as follows:

$$\text{Ag} \mid \text{AgCl} \mid \text{KCl aqueous solution of concentration } C \mid$$

▸▸ with an Ag | AgCl interface which is reactive (see section 3.3.4) :

$$\text{Ag} \rightleftharpoons \text{Ag}^+_{\text{AgCl}} + e^-$$

where Ag$^+_{\text{AgCl}}$ is an Ag$^+$ ion mobile in the AgCl solid phase.

The following equation is obtained at equilibrium:

$$\varphi_{Ag} - \varphi_{AgCl} = \frac{1}{\mathscr{F}}(\mu_{e_{Ag}} + \mu_{Ag^+_{AgCl}} - \mu_{Ag})$$

$$= \frac{1}{\mathscr{F}}(\mu_{e_{Ag}} + \mu^\circ_{Ag^+_{AgCl}} - \mu^\circ_{Ag}) + \frac{RT}{\mathscr{F}} \ln a_{Ag^+_{AgCl}}$$

$$= Cst + \frac{RT}{\mathscr{F}} \ln a_{Ag^+_{AgCl}}$$

➤➤ with an AgCl | KCl interface solution which is reactive due to Ag^+ ion exchange since Ag^+ ions can be found (in small quantity) in the solution, as a result of the AgCl solubility equilibrium. In equilibrium (see single-exchange ionic junction in section 3.3.4 with the example of AgCl considered as an ionic solid in which only Ag^+ ions are mobile), the following equation is obtained:

$$\varphi_{sol} - \varphi_{AgCl} = \frac{RT}{\mathscr{F}} \ln\left(\frac{a_{Ag^+_{AgCl}}}{a_{Ag^+_{sol}}}\right) + \frac{\mu^\circ_{Ag^+_{AgCl}} - \mu^\circ_{Ag^+_{sol}}}{\mathscr{F}}$$

The solubility equilibrium of AgCl leads to the following equation which links the activities of Ag^+ and Cl^- in the solution: $AgCl \rightleftharpoons Ag^+_{sol} + Cl^-_{sol}$

$$\ln(a_{Ag^+_{sol}} a_{Cl^-_{sol}}) = \ln K_{eq} = \frac{\mu^\circ_{AgCl} - \mu^\circ_{Ag^+_{sol}} - \mu^\circ_{Cl^-_{sol}}}{RT}$$

Finally, one obtains the following:

$$\varphi_{Ag} - \varphi_{sol} = \frac{1}{\mathscr{F}}(\mu_{e_{Ag}} + \mu^\circ_{AgCl} - \mu^\circ_{Ag} - \mu^\circ_{Cl^-_{sol}}) - \frac{RT}{\mathscr{F}} \ln a_{Cl^-_{sol}}$$

$$= Cst - \frac{RT}{\mathscr{F}} \ln a_{Cl^-_{sol}}$$

▸ One could also consider the second interface as a reactive interface by writing the chemical equilibrium as follows: $AgCl \rightleftharpoons Ag^+_{AgCl} + Cl^-_{sol}$

The equilibrium of the interface in question corresponds to the following relationship between the different electrochemical potentials:

$$\tilde{\mu}_{Cl^-_{sol}} + \tilde{\mu}_{Ag^+_{AgCl}} = \tilde{\mu}_{AgCl}$$

Therefore the internal potential difference between the two phases is expressed by the following equation:

$$\varphi_{AgCl} - \varphi_{sol} = \frac{1}{\mathscr{F}}(\mu_{AgCl} - \mu_{Ag^+_{AgCl}} - \mu_{Cl^-_{sol}}) = \frac{1}{\mathscr{F}}(\mu^\circ_{AgCl} - \mu^\circ_{Ag^+_{AgCl}} - \mu^\circ_{Cl^-_{sol}}) - \frac{RT}{\mathscr{F}} \ln(a_{Ag^+_{AgCl}} a_{Cl^-_{sol}})$$

Finally, we obtain the following again:

$$\varphi_{Ag} - \varphi_{sol} = \frac{1}{\mathscr{F}}(\mu_{e_{Ag}} + \mu^\circ_{AgCl} - \mu^\circ_{Ag} - \mu^\circ_{Cl^-_{sol}}) - \frac{RT}{\mathscr{F}} \ln a_{Cl^-_{sol}} = Cst - \frac{RT}{\mathscr{F}} \ln a_{Cl^-_{sol}}$$

which transforms as below once a reference electrode is introduced [53]:

$$E_{/Ref} = E^\circ_{AgCl/Ag} - \frac{RT}{\mathscr{F}} \ln a_{Cl^-_{sol}}$$

---

[53] *It is essential to remember the specific meaning of this equation. The GALVANI voltage $\varphi_{Ag} - \varphi_{sol}$ at the electrochemical interface terminals is presented in terms of a rigorous equation. However, it is inaccessible in experimental conditions. Once the electrode potential E is written in relation to a reference electrode, then the latter loses all rigour, since the ionic junction voltages in the electrochemical chain end up being disregarded. Consequently, as explained in appendix A.3.2, one cannot say that the chloride ion activity is a measured quantity, contrary to what this expression might suggest.*

If we base our approach on purely thermodynamic point of view, then it would be possible to build this reference electrode by placing a silver wire into a solution saturated in silver chloride, with no direct contact between the silver wire and the silver chloride precipitate. The reactive interface here would be therefore Ag | solution and the electrode's potential subsequently depends on the concentration of Ag^+ ions within the solution. However, since this concentration is extremely low (because silver chloride is barely soluble) and the equilibrium potential is a logarithmic function, the system is very sensitive to even the slightest perturbation (for example a concentration change ranging from 10^{-18} to 10^{-17} mol L^{-1} leads to a change in potential of 60 mV). Natural convection occurring in the solution or a very small current flow are both phenomena which could have a significant impact on the electrode's open-circuit potential. Finally, other electroactive species contained in very small quantities (e.g., dissolved dioxygen or another redox couple) could compete with the Ag^+/Ag equilibrium to determine the electrode potential (mixed potential).

▸▸ Calomel electrode (Hg_2Cl_2)

The calomel electrode is created by combining mercury mixed with calomel (Hg_2Cl_2, a solid mercury (I) salt barely soluble in water) with an aqueous solution of potassium chloride. Unlike in the previous chemical system, there is no risk of chlorocomplex forming, therefore a saturated solution of potassium chloride is most widely used (i.e., with a concentration close to 5 mol L^{-1} at room temperature).

Therefore the following electrochemical half-cell chain is formed:

$$Hg \mid Hg_2Cl_2 \mid KCl \text{ aqueous solution of concentration } C \mid$$

The equilibria involved in this half-cell can be described by taking into account the Hg_2Cl_2/Hg couple in the following way:

$$2\,Hg + 2\,Cl^- \rightleftharpoons Hg_2Cl_2 + 2\,e^-$$

At equilibrium, the NERNST law can be used to express the half-cell's potential with the following equation:

$$E = E°_{Hg_2Cl_2/Hg} - \frac{RT}{\mathscr{F}} \ln a_{Cl^-}$$

Calomel is crushed together with mercury for practical reasons, but also to ensure the redox system's rapidity. The resulting mixture is then put in contact, on the one hand, with mercury (in which a platinum wire is immersed) and also separately on the other hand with an aqueous solution containing potassium chloride[54]. If this electrode is set to be used in a solution containing silver salts, then an intermediate salt bridge must be introduced to prevent silver chloride precipitation. The bridge should contain a solution devoid of chloride ions, and generally with a high ionic concentration so as to minimize the two junction voltages induced by this added extension.

---

[54] A diagram of this reference electrode is given in figure 1.14 in section 1.5.1.2.

The HE is a proton sensor, and the two other electrodes are chloride ion sensors (in other words, with a potential depending on the activity of one type of ion). Whenever it is possible to fix the indicator ion concentration in the solution, these electrodes may be used as reference electrodes with no ionic junction. However, they are generally used in conjunction with a membrane (or a porous material) which separates them from the solution being studied. In this instance it is advisable to use concentrated solutions as reference solutions in order to minimize the ionic junction voltages. To give an example, a saturated solution could be used for the SCE. Similarly, it is assumed that by choosing potassium chloride (where the anion and cation have very close transport numbers) as a reference electrolyte containing chloride ions, it acts as a means of minimizing the junction voltages between the solutions. In this sense, minimizing junction voltages is particularly difficult when a hydrogen reference electrode is involved, because of the high value of the proton transport number in aqueous solution. This is another reason for avoiding the use of such an electrode when it cannot be used without an ionic junction[55].

---

[55] Appendix A.1.1 highlights the points addressed in this paragraph by giving a selection of different situations where the ionic junction voltages are numerically estimated.

QUESTIONS ON CHAPTER 3

1 - In an electrochemical system:
   ▸ at thermodynamic equilibrium, the current is always zero   true   false
   ▸ if the current is zero, then the system is always
   in thermodynamic equilibrium   true   false

2 - For the following species in their standard state:
   ▸ $\mu^\circ_{Cu} = 0 \ J \ mol^{-1}$ true   false
   ▸ $\mu^\circ_{H^+} = 0 \ J \ mol^{-1}$ true   false
   ▸ $\mu^\circ_{Cu^{2+}} = 0 \ J \ mol^{-1}$ true   false
   ▸ $\mu^\circ_{H_2} = 0 \ J \ mol^{-1}$ true   false

3 - Strictly speaking, if you have an aqueous solution containing ions,
   it is possible, for each ion individually, to measure:
   ▸ its concentration   true   false
   ▸ its activity   true   false

4 - What is the ionic strength (including the appropriate unit) of the following aqueous solutions?
   ▸ containing NaCl with a concentration of $0.1 \ mol \ L^{-1}$ ....
   ▸ containing $Cu(NO_3)_2$ with a concentration of $0.1 \ mol \ L^{-1}$ ....

5 - Based on the simplified DEBYE-HÜCKEL model, if you take the mean activity coefficient of a solute in a solution containing only NaCl, and compare it to the mean activity coefficient in a solution with the same ionic strength containing only $Cu(NO_3)_2$ then the former coefficient is

          larger   equal   smaller

6 - For a metal, it is possible to measure:
   ▸ the electrochemical potential of free electrons   true   false
   ▸ the chemical potential of free electrons   true   false
   ▸ the GALVANI potential   true   false
   ▸ the VOLTA potential   true   false

7 - On both sides of a single-exchange junction (i.e., with interfacial reaction involving only one species) between two media in which the species studied share the same standard chemical potential, identical concentrations of exchangeable species are always seen in thermodynamic equilibrium, when the latter is:
   ▸ an ion   true   false
   ▸ a neutral species   true   false

8 - The thermodynamic equilibrium of an interface involving the Cu^{2+}/Cu couple can be illustrated in the following equation: $\varphi_{metal} - \varphi_{electrolyte} = Cst + \dfrac{RT}{2\mathscr{F}} \ln \dfrac{a_{Cu^{2+}}}{a_{Cu}}$

   whereby the constant is the standard potential
   of the couple in question, relative to SHE   true   false

9 - Fill in the missing numbers (with the appropriate sign) in the equations below, which characterize the thermodynamic equilibrium of the following electrochemical chain: Cu' | Pt, H_2 | aqueous solution containing H_2SO_4 and $CuSO_4$ | Cu

Cu is chosen as the working electrode and Cu' as the counter-electrode ($U = \varphi_{Cu} - \varphi_{Cu'}$). The sign for the GIBBS energy of reaction corresponds to the overall reaction, written as: $Cu^{2+} + H_2 \rightleftharpoons Cu + 2\,H^+$

$$\Delta_r G = \ldots \mathscr{F} U = \ldots \mu_{Cu} \ldots \mu_{H^+} \ldots \mu_{Cu^{2+}} \ldots \mu_{H_2}$$

10 - Complete the simplified POURBAIX diagram for iron below. It has been plotted for an overall iron element concentration equal to C_0, ($[Fe^{3+}] + [Fe^{2+}] = C_0$) in aqueous solution. You must indicate the following:

▶ the areas of either thermodynamic stability or predominance for the following species: Fe, $Fe(OH)_3$, Fe^{2+}, Fe^{3+}

▶ the point symbolizing the potential (vs SHE) of a piece of iron immersed in a solution of $pH = pH_1$, containing Fe^{2+} ions with the concentration C_0

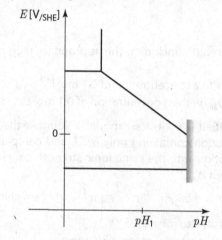

In addition:

▶ where would you locate the point symbolizing the potential (vs SHE) of a piece of iron that is immersed in a solution of $pH = pH_1$ containing Fe^{2+} ions with a concentration of $C_0/100$, in relation to the previous point?

      on the right   in the same place   above   below

▶ what reaction occurs in the previous system if a $0\ V_{SHE}$ potential is imposed at this metal interface?

       ....

▶ a piece of iron is stable in a solution
containing Fe^{3+} ions with the concentration C_0 true   false

11 - If a reference electrode Ag, AgCl | KCl 1 mol L^{-1} | has been stored with its tip immersed in distilled water, then calibration would show that its potential after storage

       has increased   has not changed   has diminished

4 - CURRENT FLOW:
A NON-EQUILIBRIUM PROCESS

The purpose of this chapter is to provide the tools to describe the relationships between voltage and current, in quantitative terms, as a function of time in electrochemical systems.

When a current flows through a system, movements of at least one type of charged species are observed on the macroscopic scale. Consequently the system is in a non-equilibrium state.

4.1 - MASS BALANCES

Once the relationship has been defined between the current and the quantities characterizing mass transport, then one can write mass balances in volume and at the interfaces to describe the electrochemical systems in which a current flows in quantitative terms.

4.1.1 - DEFINITIONS FOR THE MACROSCOPIC QUANTITIES RELATED TO THE CURRENT

Since all elementary charged species have a mass, then by definition the notion of current is related to the notion of overall mass transport.

4.1.1.1 - MOLAR FLUX

Any movement of substance, whether relating to a charged species or not, can be quantified by the amount of substance (mol) flowing through a surface per time unit. This represents the molar flux or flow rate (mol s^{-1}). Here we define an associated local macroscopic quantity: the molar flux density which depends on the position. This value is an average taken around the point in question, and calculated for a large volume when compared to atomic dimensions (greater than tens of nm^3) [1]. The latter is defined in relation to the amount of species i, dn_i, flowing through a surface of elementary area dS during an elementary time interval dt. It is a vector, $\mathbf{N}_i$, which is collinear with the local velocity of the species in question, and which is oriented in the same direction (see the development below). Its modulus is given by the following equation:

$$\|\mathbf{N}_i\| = \frac{1}{dS} \frac{dn_i}{dt}$$

---

[1] Many authors in the field of electrochemistry wrongly omit the term density when using the term molar flux, whereas it is indeed a matter of molar flux density. This may lead to misunderstanding since the terms flux and flow rate are synonyms in many other scientific domains, yet in this context they have different meanings.

with: $\mathbf{N}_i$ the molar flux density, with the modulus in   [mol m^{-2} s^{-1}]
       dn_i the variation in the amount of species i [mol]
       dS the elementary surface area   [m^2]
       dt the elementary time interval   [s]

▼ The molar flux density and local velocity vectors are collinear and in the same direction. The proportionality between these two vectors has a physical meaning. The example given below sets out this relationship for a system with unidirectional geometry, whereby one only needs to take into account the spatial variations in one direction [2]. The diagram in figure 4.1 illustrates the volume of this type of system whereby $\boldsymbol{\omega}_i$ is the local velocity of species i, and the surface (S) is chosen with its normal, $\mathbf{n}$, parallel to the velocity vector.

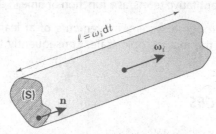

Figure 4.1 - Diagram of an elementary volume dV crossed by species i during a dt time interval

The amount of particles crossing (S) during the time interval dt is equal to the amount of particles contained in a cylinder volume with a base of area S and a length $\omega_i dt$:

$$N_i = \frac{1}{S}\frac{dn_i}{dt}$$

$$dn_i = C_i\, dV = C_i\, \omega_i\, dt\, S$$

For species i, we end up with the following equation, which can be applied to all geometric configurations:

$$\mathbf{N}_i = C_i \boldsymbol{\omega}_i$$

with: $\mathbf{N}_i$ the molar flux density, with the modulus in   [mol m^{-2} s^{-1}]
       C_i the local concentration   [mol m^{-3}]
       $\boldsymbol{\omega}_i$ the local average velocity, with the modulus in   [m s^{-1}] ◢

4.1.1.2 - CURRENT DENSITY

The molar flux density of a charged species i is also related to a charge flux density which is most commonly called current density:

$$\mathbf{j}_i = z_i \mathscr{F} \mathbf{N}_i$$

with: $\mathbf{j}_i$ the current density of species i, with the modulus in   [A m^{-2}]
       $\mathbf{N}_i$ the molar flux density of species i, with the modulus in [mol m^{-2} s^{-1}]
       z_i the algebraic charge number of species i
       $\mathscr{F}$ the FARADAY constant   [C mol^{-1}]

---

[2] *Unidirectional geometry is the simplest form of geometry. It belongs to the 1D category since the only parameter is the distance from a plane. Other usual 1D systems have cylindrical or spherical geometric forms, where the only spacial parameter is the radius (distance to the cylinder axis or the center of the sphere).*

Most often, there are several types of species in movement at a given time: the overall current is created by the movements of all the charge carriers.

The current density is defined as the vectorial sum of the current densities of each charge carrier:

$$\mathbf{j} = \sum_i \mathbf{j}_i = \sum_i z_i \mathscr{F} \mathbf{N}_i$$

When using this definition, one must pay attention to the fact that the current density is a local quantity. Consequently, when writing the sum one should only take into account the species actually present at the point in question. In an electrochemical system, the nature of the charge carriers may vary from one point to another within the device. This frequently occurs in large-sized cells with different species at the anode and the cathode. Yet this is always the case in the area surrounding a metal|electrolyte interface, as shown in the example below.

▶ Let us take the example of a unidirectional system with a planar interface between a copper electrode and an aqueous solution containing copper sulphate. The overall current density vector is the same at all points in space [3] but it is not expressed in the same way in the metallic phase compared to in the solution. In the metal, the only mobile species are electrons, whereas in the solution there are two types of charge carriers, Cu^{2+} and SO_4^{2-} ions. This gives the following equation:

$$\mathbf{j}_{metal} = \sum_i \mathbf{j}_i = \mathbf{j}_e \qquad \mathbf{j}_{electrolyte} = \sum_i \mathbf{j}_i = \mathbf{j}_{Cu^{2+}} + \mathbf{j}_{SO_4^{2-}}$$

It would be wrong to indicate that the overall current density equals the vectorial sum of the electron and ion current densities, since these species cannot coexist in the same space zone:

$$\mathbf{j} = \sum_i \mathbf{j}_i = \cancel{\mathbf{j}_e + \mathbf{j}_{Cu^{2+}} + \mathbf{j}_{SO_4^{2-}}}$$

One way of expressing this is to suggest that the observer defining the current value locally is not ubiquitous and cannot see both electrons and ions move around simultaneously.   ◢

4.1.1.3 - TRANSPORT NUMBERS

The quantity used to compare the contributions of the various charge carriers to the overall current is the transport number of a given type of charge carrier, defined as the ratio between the overall current and the partial current attributable to the charge carrier. In any geometric situation, the different current density vectors are not necessarily collinear and the general definition of a transport number is as follows:

$$\tilde{t}_i = \frac{I_i}{I} = \frac{\mathbf{j}_i \times \mathbf{j}}{\mathbf{j} \times \mathbf{j}}$$

Transport numbers are local quantities. They have no dimension and may *a priori* be positive or negative. They obey the following equation:

$$\sum_i \tilde{t}_i = 1$$

---

[3]   *This property, which applies to unidirectional geometries, is demonstrated in section 4.1.2.*

This notion is scarcely used in general cases [4], but is most commonly found exclusively in situations where migration is the only transport phenomenon to be taken into account: the transport numbers are then denoted by t_i, and are positive numbers showing values between 0 and 1 [5].

4.1.2 - VOLUME MASS BALANCE

At the macroscopic level, even without knowing the physical origins of a current [6], several general properties can be outlined pertaining to the current flow through the volume of a conducting medium [7].

A local volume mass balance can be written for each type of species: the local variation of the amount of substance depends on the balance (itself linked to the flux) between the species either leaving or entering the volume and those being created or consumed locally, for example by one or several chemical reactions. This type of mass balance can be done for neutral or charged species [8].

As shown in figure 4.2, the following equation can be written for a unidirectional system, namely one where the local quantities depend exclusively on one spatial dimension, (in other words they are homogeneous in any section perpendicular to the x-axis) and where the **N** component on the x-axis is denoted by N_x:

$$\text{variation} = \text{species in} - \text{species out} + \text{local production}$$
$$\mathrm{d}n_i = N_{x,i}(x)\,S\,\mathrm{d}t - N_{x,i}(x+\mathrm{d}x)\,S\,\mathrm{d}t + w_i\,\mathrm{d}V\,\mathrm{d}t$$

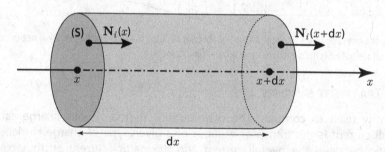

Figure 4.2 - Diagram of a volume with a molar flux of species i flowing through it

---

[4] This notion is explained in general terms via an example in appendix A.4.1. It illustrates how in a system where migration and diffusion must be taken into account, the most common transport number, t_i, has no physical meaning whereas the electrochemical transport number, $\tilde{t}_i$, expresses the ratio between the current density of the species in question and the overall current density.

[5] This typical pure migration conditions are described in section 2.2.4.3.

[6] Section 4.2.1 describes the different mass transport processes and how to write the corresponding equations. The main mechanisms involved at the microscopic level are presented in section 4.2.2.

[7] Section 4.1.3 describes what occurs at the interfaces in terms of mass balance.

[8] Expressing mass balances is also an essential tool in chemical engineering, for example with mass or energy balances for open reactors.

By writing that $dV = S\,dx$, and using a TAYLOR expansion of the output molar flux density $N_{x,i}(x + dx)$, one obtain the following equation:

$$\frac{1}{dV}\frac{dn_i}{dt} = \frac{1}{dx}\left(N_{x,i}(x) - N_{x,i}(x) - \frac{dN_{x,i}}{dx}\,dx\right) + w_i$$

Therefore, the following equation emerges for a species i in a system with unidirectional geometry:

$$\frac{\partial C_i}{\partial t} = -\frac{\partial N_{x,i}}{\partial x} + w_i$$

with: C_i the local concentration of species i [mol m^{-3}]
          $N_{x,i}$ the algebraic component on the x-axis   [mol m^{-2} s^{-1}]
                    of the molar flux density
          w_i the local production rate of species i per   [mol m^{-3} s^{-1}]
                    unit volume (algebraic)

Generalising the equation to cover all geometric configurations gives the following[9]:

$$\frac{\partial C_i}{\partial t} = -\ \text{div}\ \mathbf{N}_i + w_i$$

A so-called steady state can be reached in certain conditions, constituting a highly particular and important case from an experimental point of view[10]. One of its specific features notably includes the fact that the various concentrations are not time-dependent. Therefore, the volume mass balance at steady state is written in the following way:

$$\text{div}\ \mathbf{N}_i = w_i$$

Generally speaking, as defined in homogeneous kinetics, the production rate per unit volume and the reaction rates are linked whether the system is at steady state or not. In fact, the rate of a chemical reaction (v in mol m^{-3} s^{-1}) is defined by the following equation:

$$v = \frac{1}{v_i}\frac{1}{V}\frac{dn_i}{dt}$$

with: V the volume of the system in question   [m^3]
          v_i the algebraic stoichiometric number of species i
          dn_i the variation in the amount of species i [mol]
          dt the time interval   [s]

dn_i/dt represents the production rate of species i by the reaction concerned. When there are several reactions occurring simultaneously (see example below) then this gives the following equation:

$$w_i = \sum_{\text{reactions}} v_{i_r} v_r$$

---

[9]   div is the divergence operator: $\text{div}\ \mathbf{X} = \dfrac{\partial X_x}{\partial x} + \dfrac{\partial X_y}{\partial y} + \dfrac{\partial X_z}{\partial z}$.

[10] This equation is often applied in electrochemistry to quasi-steady systems (see section 1.6.4).

If one write the local volume charge balance in the bulk of the conducting medium (that is, located at a distance from the interfaces with the surrounding media) based on the different volume mass balances, then you end up with the following equation:

$$\frac{\partial}{\partial t}\left(\sum_i z_i \mathscr{F} C_i\right) + \text{div } \mathbf{j} = 0$$

In fact the overall local charge production balance in the bulk of the solution is always zero.

▶ For example, when certain olefins, R, are reduced in an organic medium, the charge transfer step occurring at the electrode produces an anion radical, which further reacts once in the electrolyte. One of the mechanisms that can be seen here involves three successive reactions in the solution.

After the cathodic electron transfer reaction occurs, producing the radical anion denoted by $R^{\bullet-}$:

$$R + e^- \longrightarrow R^{\bullet-}$$

the anion radical can then be seen to dimerise before two protonation reactions of the dimer formed occur. This is illustrated below with the respective volume reaction rates indicated by v_1, v_2 and v_3:

$$2\,R^{\bullet-} \xrightarrow{v_1} R_2^{2-}$$

$$R_2^{2-} + H^+ \xrightleftharpoons{v_2} R_2H^-$$

$$R_2H^- + H^+ \xrightleftharpoons{v_3} R_2H_2$$

Let us assume that before the olefins are reduced, the only ions contained in the solution are the protons and the nitrate counter-ions.

During the reaction, assuming that the anode and cathode are separated far enough apart to prevent any anodic phenomena from interfering, then the local volume mass balance for the six different species in the solution is as follows:

Volume mass balance	value of $\times z_i \mathscr{F}$
$\dfrac{\partial [R^{\bullet-}]}{\partial t} = -\text{div }\mathbf{N}_{R^{\bullet-}} - 2v_1$	$\times -\mathscr{F}$
$\dfrac{\partial [R_2^{2-}]}{\partial t} = -\text{div }\mathbf{N}_{R_2^{2-}} + v_1 - v_2$	$\times -2\mathscr{F}$
$\dfrac{\partial [R_2H^-]}{\partial t} = -\text{div }\mathbf{N}_{R_2H^-} + v_2 - v_3$	$\times -\mathscr{F}$
$\dfrac{\partial [H^+]}{\partial t} = -\text{div }\mathbf{N}_{H^+} - v_2 - v_3$	$\times +\mathscr{F}$
$\dfrac{\partial [R_2H_2]}{\partial t} = -\text{div }\mathbf{N}_{R_2H_2} + v_3$	$\times 0$
$\dfrac{\partial [R]}{\partial t} = -\text{div }\mathbf{N}_{R}$	$\times 0$
$\dfrac{\partial [NO_3^-]}{\partial t} = -\text{div }\mathbf{N}_{NO_3^-}$	$\times -\mathscr{F}$

The overall volume charge balance with the multiplying coefficients indicated above therefore gives the following:

$$\frac{\partial}{\partial t}\left(\sum_i z_i \mathscr{F} C_i\right) = - \operatorname{div} \mathbf{j} + \mathscr{F}\left[+2v_1 - 2(v_1 - v_2) - (v_2 - v_3) - (v_2 + v_3)\right] = - \operatorname{div} \mathbf{j}$$

Electromagnetism has its own equivalent of this property, as shown in the following equation:

$$\frac{\partial \rho_{ch}}{\partial t} + \operatorname{div} \mathbf{j} = 0$$

with: ρ_{ch} the charge density $[\,C\,m^{-3}\,]$
       $\mathbf{j}$ the current density, with the modulus in $[\,A\,m^{-2}\,]$

By applying these volume mass balances to the conducting media within electrochemical systems, and more generally to electrochemical chains, an important property emerges relating to the current, as demonstrated below. The following equation shows a consequence of the fact that electroneutrality applies throughout the medium on the macroscopic scale in the bulk of any conductor, and after a very short transient state [11]:

$$\sum_i z_i C_i = 0 \quad \text{and thus} \quad \operatorname{div} \mathbf{j} = 0$$

After the transient period, the molar flux perpendicular to the interface between the conductor and the insulating media equals zero. By applying the OSTROGRADSKY theorem [12], it is possible to deduce that the charge flow rate is the same in all sections of the conductor, whether perpendicular to the current density or not, as shown in figure 4.3:

$$\iint_{(S)} \mathbf{j}.\mathbf{n}\,dS - \iint_{(S')} \mathbf{j}'.\mathbf{n}'\,dS + 0 = 0$$

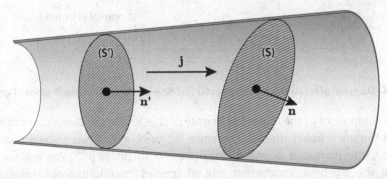

Figure 4.3 - Diagram of a conductor volume with a current flowing through it

Note that in the particular case of unidirectional geometry, the current density vector is the same at all points inside an electrochemical chain:

$$\operatorname{div} \mathbf{j} = \frac{\partial j}{\partial x} = 0 \quad \Rightarrow \quad \mathbf{j} = \text{Cst}$$

---

[11] This question is addressed in section 3.1.1.1: the order of magnitude of the time needed for electroneutrality to be established in an electrolyte is one nanosecond.

[12] $\oiint \mathbf{j}.\mathbf{n}\,dS = \iiint \operatorname{div}\mathbf{j}\,dV$

Therefore, because electroneutrality prevails throughout the conducting material's volume, the current can be clearly defined (except for the sign) since it is the same in all sections. This particular property relating to the current in a conducting medium can be extended to an electrochemical cell:

$$I = \iint\limits_{(S)} \mathbf{j} \cdot \mathbf{n} \, dS = \iint\limits_{(S')} \mathbf{j'} \cdot \mathbf{n'} \, dS$$

Remember that the current sign depends on the orientation of the normal to the interface. The common convention used by electrochemists involves orientating the normal from the metal towards the electrolyte. This implies that the current is positive at the interface with an anode and negative at a cathode interface [13].

4.1.3 - INTERFACIAL MASS BALANCE

4.1.3.1 - GENERAL CASE

In order to write the mass balance for the species i at the interface level, one needs to consider a small volume with a very low width of $2\,\delta x$, which includes the interfacial zone. This volume's limiting surfaces are respectively situated in the two extreme phases, denoted by α and β in figure 4.4 which illustrates a system with unidirectional geometry.

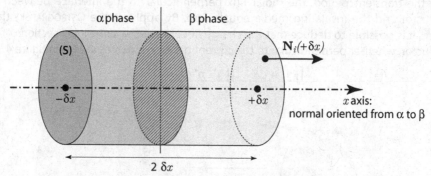

Figure 4.4 - Diagram of an interface with a molar flux flowing through in unidirectional geometry

$N_{n,i}$ is the component of the molar flux density of species i in the direction normal to the surface at the point being studied. Therefore, $N_{n,i} = \mathbf{N}_i \times \mathbf{n}$. $N_{n,i}$ is a scalar quantity. By convention, the normal is oriented from phase α to phase β [14]. We will use w_{S_i} to symbolize the algebraic production rate of species i per unit area. This species is produced or consumed locally by heterogeneous chemical reactions occurring in the interfacial zone. Γ_i is the surface concentration of species i, in mol m^{-2} [15], namely the

---

[13] This sign convention, illustrated in figures 1.5 and 1.7, is described in more detail in section 1.4.1.3.

[14] The opposite convention could also be chosen but the signs of the ensuing expressions would then be different.

[15] This notion is complex and a more detailed presentation would need to go beyond the scope of this document. In particular, in order to describe the thermodynamics of interfaces one would need to refer to the notion of surface excess as defined by the GIBBS model. We can still say that Γ_i is the integral of the volume concentration over a distance equal to the thickness of the interfacial zone

amount of substance per unit surface in a small volume with a width of $2\,\delta x$ defining the interfacial zone.

In unidirectional geometry, the variation dn_i of the amount of species i in the volume $2\,S\delta x$ between t and $t+dt$ can be expressed as:

$$\begin{aligned} \text{variation} &= \text{in} & - & \text{out} & + & \text{local production} \\ dn_i &= N_{n,i}(-\delta x)\,S\,dt & - & N_{n,i}(+\delta x)\,S\,dt & + & w_{S_i}\,S\,dt \\ &= d\Gamma_i\,S \end{aligned}$$

One therefore has: $\qquad N_{n,i_\beta}(+\delta x) - N_{n,i_\alpha}(-\delta x) = -\dfrac{d\Gamma_i}{dt} + w_{S_i}$

The following equation applies to unidirectional geometry when the thickness tends towards zero:

$$(N_{n,i})_{\text{«}x=0\text{»}_\beta} - (N_{n,i})_{\text{«}x=0\text{»}_\alpha} = -\frac{d\Gamma_i}{dt} + w_{S_i}$$

Generalising the equation to cover all geometric configurations gives the following[16]:

$$(N_{n,i})_{\text{interface}_\beta} - (N_{n,i})_{\text{interface}_\alpha} = -\frac{\partial \Gamma_i}{\partial t} + w_{S_i}$$

with : Γ_i the surface concentration of species i $\qquad\qquad\qquad$ [mol m^{-2}]
$\qquad\quad$ $N_{n,i}$ the normal component of the molar flux density of species i $\quad$ [mol m^{-2} s^{-1}]
$\qquad\quad$ w_{S_i} the production rate of species i per unit area (algebraic) $\qquad$ [mol m^{-2} s^{-1}]

Keep in mind that the signs indicated above take into account the precise definition of the orientation of the normal to the interface, namely from phase α to phase β.

The notation "$x=0$" is meant to emphasise that it is not a mathematical notation corresponding to a surface. It refers to an interfacial zone. In particular, the passage to the limit corresponds to a very small but non-zero value for δx. However, in order to simplify the equations throughout the rest of the document we will simply write $x=0$ for the quantities related to the interface in unidirectional geometry. Similarly, again to simplify the equations in what follows, we will no longer indicate the index n as a reminder that equations are written for the normal components of flux and current densities.

As in the case of homogeneous reactions in volume mass balances there is a link between the production rate per unit area and the heterogeneous reaction rates:

$$w_{S_i} = \sum_{\text{reactions}} v_{i_r}\,v_{S_r}$$

where v_{i_r} is the algebraic stoichiometric number of species i in reaction r.

---

along the normal to the surface. Based on this definition, the value of Γ_i depends on the choice of δx. However it can be shown that its time variation, which is the important quantity relating to equations which is outlined in this document, does not depend on δx.

[16] In particular, when the geometry is not unidirectional, the surface concentration Γ_i is not necessarily identical at all points on the surface and becomes a function of several variables (time and space). The term needed in the interfacial mass balance is therefore a partial derivative: $\partial \Gamma_i /\partial t$, instead of $d\Gamma_i/dt$ in unidirectional geometry.

The interfacial mass balances can be illustrated for instance by a system with a phase transfer reaction between an aqueous solution containing potassium chloride and a solution containing crown-ether (ligand L) in an organic solvent (figure 4.5).

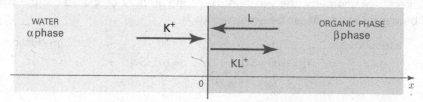

Figure 4.5 - Diagram of a phase transfer reaction

Here we have chosen to orient the x-axis in the direction from the water phase towards the organic solvent (phase α : water and phase β : organic phase). Let v_S be the rate of the following surface reaction:

$$L + K^+ \longrightarrow KL^+$$

At steady state, the variation in the surface concentration is zero and the local mass balance at the interface is expressed as follows:

$$(N_i)_{x=0_\beta} - (N_i)_{x=0_\alpha} = w_{S_i}$$

Taking into account that K^+ ions are not mobile in phase β, whereas the L and KL^+ species are not mobile in phase α, we can write the following:

$$(N_L)_{x=0_\beta} - \cancel{(N_L)_{x=0_\alpha}} = w_{S_L} = -v_S$$

$$(N_{KL^+})_{x=0_\beta} - \cancel{(N_{KL^+})_{x=0_\alpha}} = w_{S_{KL^+}} = v_S$$

$$\cancel{(N_{K^+})_{x=0_\beta}} - (N_{K^+})_{x=0_\alpha} = w_{S_{K^+}} = -v_S$$

hence $\qquad (N_{KL^+})_{x=0_\beta} = (N_{K^+})_{x=0_\alpha} = -(N_L)_{x=0_\beta} = v_S$

This correlates with the directions for the molar flux densities as those indicated in figure 4.5: N_{K^+} and N_{KL^+} have the same positive direction whereas N_L has the opposite direction.

4.1.3.2 - ADSORBED SPECIES

The same reasoning can be applied to a species adsorbed at the interface. This type of species is not mobile in either of the two phases (the adsorbed species may possibly be mobile along the interface). The fluxes normal to the surfaces are therefore zero, which in turn simplifies the general equation outlined above. The local interfacial mass balance for a species i adsorbed at the interface is:

$$\frac{\partial \Gamma_i}{\partial t} = w_{S_i}$$

In particular, at steady state the production rate of an adsorbed species is zero since its surface concentration is time-independent.

4.1.3.3 - ELECTROCHEMICAL INTERFACES

This section is restricted on describing the electrochemical interfaces between a metal and an electrolyte. Remember that there are no free mobile electrons in an electrolyte and no mobile ions in a metal. Respecting the sign conventions commonly shared by electrochemists[17], the normal to the interface is oriented towards the medium containing mobile ions, namely from the metal to the electrolyte (figure 4.6):

| METAL | ELECTROLYTE |
| phase α | phase β |

Figure 4.6 - Diagram of an electrochemical interface

The following equation applies to a species i which is mobile in the electrolyte:

$$(N_i)_{\text{interface}} = -\frac{\partial \Gamma_i}{\partial t} + w_{S_i}$$

Note that this algebraic equation is only valid for a species that is mobile in the electrolyte and if the normal to the interface is oriented from the metal to the electrolyte.

The equation reveals two terms, and therefore any molar flux can be formally divided into the sum of two terms:

▸ a term called faradic (current, flux)[18]

$$(N_i^{\text{farad}})_{\text{interface}} = w_{S_i}$$

▸ a term called capacitive[19] (or pseudo-capacitive for a neutral species)

$$(N_i^{\text{capac}})_{\text{interface}} = -\frac{\partial \Gamma_i}{\partial t}$$

At a reactive electrochemical interface, charge movements simultaneously bring into play a reorganisation of the double layer alongside the heterogeneous redox reaction itself. In simple cases, for example where no specific adsorption occurs[20] and when a supporting electrolyte is present, this step of formally splitting the molar flux densities or current densities into a faradic term and a capacitive term has a physical meaning. In fact, the two phenomena then correspond to different ion movements:

▸ the local mass balance for the ions of the supporting electrolyte is:

$$(N_i)_{\text{interface}} = -\frac{\partial \Gamma_i}{\partial t}$$

---

[17] *Stating that the current at the anode is positive involves orientating the normal to the interface from the metal to the electrolyte, as outlined in section 1.4.1.3.*

[18] *The reason why this term is chosen is because the corresponding molar flux correlates with the FARADAY law, which is outlined in its integrated form in section 2.2.2.2, and demonstrated in section 4.1.4.*

[19] *This term is chosen because most of the time the corresponding molar flux is related to movements which are required for the reorganisation of the double layer, which has a structure similar to that of a capacitor.*

[20] *See section 2.2.1.2 for a brief presentation of adsorption phenomena. A distinction is made between accumulation (linked, for example, to the double layer reorganisation) and specific adsorption when a bond to the interface is involved.*

The fraction of the current resulting from the movements of the supporting electrolyte ions is therefore exclusively capacitive;

▸ the local mass balance for electroactive ions is:

$$(N_i)_{\text{interface}} = w_{S_i}$$

The fraction of the current resulting from the electroactive ions is therefore exclusively faradic, namely produced by the interfacial redox reaction.

4.1.4 - A DEMONSTRATION OF FARADAY'S LAW

FARADAY's law, which has already been applied in this work in its integrated form (see section 2.2.2.2), can be demonstrated based on the interfacial mass balances.

In simple cases where there are no specifically adsorbed species at the electrochemical interface, the various flux densities of the electroactive species at this interface are all proportional to each other. This link looks like a mass balance taking into account the stoichiometric numbers as in any chemical reaction. Therefore, the amount of charge flowing through the system, for instance *via* the current density, can then be linked to the mass balance.

For example, let us look at a unidirectional electrochemical system with an interfacial oxidation reaction involving species that are mobile in the electrolyte. When the metal is chosen as phase α and the electrolyte as phase β, the reaction at the interface is written as follows:

$$\underset{\beta}{|v_{\text{Red}}|\,\text{Red}} \longrightarrow \underset{\beta}{|v_{\text{Ox}}|\,\text{Ox}} + \underset{\alpha}{|v_e|\,e^-}$$

Remember that in the laws of physical chemistry, stoichiometric numbers are algebraic coefficients. The general equation that is obtained as a result can equally be applied to a half-reaction occurring either in the direction of oxidation ($v_e = + n$) or reduction ($v_e = - n$)[21] provided that the reaction rate is also taken in algebraic terms following the direction of the redox reaction.

As for mobile species in the electrolyte (phase β) and for electrons in the metal (phase α), the local mass balance relating to the faradic part can be written as follows:

$$(N_i^{\text{farad}})_{\text{interface}_\beta} - (N_i^{\text{farad}})_{\text{interface}_\alpha} = + w_{S_i}$$

$$(N_e^{\text{farad}})_{\text{interface}_\beta} - (N_e^{\text{farad}})_{\text{interface}_\alpha} = + w_{S_e}$$

The production rates (denoted by w_{S_i} and w_{S_e}) are linked to the rate of the redox reaction (denoted by v_S):

$$w_{S_i} = v_i\, v_S$$

$$w_{S_e} = v_e\, v_S$$

---

[21] n is the (positive) number of electrons exchanged in the redox half-reaction in question (see section 2.1.2.1).

Finally one obtains:
$$(N_i^{farad})_{interface} = v_i \, v_S$$
$$(N_e^{farad})_{interface} = - \, v_e \, v_S$$

As for any mobile species in the electrolyte which are involved in the overall redox half-reaction, the following reaction emerges:

$$\frac{1}{v_i} (N_i^{farad})_{interface} = - \frac{1}{v_e} (N_e^{farad})_{interface}$$

The only mobile species in the metal is the electron, hence one has:

$$\mathbf{j} = \mathbf{j}_e = - \mathscr{F} \mathbf{N}_e$$

When a species i is mobile in the electrolyte, then FARADAY's law can be seen presented in the following equations:

$$j^{farad} = \mathscr{F} \frac{v_e}{v_i} (N_i^{farad})_{interface} \quad \text{or} \quad (N_i^{farad})_{interface} = \frac{v_i}{v_e \, \mathscr{F}} j^{farad}$$

with: j^{farad} the faradic current density modulus   [A m^{-2}]

$\mathscr{F}$ the FARADAY constant   [C mol^{-1}]

v_i the algebraic stoichiometric number of species i

$(N_i^{farad})_{interface}$ the molar faradic flux density at the interface   [mol m^{-2} s^{-1}]

Remember that this equation stands in algebraic terms as long as the sign convention chosen is fully respected. The interfacial mass balance is written with the normal oriented from the metal to the electrolyte. As already outlined, this boils down to considering the currents through the anode to be positive and those through the cathode to be negative. Moreover, this equation only applies to species i that are mobile in the electrolyte.

It is easy to build the integrated version of FARADAY's law (presented in section 2.2.2.2) simply by introducing the definition of the molar flux density:

$$N_i^{farad} = \frac{1}{S} \frac{dn_i}{dt}$$

One then has:
$$dn_i^{farad} = \frac{v_i}{v_e \, \mathscr{F}} j^{farad} S \, dt$$

FARADAY's law can be found again in its integrated form by making up the integral between time t and time $t + \Delta t$:

$$\Delta n_i^{farad} = \frac{v_i}{v_e \, \mathscr{F}} Q^{farad} = \frac{v_i}{v_e \, \mathscr{F}} \int_t^{t + \Delta t} I^{farad}(t) \, dt$$

▶ The diagram in figure 4.7 shows a ferric ion being reduced at the platinum electrode interface, in a system with unidirectional geometry. The aqueous solution contains Fe^{2+}, Fe^{3+}, K^+ and NO_3^- ions (KNO_3 being the supporting electrolyte).

FARADAY's law is presented in the following terms:

$$j^{farad} = \mathscr{F}(N_{Fe^{3+}}^{farad})_{x=0} = - \mathscr{F}(N_{Fe^{2+}}^{farad})_{x=0}$$

This result can be reached through two different methods, one expressing the current density in the metal (as above), the other one in the electrolyte. The first one is faster but the second one makes it easier for one to understand all of these equations. One can also note that this mathematical development illustrates the qualitative link which can be inferred from figure 4.7: the molar flux densities of the two electroactive species have opposite directions.

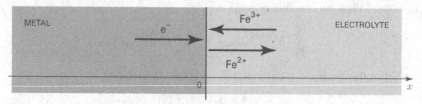

Figure 4.7 - Diagram of an interface with reduction reaction

Let us consider below the half-reaction with its actual direction of advancement (in this case reduction), with its rate denoted by v_S:

$$e^- + Fe^{3+} \longrightarrow Fe^{2+}$$

Remember that the reduction direction is chosen, therefore the following values emerge:

$$v_{Fe^{2+}} = +1 \quad v_{Fe^{3+}} = -1 \quad v_e = -n = -1$$

The following results are given by the interfacial mass balance for each species in the system (applying the same reasoning as previously in this section):

$$(N_{Fe^{3+}}^{farad})_{x=0} = -(N_{Fe^{2+}}^{farad})_{x=0} = -v_S$$

$$(N_e^{farad})_{x=0} = v_S$$

$$(N_{NO_3^-}^{farad})_{x=0} = (N_{K^+}^{farad})_{x=0} = 0$$

By definition, the overall faradic current density at any point in space is:

$$\mathbf{j} = \sum_i \mathbf{j}_i = \sum_i z_i \mathscr{F} \mathbf{N}_i$$

▸ when applied to the metal where only electrons are mobile, the following equation emerges:

$$j^{farad} = -\mathscr{F}(N_e^{farad})_{x=0} = -\mathscr{F} v_S$$

▸ when applied to the electrolyte where all the ions must be taken into account, this equation emerges:

$$j^{farad} = \mathscr{F}\left(3 N_{Fe^{3+}}^{farad} + 2 N_{Fe^{2+}}^{farad} + N_{K^+}^{farad} - N_{NO_3^-}^{farad}\right)_{x=0}$$

$$= \mathscr{F}(-3 v_S + 2 v_S + 0 - 0) = -\mathscr{F} v_S$$

FARADAY's law is obtained in both cases in the following way:

$$j^{farad} = -\mathscr{F} v_S = \mathscr{F}(N_{Fe^{3+}}^{farad})_{x=0} = -\mathscr{F}(N_{Fe^{2+}}^{farad})_{x=0}$$

If several reactions occur at an interface, the same type of reasoning demonstrates that faradic currents are additive.

Take the same example as above, but this time with a non-deaerated solution. Two simultaneous reduction reactions occur: for Fe^{3+} ions and for dissolved dioxygen:

$$e^- + Fe^{3+} \longrightarrow Fe^{2+} \qquad v_{S_1}$$

$$4 e^- + 2 H_2O + O_2 \longrightarrow 4 OH^- \qquad v_{S_2}$$

The interfacial mass balance for each of the system's charged species gives the following results (applying the same reasoning as previously in this section):

$$(N_{Fe^{3+}}^{farad})_{x=0} = -(N_{Fe^{2+}}^{farad})_{x=0} = -v_{S_1}$$

$$(N_{OH^-}^{farad})_{x=0} = 4\,v_{S_2}$$

$$(N_e^{farad})_{x=0} = v_{S_1} + 4\,v_{S_2}$$

$$(N_{NO_3^-}^{farad})_{x=0} = (N_{K^+}^{farad})_{x=0} = 0$$

For the first half-reaction you can write:

$$j_1^{farad} = \mathscr{F}(N_{Fe^{3+}}^{farad})_{x=0} = -\mathscr{F}(N_{Fe^{2+}}^{farad})_{x=0}$$

If the second half-reaction occurred alone, you would have:

$$j_2^{farad} = -\mathscr{F}(N_{OH^-}^{farad})_{x=0}$$

Therefore, when both half-reactions occur simultaneously, you end up with:

$$j^{farad} = -\mathscr{F}(N_e^{farad})_{x=0} = -\mathscr{F}(N_{Fe^{2+}}^{farad})_{x=0} - \mathscr{F}(N_{OH^-}^{farad})_{x=0}$$

namely:

$$j^{farad} = j_1^{farad} + j_2^{farad}$$

If one aims to establish the mass balance in a macroscopic volume of electrolyte in an electrochemical system, then one must take into account all the factors behind the variations. FARADAY's law only deals with variations due to an electrochemical reaction, as shown in the first example. Yet other examples can present more complicated scenarios:

▸ throughout the whole electrolyte volume, the balance of the anodic and cathodic reactions must be taken into account;

▸ when the electrolyte is circulating, as in many cases in industry, there also occurs a variation in the amount of substance, which is due to the forced convection.

4.2 - CURRENT FLOW IN A MONOPHASIC CONDUCTOR

The shape of the current-potential curves and, more generally speaking, the relationships (I, U, t) in an electrochemical system involve the kinetics of both the interfacial phenomena (in particular, charge transfers at reactive interfaces) and the mass transport inside the conducting volumes.

The latter phenomenon will be described first, in the context of a current flowing through the conducting media, before returning to phenomena linked to the current flowing through the interfaces[22]. Therefore, in this section we intend firstly to describe in more details the mass transport phenomena in the particular case of charged species.

---

[22] Section 4.3 deals with this particular aspect found in phenomena related to current flow.

CONSERVATION OF ARCHAEOLOGICAL ARTEFACTS

Document written with the kind collaboration of E. GUILMINOT,
research engineer at the Arc'Antique Laboratory in Nantes, France

Given the long periods that iron or copper artefacts remain buried under the earth or sea, it is inevitable that over time they should alter state due to the build-up of corrosion. In most cases, the corrosion process goes through chloride-containing phases. Following an archaeological excavation, these artefacts, which are known as 'active' undergo a process whereby the corrosion is reactivated, and accelerated, causing the irreversible loss of the original surface detail where the history of the artefact is engraved (inscriptions, adornments, but also deterioration marks). In the case of ferrous artefacts, damage may be more serious and lead to the partial or entire loss of the object itself. Therefore as soon as these artefacts arrive in the preservation and restoration workshops, they undergo immediate stabilizing treatment.

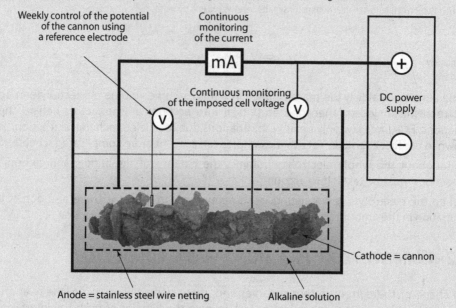

The principle of the stabilizing treatment of a metallic artefact (cannon) by electrolysis
(from a diagram kindly provided by E. GUILMINOT - Laboratoire Arc'Antique)

As a rule, when the iron artefacts emerge from long periods in sea water, they are covered with concretions or gangue (a mixture of calcite, quartz and marine organisms) which hide their original surface. The first step in treatment is to remove these concretions, which can be done using electrochemistry.

The artefact is therefore submitted to a cathodic polarisation whereby it is encased in an anodically polarised stainless steel cage, which is built to mould its shape. The cathodic potential must be carefully monitored because if the resulting hydrogen bubbling effect is too strong, it may cause cracks in the artefact. This soft cathodic treatment is meant to embrittle the gangue and thus make its removal significantly easier.

The second goal of this electrochemical treatment is to extract the chloride ions from the artefact towards the electrolyte. This removal process is an essential step to prevent the artefact from rapidly corroding again when re-exposed to air. The chloride concentration in the electrolyte is measured periodically during the operation. The electrolytic treatment process is considered complete once the chloride concentration of the electrolyte has reached sufficiently low levels.

When artefacts emerge from marine environments, the quantity of chlorides that need to be extracted can be very high: it can reach up to several kilograms of chlorides for one ton of cast iron. The baths must be regularly renewed as soon as the solution becomes saturated with chlorides.

The question of how long the treatment should last depends on the quality and origin of the artefact. If you take large size artefacts emerging from submarine environments, the average duration is from 6 months to one year for cuprous substances, 1.5 to 2 years for wrought iron and 2 to 3 years for cast iron. However, in order to stabilize certain highly damaged cast iron cannons, over 5 years of treatment are required.

State before treatment (photo J.G. AUBERT - Laboratoire Arc'Antique)

The benefit of electrolysis treatment is that the artefact can be cleaned in a uniform fashion, without seeing its original surface harmed. It can 'let the artefact tell its own tale' by revealing surface inscriptions such as the maker's name or stamp, which is the primary element used for authenticating and identifying artefacts.

Post-treatment state (photo J.G. AUBERT - Laboratoire Arc'Antique).
Cannon surfaced from the wreck of the ALCIDE dating from the beginning of the eighteenth century, exhibited in the town of Carantec in France

4.2.1 - CONDUCTION PHENOMENA: A MACROSCOPIC APPROACH

4.2.1.1 - DIFFERENT DRIVING FORCES FOR TRANSPORT

Here let us return to the classification for the various driving forces behind mass transport:

▶ migration relates to charged species moving under the influence of an electric field;

▶ diffusion relates to neutral or charged species moving under the influence of a chemical potential gradient. For example, if two solutions, in the same solvent and containing a neutral species with two different concentrations, are put in contact with each other then the system tends towards an equilibrium state: it forms a homogeneous solution with an intermediate concentration. Before reaching this equilibrium state, the system passes through various non-equilibrium states, where species can be seen to move spontaneously from the more concentrated solution to the less concentrated one. Unlike the phenomenon of migration, diffusion operates with both neutral and charged species.

Although diffusion is mostly associated with a concentration gradient, it should be noted that this phenomenon is in fact created by a chemical potential gradient. These two gradients are not identical: one can have an activity gradient without a concentration gradient. For instance, this is the case when two solutions containing the same solute in the same solvent are put in contact, whereby the solute is the only electrolyte on one side, and it is mixed with a supporting electrolyte on the other side. Initially, there is a gradient of ionic strength which results in a gradient of activity, although no concentration gradient exists.

Both of these preceding phenomena can be grouped together using the concept of electrochemical potential:

$$\widetilde{\mu}_i = \mu_i + z_i \mathscr{F} \varphi$$

The electrochemical potential gradient is therefore the driving force for both diffusion and migration, which are generally coupled together;

▶ convection relates to the overall movement of a medium when it is fluid. It can occur naturally, for instance when influenced by gravitational forces (density gradients) or uncontrolled vibrations. This natural convection, which is particularly pronounced in the case of large systems, can also be the root cause of many observations made during experiments with smaller devices, such as in analytical chemistry. In addition, in numerous experiments one uses an external device to maintain the fluid in a forced convection movement (for example with a rotating disc electrode, RDE). There is no convection in solid-state materials.

The fact of there being a temperature gradient may bring about indirect consequences on mass transport phenomena. First of all, it generally creates a chemical potential gradient and therefore gives rise to diffusion phenomena. It also entails a density gradient in the medium and consequently a natural convection movement. Moreover, one can observe a coupling of the thermal flux and the molar fluxes (see the thermodynamics of linear irreversible processes in the next section). This is what happens with the SORET effect, also called thermal diffusion, which is when mass transport is seen to occur following a temperature gradient, yet without any concentration gradient.

4.2.1.2 - THERMODYNAMICS OF LINEAR IRREVERSIBLE PROCESSES

By definition, from a thermodynamic point of view, mass transport phenomena (and especially charge transport) are irreversible processes. Here, a linear relationship can be envisaged between flux and force, yet this linearity is called into question if ever the forces applied are too strong. For instance, when mass transport occurs *via* migration, one can see a deviation from the linear behaviour when the electric field applied is either very intense [23] or when it has a very high frequency [24].

When there are several different fluxes and forces brought into play in a system, then the thermodynamics of linear irreversible processes postulates that there is a matrix relationship between the different fluxes and forces (with a symmetrical matrix: ONSAGER relations). The non-zero non-diagonal terms in the matrix signify that a coupling of the phenomena is taking place.

In the case studied here, the fluxes and forces that need to be taken into account are vector quantities, and the molar fluxes in question are relative to the mean motion of the medium. The molar flux densities of each type of species (including the solvent if applicable), $\mathbf{N}_i^{\text{rel}}$, and the heat flux density, $\mathbf{N}_T$, are created out of the gradients of electrochemical potential, $\tilde{\mu}_i$, for each type of species, and the temperature gradient. For general cases the following equations are written:

$$\mathbf{N}_i^{\text{rel}} = -\sum_k L_{ik}\, \mathbf{grad}\, \tilde{\mu}_k - L_{iT}\, \mathbf{grad}\, T$$

$$\mathbf{N}_T = -\sum_k L_{Tk}\, \mathbf{grad}\, \tilde{\mu}_k - L_{TT}\, \mathbf{grad}\, T$$

As of now, the description will be confined to systems with no temperature gradient, and containing only diagonal terms in the equations:

$$\mathbf{N}_i^{\text{rel}} = - L_{ii}\, \mathbf{grad}\, \tilde{\mu}_i$$

It should be noted that although this form for writing the equations may appear simple, since each flux density is directly proportional to a single force, it nonetheless covers implicit couplings in so far as there are migration terms (which involve the same electric field) and as electroneutrality applies. As for this notion of electroneutrality, if it is introduced into ONSAGER relations (resulting in coupling the concentrations) then all the previous equations can be rewritten to remove the potential. As a result, the electrochemical potentials of ions are replaced by terms which are based on the chemical potential of electrolytes. The latter can be measured in an experiment, unlike electrochemical potentials which are essentially theoretical concepts, which are highly useful from an educational point of view.

▶ The example given below illustrates the coupling between the molar flux densities by means of electroneutrality. It is outlined here, although the mathematical equations that one would need to firstly understand the various components of the molar flux density will be addressed only in the next two

---

[23] For an electrolytic solution this phenomenon is known as the WIEN effect.
[24] This phenomenon is known as the FALKENHAGEN effect.

paragraphs (see sections 4.2.1.3 and 4). Here we are focusing on the current flow in a unidirectional system with aqueous solution containing a 1-1 electrolyte, such as a solution of hydrogen chloride. Convection is being disregarded in this example, and solely the diffusion and migration of the different species are taken into account.

Since there are only two types of ions present, the electroneutrality in the medium means that the concentrations in the cation (H^+) and the anion (Cl^-) are equal at any point in the electrolyte volume. A concentration profile can exist, but it must be the same for both ions. Let us consider that C symbolizes this unique concentration. The molar flux densities of both types of ions are given in the following simplified form (see section 4.2.1.4):

$$\begin{cases} N_+ = -D_+ \dfrac{\partial C}{\partial x} - u_+ C \dfrac{\partial \varphi}{\partial x} \\ N_- = -D_- \dfrac{\partial C}{\partial x} + u_- C \dfrac{\partial \varphi}{\partial x} \end{cases}$$

Here, D_+ and D_-, the respective diffusion coefficients of H^+ and Cl^-, as well as u_+ and u_- their electric mobility values (see definition in section 4.2.1.4), are considered as being independent of the concentration and therefore of x. As a consequence, they are considered as constants in the equations. If we consider the volume mass balance of these species [25], then two different equations emerge to define the concentration variations with time:

$$\begin{cases} \dfrac{\partial C}{\partial t} = -\dfrac{\partial N_+}{\partial x} = D_+ \dfrac{\partial^2 C}{\partial x^2} + u_+ \dfrac{\partial}{\partial x}\left(C \dfrac{\partial \varphi}{\partial x} \right) \\ \dfrac{\partial C}{\partial t} = -\dfrac{\partial N_-}{\partial x} = D_- \dfrac{\partial^2 C}{\partial x^2} - u_- \dfrac{\partial}{\partial x}\left(C \dfrac{\partial \varphi}{\partial x} \right) \end{cases}$$

The migration term, once extracted from the first equation, then replaced in the second equation, is therefore removed to obtain the following simple equation:

$$\frac{\partial C}{\partial t} = \frac{u_- D_+ + u_+ D_-}{u_- + u_+} \frac{\partial^2 C}{\partial x^2} = D_\pm \frac{\partial^2 C}{\partial x^2}$$

The concentration field looks like it applies to a pure diffusion phenomenon (without migration) for a species (here HCl) with a mean diffusion coefficient, $D_\pm$.

However, strictly speaking it is not identical to the diffusion of a neutral molecule because in the H^+ and Cl^- case, an electric potential field is created in relation to the concentration profile.

The following equation shows the link that can be found between both fields at any time:

$$\frac{\partial}{\partial x}\left(C \frac{\partial \varphi}{\partial x} \right) = -\frac{D_+ - D_-}{u_- + u_+} \frac{\partial^2 C}{\partial x^2}$$

As explained later, if one wishes to fully describe the system in even greater detail, then one would need to include the boundary conditions [26].

---

[25] The volume mass balances are described in section 4.1.2. The system in question here has a unidirectional geometry and no chemical reaction in its volume.

[26] The fluxes at the electrochemical interfaces must be written. Various examples are given in section 4.3. In addition, an example similar to that described here, with a 1-1 electrolyte and no supporting electrolyte, is described in thorough detail in appendix A.4.1. The steady-state potential profile in the electrolyte is presented, and there is a section specifically addressing the question of whether electroneutrality applies or not.

4.2.1.3 - LINK BETWEEN MIGRATION AND DIFFUSION

In the previous description using electrochemical potentials, it is assumed that movements by migration and diffusion have identical mechanisms at the microscopic level. This hypothesis may lead to errors if ever the charge carriers are not sufficiently well identified, which is especially the case when large quantities of neutral ion pairs are involved. In fact, in this instance, the ion pairs play a part in the diffusion without contributing to migration.

The equation which presents molar flux densities as a function of the electrochemical potential implies that the proportionality coefficient for the driving forces is identical in the case of both migration and diffusion.

Another way of expressing this relationship between migration and diffusion is to use the concept of electrochemical mobility (or absolute mobility, $\tilde{u}_i$). This is the modulus of the steady-state relative velocity of the charged species when it is submitted to a unit force per mole. Let us consider a charged species moving by diffusion and migration with a ω_i^{rel} velocity relative to the average motion of the fluid which has a ω_{medium} velocity. Following the definition for electrochemical mobility, and applying the convention previously used in section 3.1 for the electrochemical potential (i.e., using the molar quantity), one obtains the following equation:

$$\omega_i^{rel} = -\ \tilde{u}_i\ \textbf{grad}\ \tilde{\mu}_i$$

The electrochemical mobility unit is therefore mol s kg^{-1} or mol m^2 s^{-1} J^{-1}. Thanks to the link between molar flux density and local velocity described in section 4.1.1.1, one ends up with the same proportionality link between flux and force (thermodynamics of linear irreversible processes) as seen previously:

$$\textbf{N}_i^{rel} = -\ C_i\ \tilde{u}_i\ \textbf{grad}\ \tilde{\mu}_i$$

One can separate the two terms of diffusion and migration by expanding the expression of electrochemical potential:

$$\textbf{N}_i^{rel} = -\ C_i\ \tilde{u}_i\ \textbf{grad}\ \mu_i - C_i\ \tilde{u}_i\ z_i\ \mathscr{F}\ \textbf{grad}\ \varphi$$
$$= -\ C_i\ \tilde{u}_i\ RT\ \textbf{grad}\ (\ln a_i) + C_i\ \tilde{u}_i\ z_i\ \mathscr{F}\ \textbf{E}$$

As will be discussed later, this equation underlines the interdependence of the coefficients for the migration and diffusion terms. Bear in mind that just because there is a connection made between the proportionality factors in the migration and diffusion components it does not mean that the corresponding molar flux densities are collinear and in the same direction. There is *a priori* no connection between the respective forces (related to the gradients of electric potential or of activity). For example, one may come across cases where the migration and diffusion currents for a species i share the same direction, and other cases where the directions are opposite [27].

---

[27] Section 4.3.1.5 gives examples describing various situations. One can even imagine situations where migration and diffusion currents are not collinear.

4.2.1.4 - EXPRESSING MOLAR FLUX AND CURRENT DENSITIES

If convection is involved, the medium moves with a mean velocity denoted by ω_{medium} and the relative molar flux density is:

$$\mathbf{N}_i^{rel} = C_i \, (\omega_i - \omega_{medium}) = - \, C_i \, \tilde{u}_i \, \mathbf{grad} \, \tilde{\mu}_i$$

Writing the definition of current densities gives:

$$\mathbf{j}_i = z_i \, \mathscr{F} \mathbf{N}_i = z_i \, \mathscr{F} C_i \, \omega_i = z_i \, \mathscr{F} (\mathbf{N}_i^{rel} + C_i \, \omega_{medium})$$

$$= - \, C_i \, z_i \, \mathscr{F} \tilde{u}_i \, RT \, \mathbf{grad} \, (\ln a_i) + C_i \, \tilde{u}_i \, z_i^2 \, \mathscr{F}^2 \, \mathbf{E} + C_i \, z_i \, \mathscr{F} \, \omega_{medium}$$

The common expression for current density does not involve electrochemical mobility but rather:

▸ the molar conductivity for the migration term

$$\lambda_i = z_i^2 \, \mathscr{F}^2 \, \tilde{u}_i$$

with the molar conductivity λ_i expressed in S m^2 mol^{-1} (or S cm^2 mol^{-1}).

There are also expressions using the electric mobility, u_i, which is the steady-state relative velocity of a charge species submitted to a unit electric field, namely 1 V m^{-1}. Its SI unit is therefore m^2 s^{-1} V^{-1}. Simple equations can be found linking the electrochemical and electrical mobility, and the molar conductivity of a given species:

$$u_i = |z_i| \, \mathscr{F} \tilde{u}_i \qquad \text{and} \qquad \lambda_i = \mathscr{F} |z_i| \, u_i$$

with : u_i the electric mobility   [m^2 s^{-1} V^{-1}]
   z_i the algebraic charge number of species i
   $\mathscr{F}$ the FARADAY constant   [C mol^{-1}]
   $\tilde{u}_i$ the electrochemical mobility   [mol s kg^{-1} or mol m^2 s^{-1} J^{-1}]
   λ_i the molar conductivity of species i [S m^2 mol^{-1}]

▸ the diffusion coefficient for the diffusion term with

$$D_i = \tilde{u}_i \, RT$$

with the diffusion coefficient D_i expressed in m^2 s^{-1} (or cm^2 s^{-1}).

This latter is known as the NERNST-EINSTEIN equation [28]. *It can also be written involving the electric mobility:*

$$\frac{u_i}{D_i} = |z_i| \, \frac{\mathscr{F}}{RT}$$

The current density of species i is then expressed as follows:

$$\mathbf{j}_i = \underbrace{- \, D_i \, z_i \, \mathscr{F} C_i \, \mathbf{grad} \, (\ln a_i)}_{\mathbf{j}_{i \, diffusion}} + \underbrace{\lambda_i \, C_i \, \mathbf{E}}_{\mathbf{j}_{i \, migration}} + \underbrace{C_i \, z_i \, \mathscr{F} \, \omega_{medium}}_{\mathbf{j}_{i \, convection}}$$

---

[28] *Some authors call this the EINSTEIN equation, preferring to save the term ' NERNST-EINSTEIN equation ' to describe the link between the molar conductivity at infinite dilution and the diffusion coefficients of an electrolyte $A_{p_+} B_{p_-}$:*

$$\Lambda^0 = (p_+ \, D_+ \, z_+^2 + p_- \, D_- \, z_-^2) \, \frac{\mathscr{F}^2}{RT}$$

with the connection between migration and diffusion (see previous section) also expressed by the following equation:

$$\lambda_i = D_i \ z_i^2 \ \frac{\mathscr{F}^2}{RT}$$

To sum up,

▸ in an ideal conducting medium, such as an electrolyte solution at infinite dilution, the activities are assumed as being equal to the concentrations and in addition the molar conductivity at infinite dilution, λ_i^0, and the diffusion coefficient at infinite dilution, D_i^0, can be used. One therefore obtains the following simplified equation for the various components of the current density of species i:

$$\mathbf{j}_i^0 = \underbrace{- \ D_i^0 \ z_i \ \mathscr{F} \ \mathbf{grad}\, C_i}_{\mathbf{j}_{i\,\text{diffusion}}^0} + \underbrace{\lambda_i^0 \ C_i \ \mathbf{E}}_{\mathbf{j}_{i\,\text{migration}}^0} + \underbrace{C_i \ z_i \ \mathscr{F} \ \boldsymbol{\omega}_{\text{medium}}}_{\mathbf{j}_{i\,\text{convection}}^0}$$

with the following equation: $\lambda_i^0 = D_i^0 \ z_i^2 \ \dfrac{\mathscr{F}^2}{RT}$

Moreover we can equally write the molar flux density as a function of D_i^0, λ_i^0, $\widetilde{u}_i^0$ or u_i^0:

$$\mathbf{N}_i^0 \ = \ - \quad D_i^0 \ \mathbf{grad}\, C_i \quad + \ z_i \ \frac{\mathscr{F}}{RT} \ D_i^0 \ C_i \ \mathbf{E} \ + \ C_i \ \boldsymbol{\omega}_{\text{medium}}$$

$$\mathbf{N}_i^0 \ = \ - \ \frac{\lambda_i^0}{z_i^2} \ \frac{RT}{\mathscr{F}^2} \ \mathbf{grad}\, C_i \quad + \ \frac{\lambda_i^0}{z_i \ \mathscr{F}} \ C_i \ \mathbf{E} \quad + \ C_i \ \boldsymbol{\omega}_{\text{medium}}$$

$$\mathbf{N}_i^0 \ = \ - \ \widetilde{u}_i^0 \ RT \ \mathbf{grad}\, C_i \quad + \ z_i \ \mathscr{F} \ \widetilde{u}_i^0 \ C_i \ \mathbf{E} \quad + \ C_i \ \boldsymbol{\omega}_{\text{medium}}$$

$$\mathbf{N}_i^0 \ = \ \underbrace{- \ \frac{u_i^0}{|z_i|} \ \frac{RT}{\mathscr{F}} \ \mathbf{grad}\, C_i}_{\mathbf{N}_{i\,\text{diffusion}}^0} \quad + \ \underbrace{\frac{z_i}{|z_i|} \ u_i^0 \ C_i \ \mathbf{E}}_{\mathbf{N}_{i\,\text{migration}}^0} \quad + \ \underbrace{C_i \ \boldsymbol{\omega}_{\text{medium}}}_{\mathbf{N}_{i\,\text{convection}}^0}$$

▸ for any conducting medium, such as a concentrated electrolyte solution, the activity coefficients, γ_i, cannot be ignored. As well as $\widetilde{u}_i$, λ_i and D_i, they depend *a priori* on the concentrations of all the other species, even if there is no coupling between the fluxes of the different species, in terms of the thermodynamics of irreversible processes. Since one rarely has an accurate idea of the link between activity and concentrations, it is difficult to measure the diffusion coefficient in an experiment. In this case, an apparent diffusion coefficient D_i^{app} can then be introduced, which depends on the concentrations of all the other species. By introducing $a_i = \gamma_i \, C_i$ into the general equation for diffusion current density, the following emerges:

$$\mathbf{j}_{i\,\text{diffusion}} \ = - \ D_i \ z_i \ \mathscr{F} \ C_i \ \mathbf{grad}\, (\ln a_i)$$

hence: $$\mathbf{j}_{i\,\text{diffusion}} \ = - \ z_i \ \mathscr{F} \ C_i \ D_i \ \mathbf{grad}\, (\ln \gamma_i + \ln C_i)$$

$$= - \ z_i \ \mathscr{F} \ C_i \ D_i \left(1 + \frac{\partial \ln \gamma_i}{\partial \ln C_i}\right) \mathbf{grad}\, (\ln C_i)$$

and finally:　　　$$\mathbf{j}_{i\,\text{diffusion}} = -\,z_i\;\mathscr{F}\;\underbrace{D_i\left(1+\frac{\partial \ln \gamma_i}{\partial \ln C_i}\right)}_{D_i^{\text{app}}}\,\mathbf{grad}\,C_i$$

It should be noted that the apparent diffusion coefficient, D_i^{app}, which depends on the various species concentrations by means of the activity coefficient, does not follow the equation involving λ_i derived from the Nernst-Einstein equation:

$$D_i^{\text{app}}\;z_i^2\;\frac{\mathscr{F}^2}{RT} \neq D_i\;z_i^2\;\frac{\mathscr{F}^2}{RT} = \lambda_i$$

Finally, one ends up with very general equations:

$$\mathbf{N}_i = -\underbrace{D_i^{app}\,\mathbf{grad}\,C_i}_{\mathbf{N}_{i\,\textit{diffusion}}} + \underbrace{\frac{z_i}{|z_i|}\,u_i\,C_i\,\mathbf{E}}_{\mathbf{N}_{i\,\textit{migration}}} + \underbrace{C_i\,\omega_{medium}}_{\mathbf{N}_{i\,\textit{convection}}}$$

$$\mathbf{j}_i = -\underbrace{z_i\,\mathscr{F}\,D_i^{app}\,\mathbf{grad}\,C_i}_{\mathbf{j}_{i\,\textit{diffusion}}} + \underbrace{\lambda_i\,C_i\,\mathbf{E}}_{\mathbf{j}_{i\,\textit{migration}}} + \underbrace{C_i\,z_i\,\mathscr{F}\,\omega_{medium}}_{\mathbf{j}_{i\,\textit{convection}}}$$

In order to simplify the notations, we will only use the symbol D_i, even if the system in question is not ideal. This is therefore an assimilation of D_i^{app} and D_i. However, we must keep in mind that in general cases these equations commonly comprise the three different types of driving forces for the moving species, yet the proportionality coefficients are not interconnected in any simple way. In addition, in an experiment it is not possible to ascertain the distribution between the migration and diffusion terms for a charged species. Only a hypothesis such as the Nernst-Einstein equation can lead to a quantitative distribution between these two terms [29]. Even if such a hypothesis is made frequently, one must be aware that it does not strictly apply in a non-ideal medium.

4.2.1.5 - General equations in a monophasic conductor

If one can describe the concentration profiles for all species and the potential profile at any time and at any point, i.e., the functions $C_i(x, y, z, t)$ and $\varphi(x, y, z, t)$, then it is possible to use the above equations to describe the time evolution of the current through a conductor.

Remember that this section is not focused on interfaces but on the volumes of the conducting media. We will not attempt at first to describe the boundary conditions [30]. At this stage the voltage in question is therefore the potential difference across the conductor outside the interfacial zones, in other words, what is usually called the ohmic drop.

---

[29] *Later in this section we give an example showing how migration and diffusion are distributed with different proportions in the anion and cation fluxes. An additional example is outlined in appendix A.4.1. In both cases the Nernst-Einstein equation is supposed to apply.*

[30] *Section 4.2.3 addresses this issue.*

In most electrochemical systems, except for poor conducting media [31], one can assume that electroneutrality applies on a macroscopic scale in the zone located away from the interfaces. The mathematical description of the system therefore combines the electroneutrality and the various volume mass balances of each species with the equations for the molar flux densities in the framework of the thermodynamics of linear irreversible processes. The result is the following system with $(n+1)$ differential equations for $(n+1)$ unknowns (n concentrations and the potential):

$$\forall i \begin{cases} \dfrac{\partial C_i}{\partial t} = -\operatorname{div}\mathbf{N}_i + w_i \\ \displaystyle\sum_i z_i\, C_i = 0 \end{cases}$$

with: $\qquad \forall i \begin{cases} \mathbf{N}_i = -D_i\ \mathbf{grad}\, C_i - \dfrac{|z_i|}{z_i} u_i\, C_i\ \mathbf{grad}\,\varphi + C_i\ \omega_{\text{medium}} \\ w_i = f(C_j) \qquad\qquad \text{homogeneous kinetic laws} \end{cases}$

In order to give a full description of the system and establish the (I, U, t) relationships, one also needs to take into account its geometry, the initial state and the boundary conditions, i.e., the equations at all the boundaries, including the electrochemical interfaces.

In all systems, one of the impacts of electroneutrality in the conducting volumes ($\sum_i z_i C_i = 0$) is that the overall convection current density is always zero, resulting in the following equation:

$$\mathbf{j} = \underbrace{-\mathscr{F}\sum_i z_i\, D_i\ \mathbf{grad}\, C_i}_{\mathbf{j}_{\text{diffusion}}}\ \underbrace{-\sum_i \lambda_i\, C_i\ \mathbf{grad}\,\varphi}_{\mathbf{j}_{\text{migration}}}\ +\ \underbrace{\mathscr{F}\ \omega_{\text{medium}}\sum_i z_i\, C_i}_{\mathbf{j}_{\text{convection}}}$$

One can also observe that if the diffusion coefficients are considered as sufficiently close to each other, then the overall diffusion current density vanishes as shown below:

if $\qquad\qquad\qquad \forall i \qquad D_i \approx D$

then $\qquad \mathbf{j}_{\text{diffusion}} \approx -\mathscr{F}\, D\sum_i z_i\ \mathbf{grad}\, C_i \approx -\mathscr{F}\, D\ \mathbf{grad}\!\left(\sum_i z_i\, C_i\right) = 0$

hence $\qquad\qquad\qquad\qquad \mathbf{j}_{\text{migration}} \approx \mathbf{j}$

Although this approximation is often faulty in numerical terms, it is nonetheless used quite frequently in electrochemistry. An example where such an approximation cannot be accepted is discussed in detail below.

To give a particularly important example of these volume equations, take the case where diffusion is the only mass transport mode of a given species at a given point in space,

---

[31] *Section 3.1.1.1 addresses this aspect. If electroneutrality does not apply, then the corresponding equation must be replaced by a more general equation: one of the MAXWELL equations if there are significant magnetic phenomena involved, or merely the POISSON equation. This may occur in poor conducting media or if the time scale considered is too short for electroneutrality to be restored (this may happen if high local ion depletion occurs in the medium).*

and where there is no local volume mass source such as a homogeneous chemical reaction. These conditions are frequently come across in electrochemistry, and indeed they will be addressed in detail in the forthcoming sections (see sections 4.3 and 4.4). The corresponding equation is known as the second FICK law [32], and is written as follows:

$$\frac{\partial C_i}{\partial t} = D_i \, \nabla^2 C_i$$

or in a system with unidirectional geometry:

$$\frac{\partial C_i}{\partial t} = D_i \, \frac{\partial^2 C_i}{\partial x^2}$$

The example given here is rather complex compared to the situations typically described in electrochemistry text books, in particular because there is no supporting electrolyte and the diffusion coefficients are significantly different. However, using an example involving simple calculations can illustrate some of the consequences which emerge as a result of the differences in diffusion coefficients. Sections 4.3 and 4.4 should be read before returning to the details of this calculation.

▶ *Presentation of the system studied*

Let us consider the electrolysis of a deaerated aqueous solution containing highly concentrated hydrogen chloride placed between two platinum electrodes. For the sake of simplicity, we will assume that water oxidation is a slow process, and therefore the only phenomenon to be observed at the anode is the oxidation of chloride to dichlorine. Protons are reduced to dihydrogen at the cathode. The faradic yield of both half-reactions is therefore 100%. The electrolysis cell has unidirectional geometry, and thanks to forced convection the composition of the solution remains homogeneous outside the two diffusion layers which are next to the electrodes. In the absence of a supporting electrolyte, mass transport is therefore ensured by migration and diffusion in both diffusion layers, and by migration and convection in the homogeneous zone, as illustrated in figure 4.8.

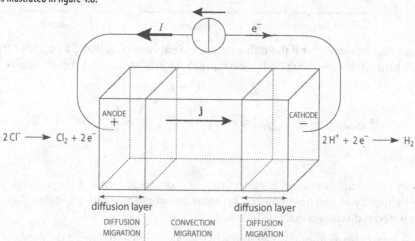

Figure 4.8 - *Diagram of an electrolysis cell showing the various mass transport modes involved*

[32] The first FICK law corresponds to the following equation for the diffusion molar flux density: $\mathbf{N}_i = - D_i \, \mathbf{grad} \, C_i$. The second FICK law is then based on the volume mass balance:

$$\frac{\partial C_i}{\partial t} = - div \, \mathbf{N}_i = D_i \, div \, (\mathbf{grad} \, C_i) = D_i \, \nabla^2 C_i.$$

Here it is also assumed that the molar conductivities are equal to their values at infinite dilution (see table 4.2 in section 4.2.2.4) and are independent of the concentrations: $\lambda_+ = \lambda_+^0 = 35.0$ mS m^2 mol^{-1} and $\lambda_- = \lambda_-^0 = 7.6$ mS m^2 mol^{-1}. It is also assumed that the NERNST-EINSTEIN equation applies. Therefore, the following values emerge for the diffusion coefficients: $D_+ = 9.2 \times 10^{-9}$ m^2 s^{-1} and $D_- = 2.0 \times 10^{-9}$ m^2 s^{-1}.

▸ **Direction of the various molar flux densities**

In the diffusion layers the overall molar flux densities for each ion are governed by the redox reactions (see FARADAY's law in section 4.4). In the homogeneous zone, since no diffusion occurs, the current is equal to the migration current which defines the direction of the ion molar fluxes (see section 2.2.4.2) as illustrated in figure 4.9.

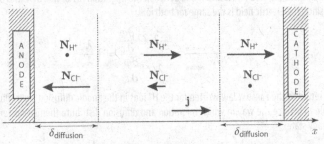

Figure 4.9 - Diagram of an electrolysis cell with the different molar flux densities of the ions in the three zones of the electrolyte (a dot represents a zero vector)

The equations outlined in the next parts of this example involve concepts that will be explained further on (see in particular section 4.3.1). However they rely on the equations for both molar flux densities and the FARADAY law, both of which have been already outlined.

▸ **Steady-state concentration profiles**

In this 1-1 electrolyte, as there are only two types of ions, the medium's electroneutrality means that the cation (H$^+$) and anion (Cl$^-$) end up with equal concentrations at all points in the electrolyte (denoted by C). The diffusion layers show a concentration profile, however this must be identical for both ions. At steady state, the molar flux densities, which are identical at all points, are given by the FARADAY law. Let us focus on the anodic diffusion layer. The H$^+$ flux is zero because only Cl$^-$ is electroactive at the anode:

$$\begin{cases} N_+ = 0 \\ N_- = -\dfrac{I}{\mathscr{F}S} \end{cases}$$

If one expresses the interfacial molar flux densities by introducing the NERNST-EINSTEIN equation, it gives the following:

$$\begin{cases} N_+ = -D_+ \dfrac{\partial C}{\partial x} - D_+ \dfrac{\mathscr{F}}{RT} C \dfrac{\partial \varphi}{\partial x} = 0 \\ N_- = -D_- \dfrac{\partial C}{\partial x} + D_- \dfrac{\mathscr{F}}{RT} C \dfrac{\partial \varphi}{\partial x} = -\dfrac{I}{\mathscr{F}S} \end{cases}$$

or:
$$\frac{\partial C}{\partial x} = -\frac{\mathscr{F}}{RT} C \frac{\partial \varphi}{\partial x}$$

hence:
$$\frac{\partial C}{\partial x} = \frac{I}{2 D_- \mathscr{F} S}$$

At the anode, the concentration profiles only depend on the diffusion coefficient of the electroactive species, which in this case is the anion. It is as if we had pure diffusion with an apparent diffusion coefficient equal to $2 D_-$.

▸ *Relative contributions of the migration and diffusion molar flux densities at steady state*

The fact that electroneutrality applies throughout the electrolyte imposes a strong constraint, which in turn creates a link between the diffusion molar flux densities of the two ions (algebraic normal components) at all points in the electrolyte:

$$\frac{N_{+_{diffusion}}}{N_{-_{diffusion}}} = \frac{-D_+\dfrac{\partial C}{\partial x}}{-D_-\dfrac{\partial C}{\partial x}} = \frac{D_+}{D_-}$$

As for the migration terms, a link is also made between the migration molar flux densities of the two ions at all points, since the electric field is the same for both ions:

$$\frac{N_{+_{migration}}}{N_{-_{migration}}} = \frac{-\lambda_+ C\dfrac{\partial \varphi}{\partial x}}{\lambda_- C\dfrac{\partial \varphi}{\partial x}} = -\frac{\lambda_+}{\lambda_-}$$

Now if we return to the FARADAY law written for the H^+ ions in the anodic diffusion layer (flux equal to zero), you are able to deduce the way in which migration and diffusion distribute themselves for each ion in this zone at steady state:

$$\frac{N_{+_{diffusion}}}{N_{+_{migration}}} = -1 \quad \text{and} \quad \frac{N_{-_{diffusion}}}{N_{-_{migration}}} = \frac{N_{-_{diffusion}}}{N_{+_{diffusion}}}\frac{N_{+_{diffusion}}}{N_{+_{migration}}}\frac{N_{+_{migration}}}{N_{-_{migration}}} = -\frac{D_-}{D_+}\left(-\frac{\lambda_+}{\lambda_-}\right) = 1$$

At steady state, the moduli of the migration and diffusion molar flux densities are therefore identical for any given species. The steady-state current is equally due to diffusion and migration. Let us stress the formal nature of such a distribution for the diffusion and migration fluxes, that is wholly based on the NERNST-EINSTEIN law. Furthermore, the current is fully transported by the electroactive species. Non-electroactive ions have no macroscopic movement at steady state. This is not the case in a transient state, since the steady-state profile has to be built up (the interfacial flux of the electroinactive ions is still zero however the transient flux density is not constant throughout the whole volume of the diffusion layer).

▸ *Quantitative links between migration and the overall current densities at steady state*

If you return to the equation for FARADAY's law for Cl^- ions in the anodic diffusion layer, you can write the following:

$$N_- = N_{-_{diffusion}} + N_{-_{migration}} = 2\,N_{-_{migration}} = -\frac{I}{\mathscr{F}\,S}$$

and finally:

$$j_{migration} = j_{+_{migration}} + j_{-_{migration}} = \mathscr{F}\,(N_{+_{migration}} - N_{-_{migration}})$$

$$= \mathscr{F}\,N_{-_{migration}}\left(\frac{N_{+_{migration}}}{N_{-_{migration}}} - 1\right) = \frac{I}{2S}\left(\frac{\lambda_+}{\lambda_-} + 1\right)$$

On the anodic side, the numerical values are therefore:

$$j_{migration} = 2.8\,j \quad \text{and} \quad j_{diffusion} = 1 - j_{migration} = -1.8\,j$$

When applying the same type of reasoning to the cathodic side, you obtain the following:

$$\frac{N_{-_{diffusion}}}{N_{-_{diffusion}}} = -1 \quad \text{and} \quad \frac{N_{+_{diffusion}}}{N_{+_{diffusion}}} = 1$$

hence:

$$j_{migration} = \frac{I}{2S}\left(\frac{\lambda_-}{\lambda_+} + 1\right)$$

on the cathodic side, the numerical values are:

$$j_{migration} = 0.6\,j \quad \text{and} \quad j_{diffusion} = 1 - j_{migration} = +0.4\,j$$

In this example, as in most cases, the migration current density has the same direction as that of the overall current density although they are not equal. These two vectors are identical when the two diffusion coefficients are equal.

4.2.2 - CONDUCTION PHENOMENA:
MECHANISMS AND ORDERS OF MAGNITUDE

Scientific literature frequently presents microscopic mechanisms for electric conduction using examples which only involve migration. In fact, once the charge carriers have been well identified, one can assume that the microscopic mechanisms briefly presented below also apply to diffusion. They describe how the mass transport is carried out at the microscopic level, under the influence of an external force, regardless of the nature of the forces involved (in migration or in diffusion).

4.2.2.1 - EXAMPLES OF CONDUCTION MECHANISMS

Depending on the materials involved, there is a range of different conduction mechanisms that can be imagined at the microscopic level. In experimental data, the influence of temperature on the conducting properties is an important criterion for determining the conduction mechanism. Some types of mechanisms are described below.

▸▸ Electronic conduction in a metal

A metal can be described as a fixed lattice of metal cations, with a gas of electrons circulating through it (band model – energy continuum). The model used to describe the movement of these conduction electrons is called the 'mean free path'. The electrons experience collisions but are not submitted to any force between two collisions. The collisions may be due to:
- impurities which create defaults in the crystalline lattice (primary phenomenon at very low temperature),
- phonons which describe the vibrations of the crystalline lattice (primary phenomenon at and over room temperature).

The electron mobility is therefore proportional to the mean time interval between two collisions, τ, which corresponds to the mean free path. This model accounts for the temperature influence in qualitative terms. When the temperature increases, the mobility decreases, as does τ. This model also takes into account how electronic mobilities depend on the nature of the metal in question by introducing different values for τ and for the amount of charge carriers per unit volume. The order of magnitude of τ is one nanosecond at very low temperature and one femtosecond at room temperature.

▸▸ Ionic conduction in an electrolytic solution

In dilute solutions, one can assume that the ions move *via* a Brownian motion, that is to say by means of the ions and solvent molecules colliding together. In the absence of any external force, the statistical mean value of ion displacement is zero. However, if there is an external force, for example when there is an external electric field, then a trend emerges in terms of the displacement direction. Collisions with solvent molecules are opposed in statistical terms to the ion movement in the direction of the external force. Very rapidly, a steady state is reached, in which the mean ion velocity is determined by the balance between the external force and the friction force exerted by the solvent. The time required for this to occur is around one picosecond in aqueous solution at room temperature (see the calculation below).

▶ In a system without convection ($\boldsymbol{\omega}_{\text{medium}} = \boldsymbol{0}$), for a solvated ion with a z_i charge number, the NEWTON's law reveals the link between the acceleration and the sum of external forces. When there is an external applied electric field, $\mathbf{E}$, the forces involved are the electric force and the friction force. The latter, which is considered in the approximation of the viscous friction model, is proportional to the $\boldsymbol{\omega}_i$ velocity of the ion in the solvent, and the following equation is obtained:

$$z_i |e| \, \mathbf{E} - \alpha \, \boldsymbol{\omega}_i = m \, \mathbf{a} = m \frac{\mathrm{d}\boldsymbol{\omega}_i}{\mathrm{d}t}$$

This first order differential equation can also be written as follows:

$$\frac{\mathrm{d}\boldsymbol{\omega}}{\mathrm{d}t} + \frac{\alpha}{m} \, \boldsymbol{\omega}_i = \frac{z_i |e|}{m} \mathbf{E}$$

It is easily solved when the external electric field is constant (with $\mathbf{A}$: the integration constant):

$$\boldsymbol{\omega}_i = \frac{z_i |e|}{\alpha} \mathbf{E} + \mathbf{A} e^{-\frac{\alpha t}{m}}$$

After a transient period with a characteristic time constant of $\tau = m/\alpha$, the ions reach a steady state with a limiting velocity proportional to the electric field:

$$\boldsymbol{\omega}_{\text{lim}} = \frac{z_i |e|}{\alpha} \mathbf{E}$$

In order to get an idea of the time needed to reach this steady state in aqueous solutions, you can evaluate α by using the electric mobility (around $10^{-7} \, \text{m}^2 \, \text{s}^{-1} \, \text{V}^{-1}$, see table 4.1 in section 4.2.2.3) which is equal to the limiting velocity in a unit electric field.

$$\tau = \frac{m}{\alpha} \approx \frac{m}{|e|} 10^{-7} = \frac{M}{\mathcal{N} |e|} 10^{-7}$$

The molar mass of a solvated ion can be taken as 100 g mol^{-1} which brings an order of magnitude of 10^{-13} s for the characteristic time constant of the transient state. ◢

The main effect observed when the temperature of the solution increases is a decrease in solvent viscosity, which coincides therefore with a decrease in the friction coefficient. Furthermore, one usually observes an increase in the electric mobility of the solvated ions [33].

In a concentrated solution the moving ions do not only collide with the solvent molecules but also with the other ions, especially those with opposite charges which also drag solvent molecules along with them. This slows down the movement by

---

[33] *For instance, an aqueous solution containing potassium chloride with a concentration of 0.02 mol L^{-1} sees an increase in conductivity from 0.25 S m^{-1} at 20 °C to 0.33 S m^{-1} at 35 °C.*

increasing the medium's friction. This phenomenon is called the electrophoretic effect, and plays a part in reducing mobility in concentrated solutions in comparison to diluted solutions. A second effect also comes into play in this reduction process: the relaxation effect. The ionic atmosphere around a moving ion is no longer symmetrical[34] because the formation/destruction of this atmosphere is not immeasurably fast. The electric dipole that is created as a result sets up a resistance to the external force and consequently slows down the ion's movement.

▸▸ Ionic conduction in a crystalline solid

In crystalline solids, the conduction mechanisms envisaged generally involve defects in the crystal lattice (vacancies or interstitial ions). The latter are created by local disorder, and their relative charge values always respect the bulk electroneutrality. For example, the relative charge of a vacancy is the opposite to that of the ion when the latter is in a normal position in the network.

The primary types of defects are:

▸ SCHOTTKY's disorders, which occur in many alkali halides: the alkali ion vacancies and the halide vacancies are equal in number;

▸ interstitial ions: when ions are in an interstitial position, such as in silver halides, electroneutrality is once again respected, because one cation vacancy is produced for every interstitial ion created (FRENKEL's disorder);

▸ as a way of improving the conduction properties, crystals are often doped with impurities with different valencies. This occurs for example in the case of zirconia doped with calcium oxide CaO or yttrium oxide Y_2O_3. These defects are called extrinsic defects. Electroneutrality is then ensured by creating either vacancies or interstitial ions.

The diagram in figure 4.10 shows the movement of an ion when conduction occurs *via* vacancies. The mechanism used here jumps from an occupied site to a vacant site.

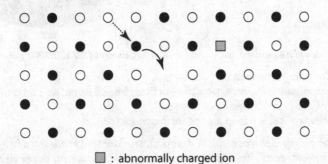

☐ : abnormally charged ion

Figure 4.10 - Diagram showing a conduction mechanism in a crystalline solid

The higher the temperature, the higher the rate of intrinsic defects is. In addition, temperature increase makes it easier to cross the potential barrier from one site to another (activation energy).

---

[34] *Contrary to the hypothesis made in the DEBYE-HÜCKEL model (see appendix A.3.2).*

ENERGY STORAGE: SUPERCAPACITORS

Document written with the kind collaboration of K. GIRARD,
R&D engineer for Batscap, based in Quimper in France

In normal operating conditions, no redox reactions occur in supercapacitors. They function based on a well-known electrochemical phenomenon: the double layer phenomenon. In electrochemistry, this phenomenon reflects the accumulation of ions in the vicinity of a charged electrode (the charge of the electrode being opposite to that of ions). Following this principle, high amounts of charge can be accumulated based on the following relationship: $q = C_{dl} S (E - E_0)$, whereby C_{dl} is the double layer capacitance, S the area of the surface developed and E_0 the system's open-circuit potential.

The thickness of the HELMOTZ double layer, which is about a few nanometers, over an extensively developed area, means that a significantly greater quantity of charge can be accumulated compared to classical capacitors. It is possible to obtain a capacitance of several thousand Farads for one unit element.

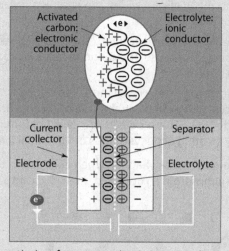 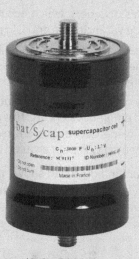

Schematic view of a supercapacitor (© Batscap) *and unit element (2.7 V, 3000 F)* (photo Batscap)

Supercapacitors are composed of:
▸ two porous electrodes (to maximize the surface) based on active carbon,
▸ a separator based on polymer or cellulose,
▸ an organic electrolyte, which is a good ionic conductor.

In the absence of any electrochemical reaction, the kinetic rate is significantly boosted as a result. This extremely high-speed operation results in a high power density, both in charge and discharge, reaching power densities of over 10 000 W/kg. On the other hand, there is a relatively low amount of embedded energy, about a few Wh/kg in such systems. They also have a high level of cyclability, clocking up some several thousand cycles. Supercapacitors can be particularly useful for applications which need to attain high power levels during a short time period with frequent recharging. For example, a tramway which runs without a catenary requires a considerable power surge to set it off at the start, but can subsequently be recharged at each stop.

▶▶ Ionic conduction in a polymer electrolyte

In a polymer electrolyte at temperatures higher than a certain value called glass transition temperature, the polymer chains behave in a similar way to solvent molecules in a liquid, although no macroscopic displacement occurs. The conduction model, called the free volume model, is close to the model that can be developed in certain liquid conducting media. The ion moves as a result of deforming neighbouring polymer chains (or moving molecules in a liquid) which provide an adjacent cavity or free volume in which this ion can lodge itself.

4.2.2.2 - CONDUCTIVITY MEASUREMENTS

Whenever diffusion can be disregarded, the overall current density is equal to the migration current density, as presented in the following equation:

$$\mathbf{j} = \sum_i \lambda_i\, C_i\, \mathbf{E} = \sigma \mathbf{E}$$

One method for obtaining the overall conductivity parameter is to impose a potential difference across the conductor, with a well-known geometry, and then to measure the resulting current. The ratio between these two quantities enables us to determine the resistance, as well as to calculate the conductivity through the conductor's geometric parameters.

For instance, the following equation shows the voltage, U, between two sections separated by a ℓ distance in a homogeneous unidirectional conductor and with an equipotential section of area S:

$$U = R\,I \quad \text{with} \quad R = \frac{\rho\,\ell}{S} = \frac{\ell}{\sigma\,S}$$

▼ Transparent electronic conductors are required in certain applications. A thin layer of an electronically conducting oxide deposited on a glass substrate is then used.

Given this rather particular geometric configuration, as shown in figure 4.11, the parameter that defines the conductor's properties combines both the material's conductivity (or resistivity) and its thickness:

$$R = \frac{\rho\,L}{S} = \frac{\rho}{e}\frac{L}{\ell}$$

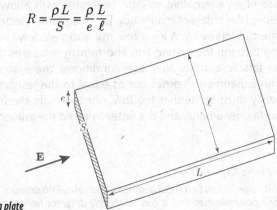

Figure 4.11 - Diagram of a conducting plate

The sheet resistance of a given thin conductor, $R_\square$, expressed in Ω per square, is the resistance of a square plate sample of the conductor ($\ell = L$).

For example, for a thin layer of SnO_2:F (tin oxide doped with fluorine with a conductivity of $\sigma = 2.5 \times 10^5$ S m^{-1}, see table 4.1) and a thickness $e = 5\,000$ Å, the square resistance is the following:

$$R_\square = \frac{1}{\sigma e} = \frac{1}{5 \times 10^{-7} \times 2.5 \times 10^5} = 8\ \Omega_{/\square}$$

It is relatively easy to apply this technique to a metal (direct potential and current). When dealing with electronic conductors with lower conductivity (e.g., semiconductors), then some improvements are needed, such as the 'four-points device' or developing samples with a particular geometry.

The experimental conductivity values show dramatically different orders of magnitude: from 10^7 S m^{-1} for the best metallic conductors to 10^{-15} S m^{-1} for insulators[35].

For an ionic conductor (or similarly for a semiconductor), applying this method is, by definition, a much more complicated affair: measuring currents or potential differences requires metallic contacts to be put in place with the conductor in question. A minimum of two electrochemical interfaces must therefore be introduced, which generally leads to deviations from OHM's law. It is assumed that in usual experimental conditions (see below) one can keep the influence of these interfacial phenomena to a minimum in the relationship between U and I.

In addition, for precise conductivity measurements, one always takes the precaution of calibrating the cell used with a reference ionic solution. This is most often a decimolar KCl solution, at a controlled temperature. Knowing the value of this electrolyte's conductivity (see data tables for different temperatures), one can determine the proportionality factor, called the cell constant. One can then also measure the conductivity of a different electrolyte in the same conductivity range using the same conductivity cell. It should be noted that if the value of the cell constant obtained is very different from that given by the manufacturer[36] or from that deduced from the cell geometry, then it is not enough to simply correct the proportionality factor. In fact it should be seen as a sign that the measuring conditions are not adequately suitable (inappropriate choice for the selected frequency, damaged electrodes, modified side-effects, etc.).

The use of an alternative voltage generator with a low amplitude and an appropriate frequency (the orders of magnitude for an aqueous electrolytic solution at room temperature are the following: $\Delta V \approx$ a few mV and $f \approx$ a few kHz) minimises the current values flowing through the system and enormously increases the capacitive current compared to the faradic current. In these conditions, the system's impedance as determined through experiments comes out as equal to the resistance of the ionic conductor. The frequency must be neither too low, nor too high, the former to avoid the influence of electrode phenomena, and the latter to avoid the influence of the electrolyte's dielectric

---

[35] See table 4.1 in section 4.2.2.3.

[36] This value relates to the cell geometry and takes into account the possible side-effects. For instance for a parallelepiped cell, it can be slightly different from the ℓ/S ratio. This is what is sometimes called primary current distribution, which can be obtained by solving the LAPLACE equation in the particular system's geometry, assuming the electrolyte conductivity is constant. It is usually determined by an experimental measurement.

properties. In addition, using metallic electrodes with a very high roughness, opens up the choice range for the measurement frequency[37].

Since the quantity measured, namely the electric conductivity, depends significantly on temperature, then it is better to use a temperature-controlling device when making precise measurements.

When trying to differentiate the contributions of the various ions, it is not enough to simply have a conductivity measurement. One needs to carry out additional experiments to determine the transport numbers.

4.2.2.3 - ORDERS OF MAGNITUDE FOR CONDUCTION PARAMETERS

Table 4.1 (see the next two pages) gives the values for the mass transport quantities typically seen for various types of charge carriers in several typical examples of media.

4.2.2.4 - MODELS FOR SOLUTIONS AT INFINITE DILUTION

In a simple infinitely diluted solution, the parameters for charge transport can be linked to some of the solvent's properties. In fact, the movement of a solvated ion in the solution can be viewed as similar to the displacement of a rigid sphere in a viscous medium.

Based on the STOKES law (viscous friction hypothesis), for a sphere with r_i radius moving with a ω_i^{rel} relative velocity, through a medium with η viscosity, the friction force is of the following type:

$$\mathbf{f}_{fr} = - 6\,\pi\,r_i\,\eta\,\omega_i^{rel}$$

The balance at the microscopic level between the friction force and a unit external force (namely with a modulus equal to 1 N) applied to an ion i which is considered to be a sphere with an radius r_i, results in a movement with a steady-state velocity. The velocity modulus, once divided by the AVOGADRO constant, is equal to the electrochemical mobility:

$$\widetilde{u}_i^{\,0} = \frac{1}{6\,\pi\,\eta\,\mathcal{N}\,r_i}$$

Depending on the viscosity η of the solvent and on the radius r_i of the solvated ion, one can therefore write equations for the constants which characterise the following:

▸ migration: $\quad \lambda_i^{\,0} = \dfrac{z_i^2\,\mathscr{F}^2}{6\,\pi\,\eta\,\mathcal{N}\,r_i}$

▸ diffusion (derived from the STOKES-EINSTEIN equation): $\quad D_i^{\,0} = \dfrac{RT}{6\,\pi\,\eta\,\mathcal{N}\,r_i}$

---

[37] In this context, platinised platinum (the same as that used for hydrogen reference electrodes, see section 1.4.1.2) is often used. When the contact area with the electrolyte increases, the double layer capacity increases substantially, when compared to the system's dielectric characteristics. The reason for this is that mean roughness is larger than the thickness of the double layer. In terms of impedance measurements, this leads to a broader frequency range in which the system's impedance is equal to the electrolyte's resistance.

Table 4.1 – Concentration, electric and electrochemical mobilities, electric conductivity, molar conductivity and diffusion coefficient in some media at 25°C (unless otherwise specified)

		C concentration [mol L⁻¹]	u electric mobility [m² s⁻¹ V⁻¹]	$\tilde{u}$ electrochemical mobility [mol s kg⁻¹]	$\sigma = F\,C\,u$ electric conductivity [S m⁻¹]	λ molar conductivity [S m² mol⁻¹]	D diffusion coefficient [m² s⁻¹]
Metals							
Ag	electrons	97	6.6×10^{-3}	6.9×10^{-8}	6.2×10^{7}	6.4×10^{2}	1.7×10^{-4}
Cu	electrons	140	4.4×10^{-3}	4.5×10^{-8}	5.9×10^{7}	4.2×10^{2}	1.1×10^{-4}
Au	electrons	98	4.8×10^{-3}	5.0×10^{-8}	4.6×10^{7}	4.6×10^{2}	1.2×10^{-4}
Intrinsic semiconductors							
Si	electrons	2×10^{-11}	0.13	1.5×10^{-6}	3×10^{-2}	1.4×10^{4}	3.6×10^{-3}
	holes	2×10^{-11}	0.05	5.2×10^{-7}	1×10^{-2}	4.8×10^{3}	1.3×10^{-3}
C diamond (insulating)	electrons	$\approx10^{-20}$	0.18	1.9×10^{-6}	$\approx2\times10^{-16}$	1.7×10^{4}	4.6×10^{-3}
	holes	$\approx10^{-20}$	0.12	1.2×10^{-6}	$\approx10^{-16}$	1.1×10^{4}	3.1×10^{-3}
Extrinsic semiconductors							
As-doped Si	holes << electrons	8×10^{-5}	1.3×10^{-4}	1.3×10^{-9}	1.0	1.3×10^{1}	3.3×10^{-6}
F-doped SnO_2		0.7	3.7×10^{-3}	3.8×10^{-8}	2.5×10^{5}	3.6×10^{2}	9.5×10^{-5}

Aqueous solutions						
H^+	0.1	3.6×10^{-7}	3.7×10^{-12}	3.5	3.5×10^{-2}	9.2×10^{-9}
Na^+	0.1	5.2×10^{-8}	5.3×10^{-13}	0.5	5.0×10^{-3}	1.3×10^{-9}
OH^-	0.1	2.1×10^{-7}	2.2×10^{-12}	2.0	2.0×10^{-2}	5.5×10^{-9}
Cl^-	0.1	7.9×10^{-8}	8.2×10^{-13}	0.8	7.6×10^{-3}	2.0×10^{-9}
$K^+ Cl^-$ at 20°C	0.1	—	—	1.2	1.2×10^{-2}	—
$K^+ Cl^-$ at 35°C	0.1	—	—	1.6	1.6×10^{-2}	—
Dioxygen, dissolved O_2 (25°C and 1 atm)	3×10^{-4}	—	—	—	—	2.5×10^{-9}
Molten salts						
NaCl at 850°C Na^+	26	8×10^{-8}	8.10^{-13}	2×10^{2}	8×10^{-3}	8×10^{-9}
Cl^-	26	2×10^{-8}	2.10^{-13}	5×10^{1}	2×10^{-3}	2×10^{-9}
Ionic solids						
$(ZrO_2)_{0.9}(Y_2O_3)_{0.1}$ at 800 °C oxygen vacancy	≈ 5	6×10^{-9}	6×10^{-14}	≈ 3	6×10^{-4}	5×10^{-10}
Polymer electrolytes						
$LiClO_4$ in POE at 60 °C Li^+	≈ 1	10^{-10}	10^{-15}	$\approx 10^{-2}$	10^{-5}	3×10^{-12}
ClO_4^-	≈ 1	4×10^{-10}	4×10^{-15}	$\approx 4 \times 10^{-2}$	4×10^{-5}	10^{-11}
Plasmas / gas						
Plasma He (25°C et 100 Pa) electrons	10^{-9}	3×10^{2}	3×10^{-3}	30	3×10^{7}	7
Dioxygen, O_2 (25°C and 1 atm)	4×10^{-2}	—	—	—	—	2×10^{-5}
Dioxygen, O_2 (100°C and 1 atm)	3.2×10^{-2}	—	—	—	—	3×10^{-5}
Dioxygen, O_2 (25°C and 50 atm)	2	—	—	—	—	4×10^{-7}

The following values can be found in scientific literature.

Table 4.2 - Molar conductivity at infinite dilution for different ions in an aqueous solution at 25°C

	λ_+^0 [mS m^2 mol^{-1}]		λ_-^0 [mS m^2 mol^{-1}]
H^+	34.98	F^-	5.54
D^+	25	Cl^-	7.63
Li^+	3.86	Br^-	7.81
Na^+	5.01	I^-	7.70
K^+	7.35	OH^-	19.92
Ag^+	6.19	OD^-	11.9
NH_4^+	7.34	NO_3^-	7.14
Cu^{2+}	10.7	SO_4^{2-}	16.0
Ca^{2+}	11.9	$Fe(CN)_6^{3-}$	30.3
Zn^{2+}	10.6	$Fe(CN)_6^{4-}$	44.2
Fe^{2+}	10.8	CO_3^{2-}	13.9
Fe^{3+}	20.4	HCO_3^-	4.45

Scientific literature provides many tables that actually refer to the concept of an equivalent, now commonly abandoned[38], though they do not directly give the values of molar conductivities but rather the values of the terms: $\dfrac{1}{|z_i|} \lambda_i{}^0$.

What is confusing is that the conductivity data for ions with a charge number different from 1 are often presented in á way whereby a fraction is placed in front of the ion in question. For instance, scientific literature frequently presents the value for the Ca^{2+} ion as $\lambda_{\frac{1}{2}Ca^{2+}}$, with the following definition: $\lambda_{\frac{1}{2}Ca^{2+}} = \dfrac{1}{2} \lambda_{Ca^{2+}}$

If you look at the values in the previous table for ions sharing the same charge, then you can see that the molar conductivity tends to increase with the element's atomic number, for instance with the family of alkali cations (Li^+, Na^+ and K^+). This observation is the opposite to what one would expect given the size of the ions, if you follow what classical tables usually show as corresponding to a non-solvated ion in an ionic crystal. In fact, it is important to remember that when applying the equation for evaluating the mobility or the diffusion coefficient (see the STOKES-EINSTEIN equation in section 4.2.2.4) the actual size of the solvated ion in the medium must be used. For example, the Li^+ ion, which is small, interacts more strongly with a polar solvent such as water, and its coordination sphere is big. This trend is less clear in the family of halide ions which are bigger and therefore less solvated, except in the case of the F^- ion which is small.

One can see how the values for ions with different charges, such as the Fe ions, depend on the complexation and oxidation number of the Fe element. This is not surprising, however it is more difficult to understand from the raw values, because the charge can change depending on the complexation state. It is therefore better to make a

---

[38] *This concept is mentioned in the transport number calculation described in section 2.2.4.3.*

comparison between the values for electrochemical mobility or diffusion coefficients which do not involve the charge. Therefore one can note that at 25°C the Fe^{2+} diffusion coefficient given by the NERNST-EINSTEIN equation is $7.2 \times 10^{-10}\, m^2\, s^{-1}$ whereas that of Fe^{3+} is $6 \times 10^{-10}\, m^2\, s^{-1}$. This shows that the bare Fe^{3+} ion, which is smaller, tends to have stronger interactions with water, especially because the charge is higher. Its solvation sphere is therefore slightly bigger than that of Fe^{2+}. There is a different scenario for media containing cyanide: the respective diffusion coefficients become $7.3 \times 10^{-10}\, m^2\, s^{-1}$ and $8.9 \times 10^{-10}\, m^2\, s^{-1}$. The mobility order is changed, which can be explained by the fact that because the $Fe(CN)_6^{4-}$ ion has a higher charge, then it will be less mobile than $Fe(CN)_6^{3-}$ because of the stronger interactions with water molecules.

The unusually high value that can be observed in proton mobility involves very different concepts still controversial today. In fact, the solvated proton (hydronium H_3O^+) is roughly the same size as that of K^+, but the molar conductivities or the diffusion coefficients are very different (factor of 5). Assuming that only the STOKES process is used to displace the solvated H^+, the mobility value would be small given the large size of this ion. In fact, proton mobility is mostly due to an exchange of H^+ taking place between neighbouring water molecules which are well oriented. Here a 'tunnelling effect' process could occur, given the properties of the H^+ species. The influence of mass (through the comparison between H^+ and D^+) certainly works in favour of this interpretation. This particular mode has a hopping velocity that is governed by the time it takes for the molecules to be well oriented. This means aligning the free orbital of the O atom in the water molecule in front of an H atom in H_3O^+, as shown in the diagram in figure 4.12. Modelling is actually quite a complex process because each molecule involved in this mechanism is bound to neighbouring molecules by hydrogen bonds. It is generally recognised that the physical movement of the H_3O^+ (STOKES flow model) entity occurs only in a small proportion (20% of the overall mobility).

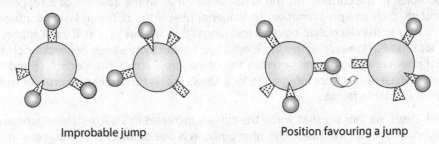

Improbable jump   Position favouring a jump

Figure 4.12 - Diagram of the tunelling conduction mechanism of protons in water

4.2.2.5 - CASE OF CONCENTRATED SOLUTIONS

Experimental studies that were carried out, in particular by KOHLRAUSH, on aqueous solutions containing only one strong electrolyte (a single type of cation and anion) produced an equation that showed how the molar conductivity is a function of the electrolyte's concentration. This is shown in the following:

$$\Lambda = \Lambda^0 - Cst \sqrt{C}$$

Λ^0 gives the value for the molar conductivity extrapolated to $C = 0$, and is called the 'molar conductivity at infinite dilution' of the electrolyte (in S m^2 mol^{-1}). This quantity represents the molar conductivity for an ideal system with no interactions. The above experimental formula can be applied to systems whose concentrations are lower than about 0.1 mol L^{-1}.

Subsequently, this empirical law has been linked to the ionic conduction model in liquid electrolytes involving electrophoretic and relaxation effects, which are found to be very strong in concentrated solutions (see section 4.2.2.1). For a 1-1 strong electrolyte solution, the molar conductivity of an ion i is shown in the following equation:

$$\lambda_i = \lambda_i{}^0 - (A + B\,\lambda_i{}^0)\sqrt{C_i}$$

with $A = 60.3$ and $B = 0.229$ at 25 °C in water, if C_i is in mol L^{-1} and λ_i^0 in S cm^2 mol^{-1}; λ_i^0 is called molar conductivity at infinite dilution of the ion i (in S m^2 mol^{-1}). A, B and λ_i^0 are temperature-dependent.

Given the values of A, B, the molar conductivity of an ion can be considered as independent of the concentration an equal, with an accuracy rate of about 5%, to its value at infinite dilution once the concentration is lower than a few 10^{-3} mol L^{-1} at 25 °C. In such conditions, the electric conductivity of the electrolytic solution is simply proportional to the electrolyte concentration.

4.2.3 - SITUATIONS IN WHICH THE OHMIC DROP
DOES NOT FOLLOW THE MACROSCOPIC OHM LAW

In numerous experimental situations, one can assume that the conductivity is identical at all points throughout the electrolyte[39]. In this instance, the ohmic drop is then proportional to the current. Yet this is not always true: in the absence of a supporting electrolyte, such an approximation no longer applies if the diffusion layers contribute significantly to the electrolyte's overall resistance[40]. OHM's law ($\mathbf{j} = \sigma\,\mathbf{E}$) still applies[41] at a local level. However, since the electrolyte conductivity varies depending on the spatial position, then the link between the ohmic drop across the electrolyte and the current, (which results from integrating local OHM's law) is no longer proportional with a constant resistance factor.

In such cases, we can say that when the current increases in steady-state experiments, the overall resistance increases. This phenomenon is due to lower concentrations in the vicinity of one electrode resulting in a local decrease in the electrolyte conductivity. On the other hand, the electrolyte conductivity in the other diffusion layer may increase at the same time. However the zone with low levels of conductivity has a greater impact on

---

[39] This is true in systems with a supporting electrolyte or in systems where the volumes of the diffusion layers are insignificant compared to the overall electrolyte volume.

[40] An example of this type is addressed in detail in appendix A.4.1: the contribution of transport by diffusion is significant in the overall electrolyte volume.

[41] An additional complication may arise if the media studied contain charge carriers with transport properties very different from each other. In fact we have seen in section 4.2.1.5 that, in such conditions, the migration current and the total current are not identical ($\mathbf{j} \neq \mathbf{j}_{migration} = \sigma\,\mathbf{E}$). The example outlined in appendix A.4.1 takes this aspect into account.

the overall resistance of the electrolyte. This goes hand in hand with a deviation from strict electroneutrality, which in turn has an immediate impact on the electrolyte's potential profiles. They are no longer linear, as one would predict when applying the LAPLACE law. When the current approaches the steady-state limiting current, the electrolyte's overall resistance greatly increases. In such operating conditions the electroneutrality can no longer be used, even for calculating the concentration profiles.

4.3 - CURRENT FLOW
THROUGH AN ELECTROCHEMICAL INTERFACE

When a current flows through a heterogeneous system it gives rise to phenomena in the different volumes (mass transport and possible chemical reactions) as well as interfacial phenomena. This section focuses on the overall characteristics (U, I, t) of an electrochemical system in relation with the characteristics of the studied interfaces. The influence of these latter on how that system is described can be found in two different ways. On the one hand they define the conditions at the volume boundaries for concentration and potential profiles, and on the other hand they determine the interfacial contributions which, added to the ohmic drop, give the overall cell voltage [42].

For certain electrochemical systems it is possible to find experimental conditions which minimise the interactions between the anode and the cathode. Both electrodes remain related to each other since they are crossed by the same current, yet the difference is that the mass transport phenomena occurring at both interfaces do not interact with each other. This type of scenario, which is typically sought after in analytical experiments, is explored in this paragraph, focused exclusively on describing one single electrochemical interface [43]. Moreover, it is worth noting that the same approach can be applied to any interface, such as for instance an ionic junction.

4.3.1 - POTENTIAL AND CONCENTRATION PROFILES AT AN INTERFACE

4.3.1.1 - POTENTIAL PROFILE

In electrochemistry, on usual time scales ($\gg 10^{-15}$ s) it can be assumed that electron exchange is immeasurably fast (no energy barrier) at the junction between two metals. In these conditions, the current flow through a metallic junction does not modify the latter in any way. In particular, the electronic junction voltage is the same as in equilibrium, when the current is zero, as represented in figure 4.13.

---

[42] Refer to section 2.2.3 and to figures 2.15 and 2.16 in section 2.2.4.2 for the qualitative description of potential profiles in an electrochemical cell, including the different contributions to the voltage across an electrochemical system with a current flowing. In addition, remember that the description of the concentration or potential profiles is on the spatial scale of the diffusion layers and not on the double-layers' scale (see section 3.3.1). In particular, while the potential profile is continuous on the scale of the double-layer dimensions, it shows discontinuities on the scale of the diffusion layers at the interfaces.

[43] Section 4.4 provides examples where both the electrodes strongly interact as a result of mass transport phenomena.

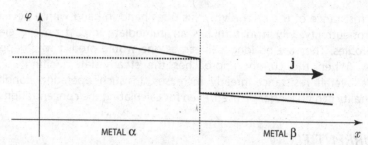

Figure 4.13 - Potential profiles at a junction between two metals with different conductivities (different slopes with $\sigma_\alpha < \sigma_\beta$) in equilibrium (dotted line) and when a current is flowing (solid line)

On the other hand, the current flow disrupts the interface in the case of electrochemical interfaces and ionic junctions. The voltage across the interface (or the junction voltage for an ionic junction) is generally different from that observed at open circuit. It depends *a priori* on the parameters of the system and deviates from a linear law, such as OHM's law. Most often it can be assumed that the double layer thickness is much lower than that of the diffusion layers[44]. Typically, one ends up with the following:

$$\left.\begin{array}{l} \delta_{\text{double layer}} \approx 10 \text{ Å} \\ \delta_{\text{diffusion layer}} \approx 10 \text{ µm} \end{array}\right\} \quad \Rightarrow \quad \delta_{\text{double layer}} \ll \delta_{\text{diffusion layer}}$$

In the following we will stay with such systems where both layers have very different scales. In this configuration, on the scale of the diffusion layer, the potential gradient (i.e., the electric field) is negligible compared to those in the double layer, as illustrated in figure 4.14.

The electric field is extremely strong in the double layer, yet it is much weaker throughout the rest of the electrolyte. However, since one may be dealing with large distances in the electrolyte, this relatively low electric field in the electrolyte may result in a non-negligible voltage across the electrolyte, corresponding to the ohmic drop.

On the other hand, the driving force behind the electrochemical reaction occurring in the interfacial zone is related to the double layer voltage (denoted by $\Delta\varphi_0$ in figure 4.14). The relationship between the double layer voltage and the current is precisely the subject of studies and models in electrochemical kinetics[45].

4.3.1.2 - CONCENTRATION PROFILES

Remember that the mass balance at an interface displays each interfacial molar flux density (and therefore also the current densities and currents of each type of charge carrier) in terms of the sum of two terms, their names being faradic and capacitive (see section 4.1.3.3).

---

[44] There are exceptions to the typical situation described here: for example in electrolytes with very low conductivity, and therefore with a very thick double layer (see section 3.3.1) or for experiments lasting for a very short duration (such as in voltammetry with a high scan rate) or in systems in which the electrodes are very close to each other. In these two last instances the diffusion layer is thinner than 10 µm.

[45] Section 4.3.2 gives a brief overview of the models found in electrochemical kinetics.

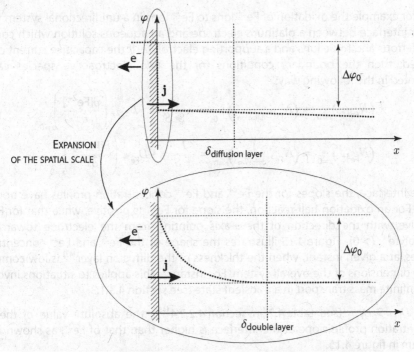

Figure 4.14 - Potential profile at an electrochemical interface with current flow
$\Delta\varphi_0$ is the voltage across the double layer.

For an electrochemical interface, the molar flux density is divided as follows:

$$(N_i)_{\text{interface}} = (N_i^{\text{capac}})_{\text{interface}} + (N_i^{\text{farad}})_{\text{interface}} = -\frac{\partial \Gamma_i}{\partial t} + w_{S_i}$$

Using FARADAY's law it is easy to introduce the current in these equations. For example, if only one half-reaction is involved at the interface, one can write:

$$(N_i^{\text{farad}})_{\text{interface}} = \frac{1}{\mathscr{F}}\frac{\nu_i}{\nu_e}\, j^{\text{farad}} = \frac{1}{\mathscr{F}\,S}\frac{\nu_i}{\nu_e}\, I^{\text{farad}}$$

When there is a supporting electrolyte, one can show that the migration currents are negligible for the electroactive species when compared to their diffusion currents [46]. In addition, the normal component of the convection flux density at the interface is inevitably zero: the electrode constitutes an impervious wall for the overall movement of the electrolyte medium. Therefore when there is a supporting electrolyte, one should only take into account the normal component of the interfacial diffusion flux density for an electroactive species.

In these conditions, when one has a single redox half-reaction at the interface, the slope for the concentration profile of each electroactive species is proportional to the faradic current.

[46] This property is widely accepted in electrochemistry, and is demonstrated in the case of an example in appendix A.4.1.

Take for example the oxidation of Fe^{2+} ions to Fe^{3+} ions in a unidirectional system with a planar interface between a platinum electrode and an aqueous solution which contains both ferrous and ferric ions and a supporting electrolyte. If the capacitive current can be ignored, then the boundary conditions for the two electroactive species can be presented in the following way:

$$\left(N_{Fe^{2+}}\right)_{x=0} = \left(N_{Fe^{2+}_{\text{diffusion}}}\right)_{x=0} = -\frac{I}{\mathscr{F}S} = -D_{Fe^{2+}}\left(\frac{\partial[Fe^{2+}]}{\partial x}\right)_{x=0}$$

$$\left(N_{Fe^{3+}}\right)_{x=0} = \left(N_{Fe^{3+}_{\text{diffusion}}}\right)_{x=0} = \frac{I}{\mathscr{F}S} = -D_{Fe^{3+}}\left(\frac{\partial[Fe^{3+}]}{\partial x}\right)_{x=0}$$

At the interface, the slopes for the Fe^{2+} and Fe^{3+} concentration profiles have opposite signs. For an oxidation half-reaction, the slope for Fe^{2+} is positive, while that for Fe^{3+} is negative, with the direction of the x-axis pointing from the electrode towards the electrolyte $(I > 0)$. Figure 4.15 illustrates the shape of the Fe^{2+} and Fe^{3+} concentration profiles, at a given instant, when the thickness of the diffusion layer[47] is low compared to the dimensions of the overall system. For instance, this applies to situations involving semi-infinite mass transport in a transient state (see section 4.3.1.3).

Since $D_{Fe^{2+}} > D_{Fe^{3+}}$ (see table 4.2 in section 4.2.2.4[48]), the absolute value of the Fe^{3+} concentration profile slope at the interface is higher than that of Fe^{2+}, as shown in the diagram in figure 4.15.

Remember not to get mixed up between the concentration profiles which are variations in space, and the concentration variations over time. Fe^{3+} is produced by the half-reaction at this interface, therefore the amount of Fe^{3+} ions increases over time (increasing function) whereas its concentration profile in space decreases. If you examine the concentration profile as shown in figure 4.15, you can see how the concept of production (or consumption) over time can be visualized through the surface (i.e., the integral) lying between the profile at instant t and the initial profile. This surface represents the amount of species produced (or consumed) per unit surface area between these two instants. In particular, the fact that just as many Fe^{3+} ions are produced as Fe^{2+} ions are consumed is depicted visually in figure 4.15 since both surfaces, denoted by + and −, are equal in absolute value[49].

---

[47] *As outlined in section 2.2.1.1, here the diffusion layer thickness depends mainly on the time factor, and the diffusion coefficient of Fe^{2+}. Section 4.3.1.3 provides more quantitative data on the different values of diffusion layer thickness in a semi-infinite transient state.*

[48] *The diffusion coefficient values can be calculated from the molar conductivities at infinite dilution (see table 4.2 in section 4.2.2.4) using the NERNST-EINSTEIN equation (see section 4.2.1.4) :*

$$D_i^0 = \lambda_i^0 \frac{1}{z_i^2}\frac{RT}{\mathscr{F}^2}$$

To get a clearer view of the effect of diffusion coefficient differences between different electroactive species, one should note that the ratio selected for plotting the concentration profiles in figure 4.15 is 2, whereas in actual fact the values shown in table 4.2 give a ratio equal to 1.2 in the case of the Fe^{3+}/Fe^{2+} couple.

[49] *This property can be stretched to apply to the differences in concentration profiles at different instants. However remember that this type of reasoning is not applicable to concentration profiles*

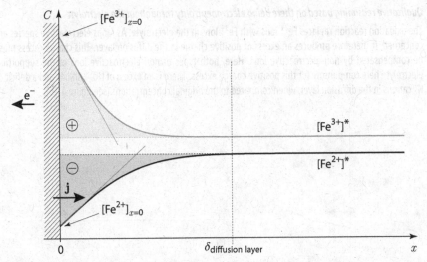

Figure 4.15 - Concentration profiles of Fe^{3+} ions (grey) and Fe^{2+} ions (black) during oxidation
$[Fe^{2+}]_{x=0}$ and $[Fe^{3+}]_{x=0}$: interfacial concentrations,
$[Fe^{2+}]^*$ and $[Fe^{3+}]^*$: bulk concentrations, far from the interface.

The charge balance of the electroactive species in the diffusion layer shows a change in charge distribution. This would also be the case even if the diffusion coefficients were identical. In the example illustrated in figure 4.15, just as in any oxidation half-reaction, the charge balance for the electroactive species reveals a positive charge excess in the diffusion layer in relation to the bulk of the solution. In this example, it results in a decreasing profile for the sum of $2\,[Fe^{2+}] + 3\,[Fe^{3+}]$ in the diffusion layer. The electro-neutrality of the solution within the diffusion layer still applies. This means that the other non-electroactive ions (belonging to the supporting electrolyte or the counter-ions if no supporting electrolyte) compensate for these charge modifications in the diffusion layer caused by the redox reaction. For these non-electroactive species, the concentration profiles also present variations in the diffusion layer [50].

▶ Let us return to the example of an oxidation reaction occurring at the interface between a platinum electrode and an aqueous solution containing ferrous and ferric nitrates with a KNO_3 supporting electrolyte. To simplify, we assume that all the ionic diffusion coefficients are close to each other. We have already dealt with the concentration profiles of the electroactive species. However, the concentration profiles of non-electroactive species, which are rarely touched upon because they are often not a subject of great interest, are not flat. We should however keep in mind that although the diffusion fluxes share the same order of magnitude as electroactive species, their relative variations appear to be negligible because these non-electroactive species are much more concentrated. The two approaches outlined below enable one to determine in qualitative terms the shapes of the concentration profiles of the non-electroactive species K^+ and NO_3^-.

---

in steady or quasi-steady state conditions. In this instance, the significance of the surface lying between the steady-state profile and the initial profile is more complex to interpret. A simple example is explored in appendix A.4.2.

[50] In addition to the next example, one can also refer to the example outlined in appendix A.4.1.

▶ *Qualitative reasoning based on there being electroneutrality throughout the electrolyte*

The oxidation reaction replaces Fe^{2+} ions with Fe^{3+} ions in the electrolyte. As far as electroactive species are concerned, it therefore produces an excess of positive charge in the diffusion layer. This charge excess must be compensated by non-electroactive ions. Here, both types of non-electroactive ions of the supporting electrolyte help compensate for this positive charge excess. There is an excess of NO_3^- anions and a deficit of K^+ cations in the diffusion layer, when compared to the original concentrations (see figure 4.16).

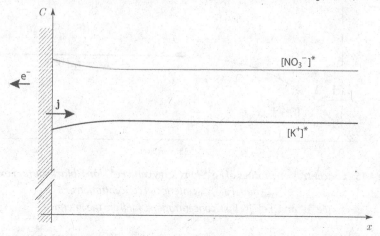

Figure 4.16 - Concentration profiles of non-electroactive ions during an oxidation half-reaction

In this figure, the concentration scale is proportionately eight times smaller than that in figure 4.15. Therefore the origin (zero concentration) cannot be seen here. The difference between the bulk concentrations of the two non-electroactive ions, $[K^+]^*$ and $[NO_3^-]^*$, is explained by the fact that the nitrate ion is also the counter-ion of the two electroactive cations.

▶ *Qualitative reasoning based on the molar interfacial fluxes*

This reasoning can only be strictly applied to interfacial molar fluxes. It can be extended to apply to the whole diffusion layer in a system with unidirectional geometry at steady state, since the molar flux densities in this case are homogeneous. In the other cases, this type of reasoning produces curves that can be considered generally correct in qualitative terms.

The interfacial conditions for the non-electroactive species are:

$$(N_i)_{x=0} = (N_{i_{\text{diffusion}}})_{x=0} + (N_{i_{\text{migration}}})_{x=0} = 0$$

In other words, since the interfacial molar fluxes are zero, then the migration and diffusion components at the interface have opposite values different from zero.

The overall current and the migration current are identical [51]. Anions migrate towards the anode and cations towards the cathode. Here anions migrate towards the interface and cations in the opposite direction. Remember that because there is a supporting electrolyte here, there is a negligible migration of electroactive species. The directions of the molar flux densities in the diffusion layer are represented in figure 4.17, in which only the directions are significant, and the sizes chosen are arbitrary.

At this interface you therefore have: $(N_{K^+_{\text{migration}}})_{x=0} > 0$ and $(N_{K^+_{\text{diffusion}}})_{x=0} < 0$

$$(N_{NO_3^-_{\text{migration}}})_{x=0} < 0 \quad \text{and} \quad (N_{NO_3^-_{\text{diffusion}}})_{x=0} > 0$$

[51] *Assuming here that the diffusion coefficients are equal, this property emerges when writing the various molar fluxes, as explained in section 4.2.1.5.*

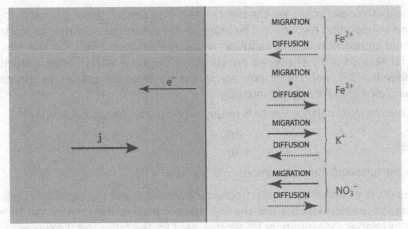

Figure 4.17 - Directions of the molar flux densities in the diffusion layer for oxidation of Fe^{2+} to Fe^{3+}

The diffusion flux of the K^+ ions is directed towards the electrode, and the concentration profile of these ions therefore is increasing. The result can be confirmed either by mathematical equation presenting the molar flux density as being proportional to the opposite of the concentration gradient or by remembering that species move by diffusion from the more concentrated zones towards less concentrated zones. The reasoning applied to the NO_3^- is symmetrical: these ions present a decreasing concentration profile in the diffusion layer.

Obviously, you end up with the same shape as the one found applying the previous reasoning (see figure 4.16) which is based on electroneutrality prevailing throughout the solution.

The two following sections outline two common experimental situations in which both electrodes can be shown to be independent as far as mass transport is concerned:

▸ a system with no convection and for a limited observation time: the system's transient state is then characterised by semi-infinite mass transport. We will represent qualitatively the shapes of the expected concentration profiles for chronoamperometry (figure 4.18) and chronopotentiometry experiments (see figure 4.19);

▸ a system with controlled stirring (forced convection): the system quickly reaches a quasi-steady state.

4.3.1.3 - EXAMPLE OF A TRANSIENT STATE: SEMI-INFINITE DIFFUSION

In a convection-free system, and for a limited observation time (when compared to the characteristic time of the system, which itself depends on the diffusion coefficients and the inter-electrode distance) the electrolyte can be separated into three zones: two diffusion layers close to the two reactive interfaces and an intermediate homogeneous zone within the electrolyte. The diffusion layers thicknesses increase with time, but they are both considered as small when compared to the inter-electrode distance. Here one refers to a transient state and each interface is defined as being in a semi-infinite mass transport condition. Both electrodes are independent, in spite of the fact that they are crossed by the same current.

To simplify the description, we choose a system with unidirectional geometriy [52], which contains a supporting electrolyte. Therefore we can ignore the migration of electroactive species in comparison to their diffusion within the diffusion layers. On the other hand, again for reasons of simplicity, we are confining ourselves to fast redox systems which have identical diffusion coefficients for all the electroactive species. In this case, the following elements are brought into play:

▸ the volume mass balance, which is simply expressed as the second FICK law:

$$\frac{\partial C_i}{\partial t} = D_i \frac{\partial^2 C_i}{\partial x^2}$$

with the following initial condition: $t = 0$, $\forall x$, $C_i = C_i^*$.

▸ FARADAY's law applied at the electrochemical interface and the equation coming from the redox reaction kinetics (for the fast system, it takes the form of the NERNST law with interfacial concentrations [53], written here for the following simple redox couple $Ox + n\,e^- \rightleftharpoons Red$):

$$\forall t \quad N_{i_{x=0}} = -D_i \left(\frac{\partial C_i}{\partial x} \right)_{x=0} = \frac{\nu_i}{\nu_e} \frac{I}{\mathscr{F}S} \quad \text{and} \quad E = E^\circ + \frac{RT}{n\mathscr{F}} \ln \frac{[Ox]_{x=0}}{[Red]_{x=0}}$$

▸ the semi-infinite boundary conditions: the concentrations tend to the initial concentrations, which are the bulk concentrations, when the distances from the interface tend towards infinite (on the scale of the system). To give an equivalent formulation, one can also write that the molar flux density tends towards zero:

$$\forall t \quad C_i \xrightarrow[x \to \infty]{} C_i^* \quad \text{or} \quad N_i = -D_i \frac{\partial C_i}{\partial x} \xrightarrow[x \to \infty]{} 0$$

▸▸ *Chronoamperometry experiment*

Figure 4.18 shows, during a chronoamperometry experiment, typical concentration profiles, at different instants, of the species consumed at the left electrode with no interaction with the right electrode. A constant potential is imposed to an electrode over time and the result in a fast redox system is that the interfacial concentration of the consumed species (here at the left electrode) is fixed [54]. In such a case, the slope of the concentration profile at the interface, namely the current, changes over time [55].

---

[52] When dealing with systems with unidirectional geometry, numerous authors would apply the term 'semi-infinite linear diffusion' to the example in question. However, in this book we prefer to use the term 'unidirectional semi-infinite diffusion' because the word linear is ambiguous. It is also often used in opposition to the 'non-linear diffusion', that is to say, it is applied in the mathematical sense. The second FICK law, as it is usually written (the same as in this document), represents a linear differential equation, and is based on the assumption that the diffusion coefficient is a constant. When this approximation does not apply, the mass balance is: $\partial C/\partial t = \partial/\partial x (D \partial C/\partial x)$ and then the differential equation is usually non-linear.

[53] The definition of a fast couple is given in section 4.3.2.6, and the impact on the laws of kinetics is outlined in section 4.3.3.2.

[54] The relationships between the interface concentrations and potential are given in appendix A.4.2 and demonstrated in a case involving equal diffusion coefficients for the two electroactive species.

[55] For the simple systems described here, the current variations over time enables one to calculate the diffusion coefficient of the species in question. This is the COTTRELL law: $I(t) = n\mathscr{F}SC^* \sqrt{D/\pi t}$.

In this chronoamperometry experiment the thickness of the diffuse layer increases with time. One can gain a good order of magnitude for this thickness by looking at the intersection between the interfacial slope of the concentration profile and the initial flat profile: $\sqrt{\pi D t}$ [56]. Even if this value is sometimes confused with the diffusion layer thickness, it is best still to distinguish between them [57]: for instance, at a distance equal to $\sqrt{\pi D t}$, the concentration is still 18% off from the initial concentration. The thickness of the diffusion layer is equal to $2.0\sqrt{\pi D t}$ for an accuracy of 1% and $2.6\sqrt{\pi D t}$ for an accuracy of 0.1%.

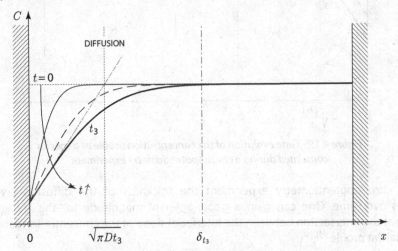

Figure 4.18 - Time evolution of the concentration profile of a species consumed in a chronoamperometry experiment

▸▸ *Chronopotentiometry experiment*

Figure 4.19 shows the results of a chronopotentiometry experiment, that is to say with a constant current density fixed over time. This involves fixing the slope at the interface (left in this case) of the electroactive species' concentration profile. The interfacial concentration then changes over time [58].

---

[56] The set of equations defining the system can be analytically integrated using the LAPLACE transform. However, the task of actually solving the equation stretches beyond the goals of this book. Here, we will only retain the orders of magnitude that emerge as a result.

[57] Let us recall that, as defined in section 2.2.1.1, the thickness of the diffusion layer matches the thickness of the volume experiencing significant diffusion phenomena. One can characterise this in quantitative terms, for example, by using a negligible difference of 0.1% (or 1% or 10%… depending on the accuracy required) for the concentration of at least one species compared to the value in the homogeneous zone where diffusion plays a minimal role.

[58] In this type of experiment it is easy to determine the time needed to obtain a near-zero interfacial concentration. In fact, if one continues the experiment, the imposed current can no longer be ensured by the main reaction. At this point another redox system is called into play, which leads to a sharp variation in the potential measured. This characteristic time can be used to determine the mass transport parameters. For a simple system, the equation involved is known as SAND's law:

$$\sqrt{\tau} = \frac{n \, \mathscr{F} \, S \, C^*}{2 \, I} \sqrt{D\pi}.$$

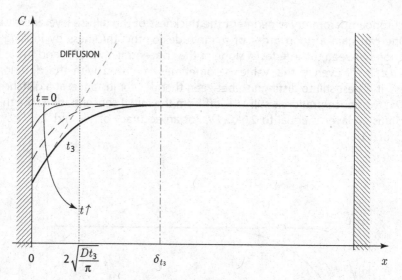

Figure 4.19 - Time evolution of the concentration profile of a species
consumed during a chronopotentiometry experiment

In this chronopotentiometry experiment the thickness of the diffusion layer also changes over time. One can gain a good order of magnitude for this thickness by looking at the intersection between the interfacial slope of the concentration profile and the initial flat profile [54]:

$$2\sqrt{\frac{Dt}{\pi}}$$

Although this value is different from the previous one, it still has the same order of magnitude. Here, at a distance equal to this value, the concentration differs by 14% from the initial concentration. The diffusion layer thickness is a multiple of the thickness previously defined, with a factor 2.2 for an accuracy of 1% and 2.8 for an accuracy of 0.1%.

4.3.1.4 - EXAMPLE OF A STEADY STATE: THE NERNST MODEL

There is another usual set of boundary conditions which emerge when, for example, forced convection is imposed within a system. This applies to a small electrolyser with an RDE in analytical chemistry, or in the case of industrial electrolysers when a system is installed which imposes forced circulation of the electrolyte or of the electrode. It can be shown in simple cases that the diffusion layer has a time-independent thickness [59]. This thickness is a function of the stirring conditions and of the transport properties of the mobile species, and is typically about 10 µm. At this point, the NERNST layer model described in section 2.2.1.1 is then used.

---

[59] This rather complicated reasoning calls for hydrodynamic calculations to be used to quantify the convection molar fluxes. Such a calculation was developed by LEVICH in the case of the RDE (some authors call it the LEVICH theory). It legitimates the use of the simplified model of the NERNST layer and, above all, provides a mathematical equation to express the thickness of that layer as a function of the system's parameters. This expression, called the LEVICH law, $\delta_i = 1.611\, D_i^{1/3} \nu^{1/6}\, \Omega^{-1/2}$, is used in appendix A.4.2.

According to this model, the concentration is equal to the initial concentration beyond a distance to the electrode equal to δ_{NERNST}. At steady state this model leads to a mathematical discontinuity in the concentration profiles at this δ_{NERNST} distance from the electrode. The actual steady-state profile does not show any angular point. Therefore the model gives incorrect results in the area surrounding δ_{NERNST}. However, it makes it easy to describe the exact characteristics of the concentration profile at the interface, and therefore it gives a correct value for the current density.

According to the NERNST model, the concentration profile of the electroactive species within the diffusion layer is described by the following system of equations:

$$\frac{\partial C_i}{\partial t} = 0 = D_i \frac{\partial^2 C_i}{\partial x^2}$$

and the following boundary conditions:

$$N_{i_{x=0}} = -D_i \left(\frac{\partial C_i}{\partial x}\right)_{x=0} = \frac{v_i}{v_e} \frac{I}{\mathscr{F} S}$$

$$C_{i_{x \geq \delta_{\text{NERNST}}}} = C_i^* \qquad \text{(fixed concentration at } \delta_{\text{NERNST}}\text{)}$$

Figure 4.20 shows, during a chronopotentiometry experiment, typical actual concentration profiles, at different instants, of the species which is consumed at the left electrode, with no interaction at the right electrode [60].

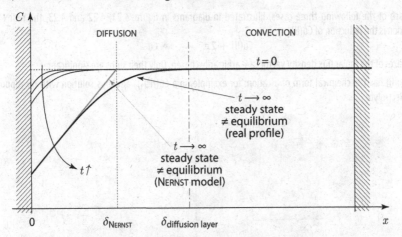

Figure 4.20 - Time evolution of the concentration profile of a species consumed in a system with forced convection

[60] As previously mentioned in note [82] of section 1.6.3, the term chronopotentiometry is used here in its etymological sense. In other words, it is not confined to cases where diffusion is the only mass transport mode for the electroactive species. Although forced convection is imposed on the system, which rapidly reaches a steady state as a result, here we are partly focusing on the transient period leading up towards that state. Therefore, we are interested in the development of the concentration profile over time when a current is imposed. In addition, the current imposed must be lower, in absolute value than a limiting value, which correlates with a steady-state interfacial concentration equal to zero. This concept of a limiting current, which has already been introduced in section 2.3.3.1 in qualitative terms when describing the current-potential curves, is outlined in quantitative terms in section 4.3.3.1.

A steady state with a non-zero current[61] is reached very quickly in these forced convection conditions. The interfacial slope is identical to that in the NERNST model as shown in figure 4.20. The NERNST layer thickness is lower than that of the diffusion layer since, at that distance from the electrode, the actual profile[62] still has a significant slope with a relative concentration difference of 11% with respect to the initial concentration. The diffusion layer thickness is equal to 1.5 δ_{NERNST} for a target accuracy of 1% and to 1.8 δ_{NERNST} for a target accuracy of 0.1%.

These few examples illustrate how one can, in simple cases, represent the concentration profiles in qualitative terms as a function of the system's specific boundary conditions. They also highlight the fact that obtaining a steady state with a non-zero current requires specific experimental conditions.

4.3.1.5 - DIRECTIONS OF THE VARIOUS CURRENT DENSITIES

One can come across a whole range of possible movement directions for electroactive species, by migration and diffusion, depending on their charge number.

The migration current direction is generally the same as that for the overall current, which in turn is directly related to the redox reaction's direction of advancement. The diffusion current direction can often be determined by difference, as illustrated in the examples below.

▶ In each of the following three cases, illustrated in diagrams in figures 4.21, 4.22 and 4.23, the redox half-reaction is the reduction of Cu(II):

$$Cu(II) + 2\,e^- \longrightarrow Cu$$

The sizes of the molar flux density vectors are arbitrarily chosen. Only their signs are significant.

▸ Cu(II) has the chemical form of a cation: for example, in a $Cu(NO_3)_2$ aqueous solution with no supporting electrolyte

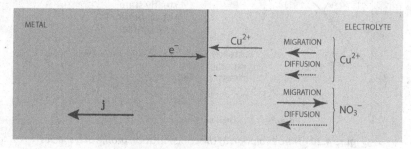

Figure 4.21 - Directions of the molar flux densities

---

[61] By using the results from the preceding example, it is possible to estimate the time needed to reach the steady state: it is the time value such that $\sqrt{Dt}$ reaches the value of the NERNST layer, 10 μm.

For a diffusion coefficient of about $10^{-5}\ cm^2\ s^{-1}$, it comes out as 0.1 s. Therefore, in usual experimental conditions, the characteristic time needed to reach the steady state is small compared to the total length of the experiment.

[62] In more precise terms, it is the concentration profile resulting from the LEVICH calculation.

▶ Cu(II) is in an anionic complexed form: for example, in a $K_2Cu(CN)_4$ aqueous solution with no supporting electrolyte

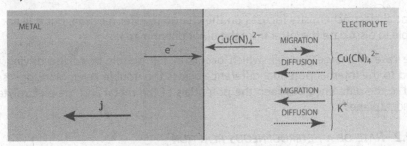

Figure 4.22 - Directions of the molar flux densities

▶ Cu(II) is in a neutral complexed form: for example, in a $CuCl_2 + KCl$ aqueous solution

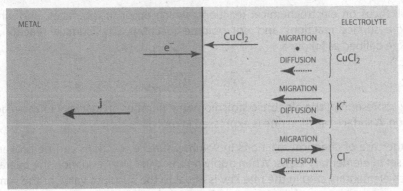

Figure 4.23 - Directions of the molar flux densities

As in the case of all non-electroactive species, the ions of the supporting electrolyte have components for migration and diffusion movement that are opposite in directions, and compensate each other perfectly in the area next to the electrode [63].

4.3.2 - KINETIC MODEL FOR A HETEROGENEOUS REACTION

4.3.2.1 - GENERAL

Many phenomena may occur when a current flows through an electrochemical interface:

▶ charge transfer reactions in the double layer,
▶ mass transport by diffusion, migration and convection in the electrolyte,
▶ homogeneous chemical or redox reactions in the electrolyte,
▶ adsorption-desorption reactions in the double layer,
▶ reactions between adsorbed species in the double layer,
▶ insertion reactions in the electrode: concentration profiles of inserted species then develop in the electrode volume,

---

[63] Section 4.3.1.2 gives an example of the concentration profile for the ions of the supporting electrolyte.

▸ interfacial diffusion along the interface (surface diffusion),
▸ crystallisation, etc.

Interfacial electrochemistry focuses on the charge transfer reaction, but it is not usually possible to dissociate the latter from the other phenomena.

In the case of charge transfer, which occurs in the double layer, the driving force is related to the internal potential difference across the double layer. More precisely, it is linked to the difference between the potentials of the metal and the electrolyte at the HELMHOLTZ plane [64].

4.3.2.2 - RATE OF A HETEROGENEOUS REACTION

Electrochemical kinetics represents a particular category of chemical kinetics, featuring a set of specific characteristics that will be outlined here.

We will focus on electrochemical reactions, which occur at interfaces, and which are therefore surface reactions and not volume reactions. The surface reaction rate is therefore defined as follows:

$$v = \frac{1}{\nu_i} \frac{1}{S} \left(\frac{\partial n_i}{\partial t} \right)_{\text{interface}}$$

with ν_i representing the algebraic stoichiometric number of species i in the reaction in question. A surface reaction rate is expressed in mol m^{-2} s^{-1}.

As in the case of chemical reactions in volume, surface reactions can be broken down into a set of elementary steps. When applying the VAN'T HOFF equation the overall order for each elementary step in the rate law is equal to the absolute value of the sum of the stoichiometric numbers of the reactants. In addition, the partial order of each reactant is equal to the absolute value of its stoichiometric number in the reaction in question. This gives the following equation:

$$v = k \prod_{i=\text{reactant}} (C_{i\,\text{interface}})^{|\nu_i|}$$

If one of the reactants is in much larger amount than the other reactants, then one can assume that its concentration is constant. The corresponding concentration is included in the rate constant, thus reducing the order of the rate law, written with a pseudo-constant.

As the concentration of free electrons in a metal is much greater than that of electroactive compounds in an electrolytic solution, then one can assume that this concentration is constant, and it no longer features in the equation for the apparent rate law of the elementary redox half-reactions at a metal|electrolyte interface. This would be different if we wanted to write kinetic laws for a redox half-reaction at an interface between a semiconductor and an electrolyte. However, here the description will be confined to metallic electrodes.

For example, the redox half-reaction of the Fe^{3+}/Fe^{2+} couple, which can be considered as an elementary step, corresponds to the following reactions:

---

[64] See a simplified description of the double layer in section 3.3.1.

$$Fe^{2+} \xrightarrow{\; v_{\text{oxidation}} \;} Fe^{3+} + e^-$$

$$Fe^{3+} + e^- \xrightarrow{\; v_{\text{reduction}} \;} Fe^{2+}$$

The rate law at a metal | electrolyte interface is:

$$v_{\text{oxidation}} = k_{\text{oxidation}} \, [Fe^{2+}]_{\text{interface}}$$

$$v_{\text{reduction}} = k_{\text{reduction}} \, [Fe^{3+}]_{\text{interface}}$$

For this mechanism, called an E mechanism [65], the rate constants are expressed in m s^{-1}.

Other redox-type elementary mechanisms exist in electrochemistry:

▸ electrosorption, where one of the redox species is in an adsorbed state,

▸ insertion, where one of the redox species is an insertion compound,

▸ deposition, where one of the redox species is a solid.

The following description is confined to systems that follow an E mechanism at a metal-lic electrode. As is usual in most documents dealing with this aspect of electrochemistry, we have taken a redox reaction involving only two species, Ox and Red, with $|v_{\text{Ox}}| = |v_{\text{Red}}| = 1$, but possibly $|v_e| = n \neq 1$. However, according to kinetic theory and mechanism, an elementary step can only involve few species and in particular the exchange of several electrons is unlikely in such a step. When a redox couple involves two or more exchanged electrons, then the overall number n is often involved in the final equation of the current-potential curve. However, all the kinetic equations must be written for each elementary step involved. This generally complex task will not be tackled in this document.

4.3.2.3 - SIMPLIFIED KINETIC MODEL OF THE E MECHANISM (SINGLE STEP)

In the case of a redox reaction, the driving force for the reaction, and consequently for the rate constants, in particular depends on the potential difference at the interface. More precisely, this driving force depends on the difference between the potential at the surface of the metal and the potential in the electrolyte at the very point where the electroactive species is located when the electron transfer occurs. The kinetic models which reflect precisely these features are complex, and therefore we are kee ping ourselves confined to equations based on common and simplified descriptions for electrochemical kinetics for the E mechanism. In this context, any changes in the redox reaction rate constant can be described using the following two kinetic parameters:

▸ the standard rate constant of the redox reaction, denoted by $k°$, which is homo-geneous to the rate constant and therefore expressed in m s^{-1} (or also cm s^{-1});

▸ the symmetry factor, denoted by α, which is a dimensionless number between 0 and 1, defined here for the reaction in the direction of oxidation [66].

---

[65] Here E stands for 'electrochemical'. In this nomenclature, a volume reaction step in the electrolyte is denoted by C. One of the most frequent complex mechanisms is the EC mechanism.

[66] Note that, in some documents, notations such as $\alpha_{\text{oxidation}}$ and $\alpha_{\text{reduction}}$ are used, with $\alpha_{\text{oxidation}} + \alpha_{\text{reduction}} = 1$, since the ratio between the two kinetic constants must be equal to the

The kinetic rate constants are then expressed as:

$$k_{\text{oxidation}} = k° \, e^{+\alpha \frac{n\mathscr{F}}{RT}(E-E°)}$$

$$k_{\text{reduction}} = k° \, e^{-(1-\alpha)\frac{n\mathscr{F}}{RT}(E-E°)}$$

Consequently, $k°$ represents the oxidation and reduction rate constant value at the standard potential $E°$.

In most cases it is difficult to give a precise physical meaning to α. However, for very simple species, such as proton electrosorption for instance, this parameter characterises the difference in curvature between the system's potential energy curve in its oxidised state (proton) and in its reduced state (adsorbed hydrogen), depending on the distance to the metal surface. Based on this model, the MORSE potential wells are represented by parabolic curves around the minimum, as shown in figure 4.24.

If one wanted to take a more comprehensive approach, then one could resort to the activated complex model for reactions in a homogeneous phase. Here we could say that the α parameter gives the link between the thermodynamic characteristic, $\Delta_r G°$, and the kinetic characteristic, namely the GIBBS energy of activation, $\Delta G^{\neq}$.

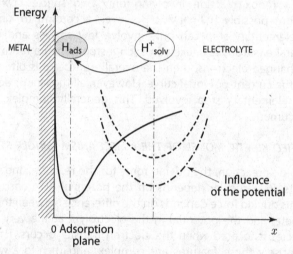

Figure 4.24 - Shape of the potential wells for the reactants and the products
of the proton electrosorption reaction: $H^+_{solv} + e^- \longrightarrow H_{ads}$

This simplified description implies in particular that the medium is ideal (with activity coefficients equal to 1) or at least it implies that the activity coefficients of the species in question can be seen as constant, while the concentrations vary due to the current flow. This is true in particular in systems with a supporting electrolyte, because the ionic strength is fixed by the supporting electrolyte and remains unchanged throughout the experiment. Here we assume that the term involving activity coefficients is constant, and furthermore that it is included in the reaction's standard rate constant.

---

thermodynamic equilibrium constant. Some authors may prefer to choose the $\alpha_{reduction}$ parameter, although here $\alpha_{oxidation}$ has been chosen.

To simplify the notations, a parameter is sometimes used, denoted by ξ, called the 'dimensionless potential' which for a given couple is defined by the following equation:

$$\xi = n \frac{\mathscr{F}}{RT} (E - E^\circ) = n f (E - E^\circ)$$

and, at 25°C: $f = \frac{\mathscr{F}}{RT} = 38.9 \; V^{-1}$ and $\frac{1}{f} = 25.7 \; mV$

we then have: $k_{\text{oxidation}} = k^\circ \, e^{+\alpha \xi}$

$$k_{\text{reduction}} = k^\circ \, e^{-(1-\alpha)\xi}$$

In electrochemistry, the notion of rate is used to quantify a system's ability to pass more or less current when equilibrium is disrupted. This property is usually determined by many parameters. For example, electrochemical reactions are generally associated with mass transport, namely to provide the interface with a reactant or to remove a product.

Numerous terms are put to use in the field of electrochemical kinetics to characterise typical situations which are limiting cases with particular shapes for the corresponding current-potential curves. In scientific literature, these terms are not always applied with the greatest rigour. In the forthcoming sections we will give a precise definition for the common terms: nernstian redox systems in section 4.3.2.4; reversible/irreversible redox reactions in section 4.3.2.5; slow/fast redox systems in section 4.3.2.6.

4.3.2.4 - RATE-LIMITING OR DETERMINING STEP

As already noted above, when there is a current flow then mass transport phenomena are automatically brought into play in addition to the electron transfer itself. The link between current and voltage generally involves all of the system's kinetic parameters. It is often interesting to consider the two following limiting cases, which are illustrated here on the E mechanism (remember we assume that $|v_{Ox}| = |v_{Red}| = 1$):

▸ the current/voltage link only involves quantities related to mass transport phenomena. In this case one says that the mass transport phenomena are rate-limiting or rate-determining, according to kinetic theory, or that the system is controlled by mass transport. In these conditions, the interfacial concentrations obey the following equation:

$$E = E^\circ + \frac{RT}{n \mathscr{F}} \ln \frac{[\text{Ox}]_{\text{interface}}}{[\text{Red}]_{\text{interface}}}$$

This equation corresponds to the NERNST law written with the interfacial concentrations in the argument of the logarithm. Beware that it is not the NERNST law in its strictest sense because, in the case here of a system carrying a current, the interfacial concentrations are necessarily different from the concentrations in the bulk of the solution (even if they may be very close in numerical terms). These parallel ways for writing the equation help to explain the origin of the term nernstian, which is often used to qualify such systems[67]. Some authors also use the terms 'local equilibrium at the interface' which is incorrect since this interface crossed by a current is not in equilibrium;

---

[67] These two concepts (control by mass transport and nernstian system) as terms are not strictly synonymous. Quasi-fast redox couples (see section 4.3.2.6) close to equilibrium conditions are nernstian, but they are not strictly controlled by mass transport.

▸ the current/voltage link only involves quantities which are related to electron transfer. One then says that the electron transfer step is rate-limiting or rate-determining, based on kinetic theory, or that the system is controlled by redox kinetics. In these conditions, the concentrations are practically constant throughout the electrolyte, and the interfacial concentrations can be taken as equal to the initial concentrations or to the bulk concentrations, [Ox]* and [Red]*:

$$[Ox]_{interface} \approx [Ox]^* \quad \text{and} \quad [Red]_{interface} \approx [Red]^*$$

Generally, these limiting cases only apply in a restricted potential range.

4.3.2.5 - REVERSIBILITY CHARACTER OF AN ELEMENTARY REACTION STEP

In kinetic theory, reversibility as a concept is defined in relation to reaction rates. In a reaction mechanism, when an elementary step can be said to occur in both directions, the rate of this step is always equal to the difference between the forward rate, $v_\rightarrow$, and the backward rate, $v_\leftarrow$:

$$v = v_\rightarrow - v_\leftarrow$$

This step is called reversible in terms of kinetic theory, which is different from the thermodynamic meaning, if and only if the overall reaction rate is very small compared to both the forward and backward rates:

$$v \ll v_\rightarrow \quad \text{and} \quad v \ll v_\leftarrow$$

An important consequence of this is that the forward and backward rates are practically equal:

$$v_\rightarrow \approx v_\leftarrow$$

Remember that the rates are not strictly equal. Therefore it would be wrong to write their difference, in other words the overall rate, as equal to zero. To give a numerical example, if the forward rate were equal to 1 000 001 and the backward rate equal to 1 000 000 (in $mol\,m^{-2}\,s^{-1}$), the overall rate, equal to 1, is different from zero. The same type of confusion is made by authors who use the term local equilibrium to refer to a reversible heterogeneous reaction[68].

This step is called irreversible, for example in the forward direction, if and only if the backward rate is negligible compared to the forward rate:

$$v_\rightarrow \gg v_\leftarrow$$

Consequently, the overall rate is almost equal to the forward rate:

$$v \approx v_\rightarrow$$

In such a case, the overall reaction is usually written using only one arrow.

Still following these two definitions, keep in mind that one can have a reaction step which is neither reversible nor irreversible.

---

[68] Similarly, the term reversible is not synonymous with nernstian. Even for an E mechanism, it is possible to find cases where nernstian systems are reversible only in a very narrow potential range close to equilibrium.

This typical case is sometimes referred to as quasi-reversible in scientific literature[69].

In electrochemistry, given that reaction rates depend on the potential, then these concepts need to be examined for each value of this parameter. A given system may present a reversible charge transfer step in one particular potential range and an irreversible one in another range. Therefore, strictly speaking, one should not use the term reversibility about a redox couple, but rather qualify the reversibility of the corresponding redox reaction at a given potential.

4.3.2.6 - RAPIDITY OF A REDOX COUPLE

The concept of reversibility as defined above is sometimes used incorrectly and mixed up with the concept of the rapidity of a couple, which is defined in this section. Even if a reversible redox reaction involves a fast redox couple in many circumstances, these two notions are nonetheless different in nature.

To define in precise terms the concept of rapidity of a redox couple for an E mechanism, one says that a redox couple is fast (respectively slow) if the standard rate constant of the redox reaction is very high (respectively very low) compared to the mass transport rate constant (denoted by m and expressed in m s^{-1}):

$$\text{fast redox couple}: \frac{k^\circ}{m} \gg 1 \qquad \text{slow redox couple}: \frac{k^\circ}{m} \ll 1$$

In the simple examples considered here, the mass transport rate constant is the ratio of the diffusion coefficient of the species in question to the diffusion layer thickness of the latter[70]:

$$m = \frac{D}{\delta}$$

Rigorously speaking, one should take into account all the mass transport rate constants of the various electroactive species.

The concept of fast or slow couples is therefore independent of the potential applied, since it is intrinsic to the system. However it does depend on other experimental parameters through the mass transport rate constant. The latter parameter is in fact a function of the quantities specific to the mass transport of the species in question (diffusion coefficient or electrochemical mobility), but it also depends on other characteristics in the system which vary according to each type of experiment, as illustrated in the examples below.

---

[69] This terminology is rather clumsy since general cases are no more quasi-reversible than quasi-irreversible.

[70] In process engineering this parameter is called the mass transfer constant. It is denoted by some authors by k_m (a notation which has the advantage of underlining the parallel shown with the reaction rate constants denoted by k). To be precise, this quantity is not based on the diffusion layer thickness as defined in this document, but rather the value calculated from the interfacial slope of the concentration profile (see section 4.3.1.4). For example, in the case of an experiment involving forced convection, one should use the thickness of the NERNST layer in order to define the mass transport rate constant.

When dealing with the same redox couple, it can behave either like a slow system or a fast system, depending on the experimental conditions chosen:

▶ in a steady-state experiment using a rotating disc electrode, the mass transport rate constant, m, depends on the rotation speed of the electrode and on the solvent viscosity;

▶ in a voltammetry experiment, m depends on the sweep rate used;

▶ in a chronoamperometry or chronopotentiometry experiment, m depends on the time elapsed;

▶ in a thin-layer cell, m depends on the inter-electrode distance, etc.

To fix some orders of magnitude, let us imagine a redox couple with an E mechanism in an aqueous solution, with usual values for the diffusion coefficients, namely about $10^{-5}\ cm^2\ s^{-1}$:

▶ for an experiment using an RDE (with a rotation speed of around 1000 rpm[71]), the redox system will be considered as:

$$\text{fast if} \qquad k° \gg 10^{-2}\ cm\ s^{-1}$$
$$\text{slow if} \qquad k° \ll 10^{-4}\ cm\ s^{-1}$$

▶ for a voltammetry experiment, a redox couple with a standard reaction rate constant of $k° = 10^{-3}\ cm\ s^{-1}$ will be fast or slow, depending on the sweep rate:

$$\text{fast if} \qquad v_{sweep} \ll 0.1\ mV\ s^{-1}$$
$$\text{slow if} \qquad v_{sweep} \gg 1\ V\ s^{-1}$$

▶ for a chronoamperometry experiment without convection a redox couple with a standard reaction rate constant of $k° = 10^{-3}\ cm\ s^{-1}$ will be considered, depending on the length of the experiment, as:

$$\text{fast if} \qquad t \gg 1000\ s$$
$$\text{slow if} \qquad t \ll 0.1\ s$$

Remember that when defining in full the complex phenomenon that is associated with current flow, it is not enough to only use the concept of the rapidity of a redox couple. In particular, if the characteristics related to current flow actually depend on the $k°/m$ ratio, they also vary as a function of the electrode potential and of the concentrations in the solution.

4.3.3 - POLARISATION OF AN ELECTROCHEMICAL INTERFACE AT STEADY STATE

The polarisation of an electrode is a function of the current flowing through it, and the product πI is usually positive[72]. Here we will give some quantitative equations but keep to simple cases. These examples highlight the key phenomena involved, and their impact on the shape of the steady-state current-potential curves.

---

[71] rpm: revolution per minute.

[72] This property is illustrated in section 2.3.1 which outlines how to plot a current-potential curve in qualitative terms.

Let us consider an electrode interface in a unidirectinal system where a redox reaction is occurring, that is wholly undisturbed by the phenomena occurring at the other electrode. Steady-state conditions prevail here, and moreover, migration phenomena are negligible. This case applies, for instance, to a system with a supporting electrolyte. When the latter is present, one can ignore both the ohmic drop phenomena, and also the migration of the electroactive species.

4.3.3.1 - CONCENTRATION PROFILES AND EQUATION FOR THE LIMITING CURRENTS

The steady-state current-potential curves reflect the influence of the mass transport kinetics of the electroactive species by showing current plateaus, whatever the redox kinetics is. The concentration profiles of the electroactive species account for this particular situation, as illustrated in figure 4.25 for the Fe^{3+}/Fe^{2+} couple.

Here one can recognise several properties that are typical of the concentration profiles of electroactive species at steady state [73]: the concentration profiles are linear for a Fe^{3+}/Fe^{2+} couple with a supporting electrolyte and unidirectional geometry. To simplify, we assume that the diffusion coefficients are equal. The diffusion layer thickness is fixed since the steady state has been reached, and it is identical for both electroactive ions. In the homogeneous solution (such as a stirred solution for instance) the concentration of Fe^{2+} ions is set higher than the concentration of Fe^{3+}. The Fe^{3+}/Fe^{2+} couple chosen in this example is fast, however this feature has no impact on the shape of the concentration profiles plotted here, nor so on the equations written in this section (only the shape of the current-potential curve will be affected).

In both anodic and cathodic domains of the current-potential curve, the concentration profiles for the operating points with a current equal to the limiting current all show that the interfacial concentration of the species consumed is negligible compared to the concentration in the bulk of the solution, i.e., the reductant Fe^{2+} in oxidation and the oxidant Fe^{3+} in reduction. The mathematical equations for the limiting currents indicated below are derived themselves from this peculiarity in the concentration profile. To obtain general equations we consider possible different diffusion coefficients for the oxidant and the reductant and any stoichiometric numbers, thus ignoring the simplification introduced in figure 4.25.

In such a system with unidirectional geometry, where diffusion is the only transport mode for the electroactive species, the molar flux density at steady state is homogeneous and proportional to the slope of the linear concentration profile:

$$N_{i_{0 \leq x \leq \delta_i}} = N_{i_{x=0}} = - D_i \frac{C_i{}^* - C_{i_{x=0}}}{\delta_i}$$

Remember that FARADAY's law produces the relationship to the current. Here, for a redox reaction with a 100% faradic yield, the slope of the concentration profile is also proportional to the current:

$$N_{i_{x=0}} = \frac{v_i}{v_e} \frac{I}{\mathscr{F} S} = - D_i \frac{C_i{}^* - C_{i_{x=0}}}{\delta_i}$$

[73] These characteristics correspond to the NERNST model described in section 4.3.1.4.

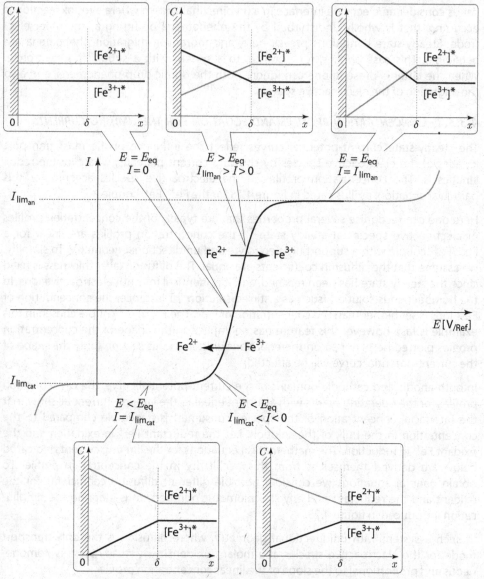

Figure 4.25 - Shape of the concentration profiles for various points
on the steady-state current-potential curve for a Fe^{3+}/Fe^{2+} system

In the particular case of where the interfacial concentration of the consumed species is
zero, or more precisely is negligible compared to the bulk concentration, the following
equation emerges for the limiting current:

$$I_{lim} = -\frac{\nu_e}{\nu_i} D_i \mathscr{F} S \frac{C_i^*}{\delta_i}$$

Using the kinetic mass transport constant, m_i, which is the ratio of the diffusion
coefficient to the diffusion layer thickness:

$$m_i = \frac{D_i}{\delta_i}$$

one obtains the following expression of the limiting current:

$$I_{\text{lim}} = -\frac{\nu_e}{\nu_i}\,\mathscr{F}\,S\,m_i\,C_i{}^*$$

with: m_i the mass transport rate constant of the consumed species i [m s^{-1}]
   $C_i{}^*$ the concentration of the consumed species i in the bulk electrolyte   [mol m^{-3}]
   $\mathscr{F}$ FARADAY's constant   [C mol^{-1}]
   ν_i the algebraic stoichiometric number of the consumed species i
   ν_e the algebraic stoichiometric number of the electron
   I_{lim} the algebraic limiting current   [A]
   S the area of the interface surface   [m^2]

Remember that for this equation, the reactant is the reductant in an oxidation reaction, while the reactant is the oxidant in a reduction reaction. Therefore in the previous example, i.e., the Fe^{3+}/Fe^{2+} couple, one has the following:

$$I_{\text{lim cat}} = -\frac{\nu_e}{\nu_{\text{Ox}}}\,\mathscr{F}\,S\,m_{\text{Ox}}\,[\text{Ox}]^* = -\,\mathscr{F}\,S\,m_{Fe^{3+}}\,[Fe^{3+}]^*$$

$$I_{\text{lim an}} = -\frac{\nu_e}{\nu_{\text{Red}}}\,\mathscr{F}\,S\,m_{\text{Red}}\,[\text{Red}]^* = \mathscr{F}\,S\,m_{Fe^{2+}}\,[Fe^{2+}]^*$$

Given the sign conventions used, once again one finds that the anodic limiting current is positive, while the cathodic one is negative.

The general link between the current and the concentration profiles also describes the interfacial concentrations at any working point. Therefore, based on the equations outlined above (FARADAY's law and limiting current) one has:

$$\frac{I}{I_{\text{lim}}} = \frac{C_i{}^* - C_{i_{x=0}}}{C_i{}^*}$$

hence the following equation for the interfacial concentration:

$$C_{i_{x=0}} = C_i{}^*\left(1 - \frac{I}{I_{\text{lim}}}\right)$$

This algebraic equation must involve the anodic limiting current for the reductant and the cathodic limiting current for the oxidant.

Therefore, in the case of the Fe^{3+}/Fe^{2+} couple, one has:

$$[Fe^{2+}]_{x=0} = [Fe^{2+}]^*\left(1 - \frac{I}{I_{\text{lim an}}}\right)$$

$$[Fe^{3+}]_{x=0} = [Fe^{3+}]^*\left(1 - \frac{I}{I_{\text{lim cat}}}\right)$$

One then finds for example that the interfacial concentration in Fe^{2+} is less than the bulk value, in the case of an anodic current ($I > 0$). On the other hand, the Fe^{3+} concentration is larger in the zone next to the interface than in the bulk of the electrolyte ($I/I_{\text{lim}_{\text{cat}}} < 0$).

SCANNING ELECTROCHEMICAL MICROSCOPE

Document written with the kind collaboration of Dr Rob SIDES, Applications Specialist,
AMETEK - Princeton Applied Research, Oak Ridge, in USA

In a traditional electrochemical measurement, a potentiostat measures an average current over the entire electrode/electrolyte interface. Rarely is a sample homogenous. Samples often consist of local sites of passivate/active nature or sites of anodic/cathodic character. This need to investigate localized electrochemistry led to the development of the Scanning ElectroChemical Microscope (SECM). Since the imaging mechanism is based on electrochemistry, the applications of a SECM are as varied as the applications offered by electrochemistry itself. Some key applications include biological sensors, reaction kinetics, porous membrane study, fuel cell catalysts, and corrosion mechanisms.

The SECM integrates a positioning system, a bipotentiostat, and an ultramicroelectrode tip. The positioning system moves the tip close to the surface of the sample. The bipotentiostat polarizes both the sample and the tip independently and measures both resulting currents. The tip is an ultramicroelectrode with a specific tapered polish and active radius lower than 100 microns. The positioning system scans the measurement tip and charts position with measured electrochemical parameters, creating a data map of the local current.

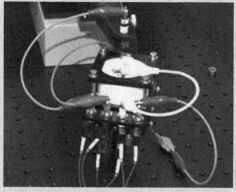

AMETEK – Princeton Applied Research SCEM

One particular SECM experiment is the approach curve in feedback mode. In this experiment, a redox active salt, the mediator, is introduced into the electrolyte. A single potentiostat polarizes the tip to cause an electrochemical reaction; however, the sample itself is not polarized. The resulting current is recorded as the tip is moved closer towards the sample. When the tip is positioned appropriately close to the sample, a local response is seen. If the specific location on the sample is conductive, the resulting nernstian response observed at the surface sample causes the current to increase when compared to the 'bulk current' (i.e., when the tip is far from the substrate). This is called 'positive feedback'. If the specific location on the sample is insulating, then mass transport to the electrode of the tip is hindered, and the current decreases when compared to the bulk current. This is called 'negative feedback'. A range of intermediate types of behavior may also occur with different samples. The quantitative analysis of such approach curves allows for a very accurate analysis of local surface kinetics to be carried out.

For imaging experiments, the same kind of approach curve is also used in order to position the tip in the sample's local imaging zone, without actually making contact with it. In surface imaging experiments the tip-to-sample distance is typically 2 to 4 times the probe diameter. Given that the tip measures 10s microns, it is important to establish a non-visual means by which to identify this zone.

In generator-collector mode, the SECM uses the bipotentiostat's second channel to polarize the sample and to actively control the redox reactions within the system. For example, if the redox species exists in the bulk solution in the oxidation state, Ox, then the sample can be polarized so as to electrochemically convert species Ox into species Red. If the tip is polarized to a more positive value, sufficient to cause an electro-chemical reaction with Red, no bulk current is initially generated at the tip, since the initial Red concentration is negligible. However, when the tip enters into an area where products resulting from the electrochemical reaction at the sample (Red) can be found, then a current is generated at the probe. This provides an imaging mechanism, which enables the probe to determine in spatial terms where exactly on the surface of the sample the electrochemical reactions are occurring most efficiently. A high current at the tip is due to a greater concentration of Red at the probe, which is in turn due to a more active Ox $\rightarrow$ Red transition in that position.

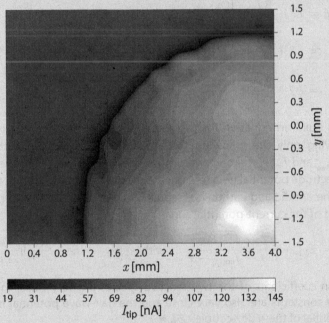

Image generated using Generator-Collector Mode by Princeton Applied Research SECM
Sample is epoxy/gold in the presence of ferricyanide. Note the high current (activity) over the gold.

Sensor evaluations or fuel cell catalyst evaluations commonly use the oxygen reduction reaction and do not require the use of any external salt. The tip can use electro-chemistry to detect products as they diffuse through porous membranes. Corrosion products may be able to undergo further electrochemical reactions, or could additionally benefit from using an ion-selective electrode as the tip. Many different applications can benefit from the ability to both control and monitor electrochemical reactions, with the added dimension of being able to provide spatial resolution thanks to the SECM.

In the three following sections, we focus on redox couples following the E mechanism (with $|v_{Ox}| = |v_{Red}| = 1$).

4.3.3.2 - FAST REDOX SYSTEMS

Remember that such a redox couple is called fast if its standard redox rate constant is very large compared to the mass transport constants of the electroactive species. Fast redox couples can be shown to be nernstian and reversible systems, in a potential range surrounding the standard potential. As for large interface polarisations in this potential range, the rate limitation due to mass transport phenomena gives rise to a limiting current at steady state. With these fast systems, when the polarisation values are very high, such systems remain nernstian, yet their reaction is no longer reversible at these potentials. In spite of this, given that the current has reached its limiting value, this change does not modify the current-potential curves in any way.

By bringing into play the nernstian character of these systems, combined with the equations for both the interfacial concentrations as a function of the current and the limiting currents given in section 4.3.3.1, one then obtains the following equations:

$$E = E° + \frac{RT}{n\,\mathscr{F}} \ln \frac{[Ox]_{x=0}}{[Red]_{x=0}}$$

$$= E° + \frac{RT}{n\,\mathscr{F}} \ln \frac{[Ox]^*}{[Red]^*} + \frac{RT}{n\,\mathscr{F}} \ln \frac{I_{lim\,an}}{-\,I_{lim\,cat}} + \frac{RT}{n\,\mathscr{F}} \ln \frac{I - I_{lim\,cat}}{I_{lim\,an} - I}$$

$$= E° + \frac{RT}{n\,\mathscr{F}} \ln \frac{m_{Red}}{m_{Ox}} + \frac{RT}{n\,\mathscr{F}} \ln \frac{I - I_{lim\,cat}}{I_{lim\,an} - I}$$

One can also build an equation for the steady-state current-potential curve of a fast system involving only the following three quantities: anodic and cathodic limiting currents, $I_{lim\,an}$ and $I_{lim\,cat}$, and half-wave potential, $E_{1/2}$. The half-wave potential is simply the electrode potential (*vs* a reference electrode) when the current is equal to the half-sum of the anodic and cathodic algebraic limiting currents. It also represents the inflexion point of the current-potential curve [74]:

$$E = E_{1/2} + \frac{RT}{n\,\mathscr{F}} \ln \frac{I - I_{lim\,cat}}{I_{lim\,an} - I} \quad \text{with} \quad E_{1/2} = E° + \frac{RT}{n\,\mathscr{F}} \ln \frac{m_{Red}}{m_{Ox}}$$

If the diffusion coefficients of the two reactive species are very close, then their mass transport rate constants are also very close and the half-wave potential is equal to the standard potential of the redox couple: $E_{1/2} \approx E°$

This general equation can also be applied to studying a mixture of oxidised and reduced species. Here one would rather introduce the overpotential, η which is equal to:

$$\eta = E - \left(E° + \frac{RT}{n\,\mathscr{F}} \ln \frac{[Ox]^*}{[Red]^*} \right) = \frac{RT}{n\,\mathscr{F}} \ln \frac{I_{lim\,an}}{-\,I_{lim\,cat}} + \frac{RT}{n\,\mathscr{F}} \ln \frac{I - I_{lim\,cat}}{I_{lim\,an} - I}$$

---

[74] The analytic equations for the current-potential curve give one the opportunity to confirm that the second derivative of the current with respect to the potential at the half-wave potential is zero.

In the following one can see again the shapes for the current-potential curves previously described in qualitative terms as being a function of the limiting currents:

▶ if both limiting currents are different from zero, then this leads to the shape in figure 4.26. In such a case, the point for $I = 0$ corresponds to an equilibrium state.

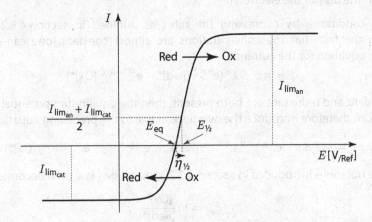

Figure 4.26 - Steady-state current-potential curve in a solution containing Ox and Red

▶ if one of the limiting currents is zero, for example if the initial solution contains no oxidant species: $I_{\text{lim}_{cat}} = 0$, then this leads to the shape in figure 4.27. In this case, $I = 0$ does not correspond to an equilibrium state: the potential at this point is a mixed potential.

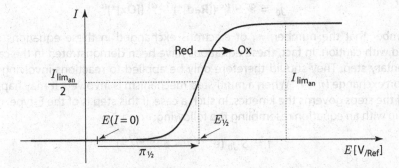

Figure 4.27 - Steady-state current-potential curve in a solution containing only Red

In the particular case of rate control by mass transport, the overpotentials or polarisations are sometimes called concentration overpotentials or concentration polarisations.

4.3.3.3 - SLOW REDOX SYSTEMS

Remember that a redox couple involving the E mechanism is called slow if its standard redox reaction rate constant is very small compared to the mass transport rate constants of the electroactive species.

Slow redox couples can be shown to be systems where the relationship between current and potential is controlled by the electron transfer phenomena in a potential range around the standard potential. The concentration profiles near the interface are not marked and the interfacial concentrations can be considered as equal to the concentrations in the bulk of the electrolyte.

In these conditions, by expressing the rates as outlined in section 4.3.2.3, and by including the fact that the concentrations are almost constant, one can obtain the following equation for the current:

$$I \approx n \mathcal{F} S k^\circ (e^{+\alpha\xi} [Red]^* - e^{-(1-\alpha)\xi} [Ox]^*)$$

If the oxidant and reductant are both present, then the equilibrium potential is defined, and one can therefore introduce the overpotential, η to the preceding equation:

$$I = n \mathcal{F} S k^\circ (e^{+\alpha n f \eta} e^{+\alpha \xi_{eq}} [Red]^* - e^{-(1-\alpha) n f \eta} e^{-(1-\alpha)\xi_{eq}} [Ox]^*)$$

Using the notations introduced in section 4.3.2.3, the NERNST law then becomes:

$$\xi_{eq} = \ln \frac{[Ox]^*}{[Red]^*}$$

The following equation, known as the BUTLER-VOLMER equation, is obtained:

$$I = S j_0 (e^{+\alpha n f \eta} - e^{-(1-\alpha) n f \eta})$$

where j_0, the exchange current density, is linked to the standard reaction rate constant via the following equation:

$$j_0 = \mathcal{F} S k^\circ ([Red]^*)^{(1-\alpha)} ([Ox]^*)^\alpha$$

Remember that the number, n, of electrons exchanged in these equations must be treated with caution. In fact, these equations have been demonstrated in the case of an elementary step. They should therefore only be applied to reactions involving a single electron exchange ($n = 1$). When a multi-step mechanism is involved, it may happen that one of the steps governs the kinetics. In such a case, if this step is of the E-type, then one ends up with an equation resembling the following:

$$I = S j_0 (e^{+\alpha f \eta} - e^{-(1-\alpha) f \eta})$$

with:

$$j_0 = n \mathcal{F} k^\circ ([Red]^*)^{(1-\alpha)} ([Ox]^*)^\alpha$$

The total number of electrons exchanged in the overall reaction, (a value which is also used in FARADAY's law) appears in the equation for j_0 but it is not in the exponential terms in the BUTLER-VOLMER equation.

The exchange current or current density is proportional to the standard rate constant of the redox reaction. Consequently, the slower the couple, the lower the exchange current is. It should be noted that the exchange current also depends on the concentration levels in the solution. This parameter is therefore not intrinsic to the redox couple in question, unlike the standard reaction rate constant. Numerical examples of j_0 and k° values for a few experimental situations are collected together in table 4.3.

Table 4.3 - Examples of kinetic parameters

The various data sources used to compile this table often provide incomplete information regarding the experimental conditions. This means therefore that it is difficult to make comparisons. The strictly precise measurement for kinetic constants is hard to attain.

Redox couple	Conditions, at 25 °C	j_0 [A m^{-2}]	$k°$ [m s^{-1}]
H^+/H_2	on Pt, 1.2 mol L^{-1} HCl	5	
H^+/H_2	on Hg, 1 mol L^{-1} HCl	$10^{-7.5}$	
H^+/H_2	on Pb, 0.25 mol L^{-1} H$_2$SO$_4$	$10^{-7.3}$	
H^+/H_2	on Pb		6×10^{-17}
H_2O/H_2	on Pt, 0.1 mol L^{-1} NaOH	0.6	
H_2O/H_2	on Pb, 6 mol L^{-1} NaOH	4×10^{-2}	
O_2/H_2O	onPt, 1 mol L^{-1} H$_2$SO$_4$	$10^{-4.6}$	
Cl_2/Cl^-	on Pt, 1 mol L^{-1} HCl	10	
$Cd^{2+}/Cd(Hg)$	on Hg, extrapolated 1 mol L^{-1} Cd^{2+}	8×10^4	
$Cd^{2+}/Cd(Hg)$	on Hg		10^{-2}
Fe^{3+}/Fe^{2+}	on Pt		4×10^{-5}
Fe^{3+}/Fe^{2+}	on Pt, equimolar 10^{-2} mol L^{-1}	10^2	10^{-4}
$Fe(CN)_6^{3-}/Fe(CN)_6^{4-}$	on Pt, 1 mol L^{-1} KCl		10^{-3}

The shape of the current-potential curve that emerges, resulting from the addition of two exponential terms, is represented in figures 4.28 and 4.29, whereby $\alpha = 0.7$ and with two different values for j_0. When $\alpha = 0.5$ (a particular case which is not shown here [75]) then the curve is symmetrical around the point ($E = E_{eq}$; $I = 0$).

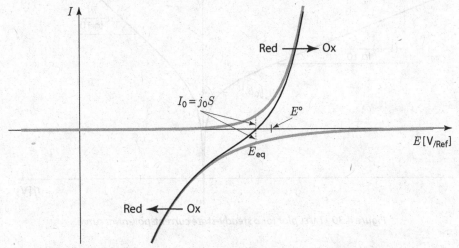

Figure 4.28 - Steady-state current-potential curve for a slow redox system
whereby $\alpha = 0.7$ and with an arbitrary value for j_0

---

[75] When the kinetic parameters are not precisely known, then it is typically assumed that $\alpha = 0.5$.

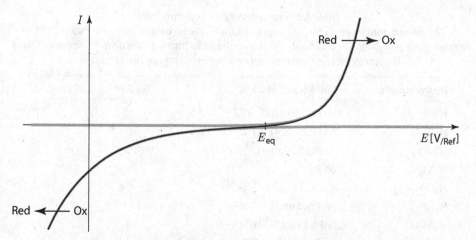

Figure 4.29 - Steady-state current-potential curve for a very slow redox system whereby $\alpha = 0.7$ and with $j_0' \ll j_0$ (in figure 4.28)

The charge transfer control conditions are limited to moderate polarisations, however if the system is slow enough, the corresponding polarisation zone can extend to over several hundred millivolts. By using the experimental data one can then easily determine the kinetic parameters, as outlined below. To do this, a TAFEL plot is often used: $\log |I| = f(\eta)$ as illustrated in figure 4.30.

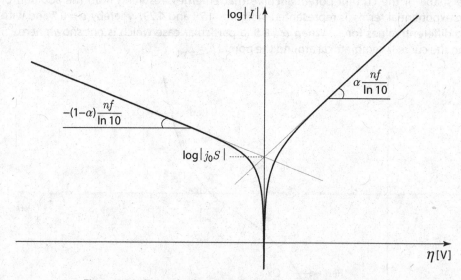

Figure 4.30 - TAFEL plot for a steady-state current-potential curve

For these types of slow systems, the redox reaction is reversible for polarisations lower than a few millivolts, however it becomes irreversible when the polarisation increases (the absolute value of the polarisation is only over 80 mV when the error on I calculated as irreversible is less than 5%, with $n = 1$ and $\alpha = 0.5$). Linear asymptotes then appear, because one of the exponentials outweighs the other, which is tantamount to irreversibility.

These asymptotes enable one to determine the kinetic parameters by using the experimental data easily extracted from the TAFEL plot (see figure 4.30):

▸ intercept $\qquad\qquad\qquad\qquad\qquad$ $\log |j_0\,S|$

▸ slope of the anodic branch $\qquad\qquad$ $\alpha\,\dfrac{n\,f}{\ln 10}$

▸ slope of the cathodic branch $\qquad$ $-(1-\alpha)\,\dfrac{n\,f}{\ln 10}$

For aqueous solution systems, with concentrations in the electroactive species of about 10^{-2} mol L^{-1}, this simple data processing is possible if the standard redox reaction rate constant is lower than 10^{-4} cm s^{-1}. This limit corresponds to an exchange current density lower than 10^{-4} A cm^{-2} = 100 μA cm^{-2} = 1 A m^{-2}.

For even greater polarisations, the slow systems can no longer be regarded as controlled by charge transfer. This is a mixed zone, where charge transfer is irreversible, and where both mass transport and charge transfer are simultaneously brought into play. The concentration profiles have steep slopes, in connection to currents which share the same order of magnitude as the limiting currents. The current-potential curve shape is almost identical for all slow systems, though showing a more or less pronounced shift with respect to the standard potential of the redox couple. Therefore the influence of the standard reaction rate constant is only shown in the potential range where these dramatic current shifts can be seen, as illustrated in figure 4.31 for two different slow redox couples. To simplify the diagram, the same standard and equilibrium potentials and the same limiting currents in oxidation and reduction have been chosen. For example, this figure depicts the current-potential curves obtained with the same solution containing Fe^{3+} and Fe^{2+} ions reacting at two electrodes with different metallic natures, which would therefore lead to different reaction kinetics.

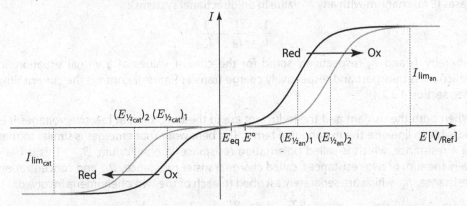

Figure 4.31 - Overall steady-state current-potential curves for two slow redox systems with the same mass transport parameters, the same concentrations, and therefore the same limiting currents. The standard reaction rate constants are different: $k°_1$ (black) > $k°_2$ (grey).

Let us stress the fact that, as indicated in figure 4.31, the comparison between the current-potential curves for two different slow systems results in horizontal translational shifts in opposite directions for both the anodic and cathodic branches. There are no changes in the slope and curvature of the current-potential curves plotted for two different slow couples, quite unlike the changes observed when moving from a fast to a slow couple.

The half-wave potentials of these slow systems are expressed by the following equations:

▶ in oxidation :
$$E_{1/2\,\text{an}} = E° - \frac{1}{\alpha}\,\frac{RT}{n\,\mathscr{F}}\,\ln\frac{k°}{m_{\text{Red}}}$$

▶ in reduction :
$$E_{1/2\,\text{cat}} = E° + \frac{1}{1-\alpha}\,\frac{RT}{n\,\mathscr{F}}\,\ln\frac{k°}{m_{\text{Ox}}}$$

Note that the definition of the half-wave potential is not the same here as for fast couples, despite the fact that they share identical notations and terms. For a slow couple, two half-wave potentials are defined, one for the anodic branch and the other for the cathodic branch, with each of them at the operating point having a current equal to half the value of the corresponding limiting current.

4.3.3.4 - GENERAL CASE

When defining the overall current-potential curve in general cases, which are sometimes called mixed control or quasi-reversible systems, first one needs to write the equation for the current, based on general rate laws involving the interfacial concentrations:

$$I = n\,\mathscr{F}\,S\,k°\,(e^{+\alpha\,\xi}[\text{Red}]_{x=0} - e^{-(1-\alpha)\,\xi}[\text{Ox}]_{x=0})$$

The latter can then be expressed in terms of current and bulk concentrations as follows:

$$C_{i_{x=0}} = C_i{}^* \left(1 - \frac{I}{I_{\text{lim}}}\right)$$

Therefore, the following relationship can be demonstrated and applied to all general cases (E mechanism with any $k°$ value in unidirectional systems):

$$\frac{1}{I} = \frac{1}{I_d} + \frac{1}{I_{ct}}$$

whereby I_d and I_{ct} respectively stand for the current values of a virtual situation, in which mass transport and respectively charge transfer limits or controls the current flow (see section 4.3.2.4) [76].

When both the oxidant and the reductant are in the bulk, then at low overvoltages it is possible to linearise the exponential terms. The behaviour that emerges is similar to that of a resistance, which is called polarisation resistance at equilibrium, R_{peq}. This is basically the sum of two resistances, called charge transfer resistance, R_{ct}, and concentration resistance, R_d, which are separately ascribed to each of the two phenomena involved:

$$R_{\text{peq}} = \underbrace{\frac{RT}{n\,\mathscr{F}}\,\frac{1}{S\,j_0}}_{R_{ct}} + \underbrace{\frac{RT}{n\,\mathscr{F}}\left(\frac{1}{I_{\text{lim an}}} - \frac{1}{I_{\text{lim cat}}}\right)}_{R_d}$$

---

[76] This general equation for a current illustrates the concept of control or limitation by one of the two phenomena. For instance, if it is mass transport that is limiting, this means that $I_{ct} \gg I_d$ and therefore that $I \approx I_d$.

When the overvoltage is sufficiently low, then it can be divided into the sum of two terms which are frequently called the activation overpotential, η_{ct}, on the one hand, related to charge transfer kinetics and on the other hand the concentration over-potential, η_d, related to mass transport kinetics:

$$\eta \approx \eta_{ct} + \eta_d$$

However, whatever the overvoltage value, the two different phenomena of mass transport and charge transfer are usually coupled together, and cannot be considered as the mere sum of two independent terms, with one being related to mass transport and the other to charge transfer. *A priori*, one cannot split the overvoltage into two such terms. As shown in figure 4.32, the stronger the overpotential, the larger the error brought by considering such a sum. The actual overpotential is represented by a black line in figure 4.32. This gives an absolute value, which is larger than that of the grey curve representing the sum of the activation and concentration overpotentials, both regarded as independent. However, when the overpotentials are very high, then the current gets very close to the limiting current. Therefore even if there is a big error in the overpotential at high values of I/I_{lim}, it has little impact on the current-potential curve.

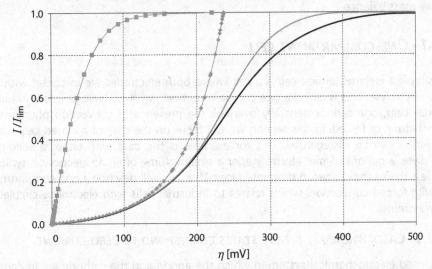

Figure 4.32 - Comparison of the current-potential curve for a redox couple whereby $k°/m$ is equal to 0.01 with curves resulting from the various virtual components
Total overpotential (black), overpotential in a situation of control by charge transfer (activation overpotential, η_{ct}, grey diamonds), overpotential in a situation of control by mass transport (concentration overpotential, η_d, grey squares) and sum of the last two terms (grey).

One can assume therefore that in the conditions indicated in figure 4.32, it is numerically acceptable to split the overpotential into an activation term and a concentration term, even if this has no real physical meaning. For overpotentials lower than 200 mV, the error is below 2% of the limiting current, whereas the maximum error is 10% for an overpotential of around 300 mV.

However, one must not conclude from this result that it is always possible to break the overpotential down into two independent phenomena added together, even if only as a rough approximation. For example, if you took the same system but in different operating conditions, in particular when using transient electrochemical methods, this type of assimilation would lead to totally wrong results.

4.4 - COMPLETE ELECTROCHEMICAL SYSTEMS WITH A CURRENT FLOWING

When dealing with analytical applications in electrochemistry, the focal point is generally the working electrode's interface: the descriptions outlined in the previous section are adequate. However, in many electrochemical applications, it is not possible to consider the kinetic and mass transport phenomena to an electrode as being separate from the other electrode's phenomena. Various examples will be given here to illustrate this point, the first describing two simple cases where both electrodes are in the same electrolyte, therefore constituting a one-compartment cell. We will then study various cases involving electrolysis cells with two compartments. In addition, we will outline the basic principle and the suitable conditions relating to mass balances known as the HITTORF mass balance.

4.4.1 - ONE-COMPARTMENT CELL

The simplest electrochemical cell is a cell where both electrodes are in contat with the same electrolyte. This is called a one-compartment cell, with no ionic junction. In industrial-size cells, one cannot generally overlook the presence of convection phenomena, either natural or forced. In this section we will focus on the case of a closed cell, where convection can be disregarded. This for example is the case with either a solid-type electrolyte, a gel or polymer electrolyte, or a small volume of liquid electrolyte, typically with less than 1 mm between both interfaces. Then we will describe a one-compartment cell with forced convection, which relates to industrial cells with electrolyte circulation (open systems).

4.4.1.1 - CASES WHERE ALL STEADY STATES CORRESPOND TO ZERO-CURRENT

In a closed electrochemical system in which the anode and the cathode are in contact with a common electrolyte, the current flow most often involves two different redox couples, one at the anode and the other at the cathode. The electrolyte changes composition over time as shown in the example in figure 4.33. This diagram depicts a chronoamperometry experiment, whereby the left electrode potential is controlled and involves a rapid and nernstian redox system with a supporting electrolyte, with no convection, and in unidirectional geometry. Therefore, the interfacial concentration is fixed for the electroactive species at the left electrode, whereas their molar flux density at the right electrode is zero since here they are not electroactive.

Firstly, let us note the fact that shortly after the start of the experiment, the concentration profile follows the same pattern as that of chronoamperometry in semi-infinite geometric conditions (see figure 4.18 in section 4.3.1.3). The specific condition of there being zero flux in the zone located far away from the left interface leads to a constant concentration in that same area. However, if the experiment is carried on further in time, then there is a change in concentration at the right interface, and the concentration of the whole electrolyte becomes non-homogeneous. In this kind of time range, it is no longer possible to view the phenomena at the two electrodes as independent[77].

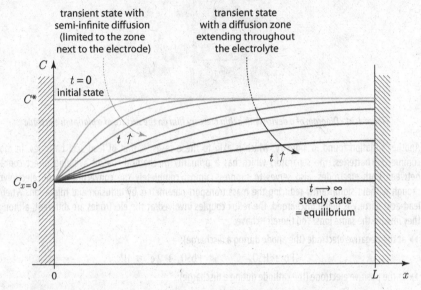

Figure 4.33 - Evolution of the concentration profile in chronoamperometry

The current decreases to the point of reaching zero. Here in such a system it is impossible to attain a steady state with a current that is anything other than zero. In a potentiostatic experiment, the only steady state that can be observed is therefore the equilibrium state with zero-current. This equilibrium state eventually settles once the transient current flow has homogenized the electrolyte composition. This composition correlates with the potential imposed at the working electrode, according to the NERNST law.

▶ A similar situation occurs when a thin film of a conducting material is being deposited on the electrode. The example shown in the diagram in figure 4.34 illustrates a thin film of mercury (an electronic conductor) that is deposited on a platinum electrode, whereby only the mercury is in contact with a liquid electrolyte containing Au^{3+} ions. By reduction at the interface between mercury and the electrolyte, these ions produce gold metal in the form of an amalgam:

$$Au^{3+} + 3\,e^- \longrightarrow Au(Hg)$$

The film is gradually transformed into an amalgam. For the Au(Hg) species, the platinum | mercury interface, which is an electronic junction, is non-reactive while the mercury | electrolyte interface is reactive.

---

[77] In a motionless electrolyte (namely without convection), such as a gel for example, one can estimate the time at which an interaction occurs between the electrodes in terms of mass transport. This is done by calculating the moment when the diffusion layer in a semi-infinite diffusion experiment (thickness of about $\sqrt{Dt}$ see section 4.3.1.3) becomes approximately as thick as the inter-electrode distance. Therefore, in a system containing an electroactive species with a diffusion coefficient of about 10^{-5} cm² s⁻¹ and an inter-electrode distance of 3 mm, it would take an experiment lasting around 3 h for the diffusion layer to reach a thickness matching the inter-electrode distance.

As in the case described above in figure 4.33, here the steady state can only correspond to an equilibrium state with a non-zero, homogeneous gold concentration in the amalgam.

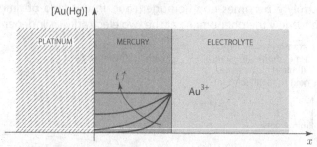

Figure 4.34 - Diagram of a device with a thin mercury film on the surface of a platinum electrode

▶ Another situation found in industry, which is akin to these cases, is that of lead-acid batteries. In most commercial batteries, the separator, which has a primarily mechanical role of avoiding direct contact between both electrodes, also serves to suppress almost completely the convection in the electrolyte, though without significantly reducing the mass transport parameters by diffusion and migration. When a lead-acid battery is being discharged, the redox couples involved at the electrodes are different, although they involve the same ions. You therefore have:

▶▶ at the negative electrode (the anode during a discharge):

$$Pb + HSO_4^- \longrightarrow PbSO_4 + 2\,e^- + H^+$$

▶▶ at the positive electrode (the cathode during a discharge):

$$PbO_2 + HSO_4^- + 3\,H^+ + 2\,e^- \longrightarrow PbSO_4 + 2\,H_2O$$

As before, the whole electrolyte is the seat for diffusion and migration phenomena after a transient period. Unlike the previous cases, both interfaces are reactive for the electrolyte ions. However, the fact that the two electrode half-reactions are different means that different stoichiometries appear in the equations, with therefore different slopes on each side as a result of applying FARADAY's law. Here again, one cannot obtain a non-zero steady-state current.

To illustrate this, figure 4.35 shows the typical changes in concentration profile that emerges over time when a lead-acid battery is being discharged at a constant current.

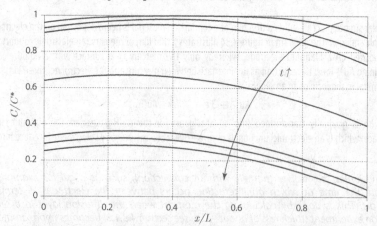

Figure 4.35 - Concentration profiles of the ions of the electrolyte, $[H^+] = [HSO_4^-]$,
at various instants while a lead-acid battery is being discharged at a constant current
The results are obtained by numerically solving the migration and diffusion equations for $t_+ = 0.75$

4.4.1.2 - OBTAINING NON-ZERO-CURRENT STEADY STATES

For such a type of closed electrochemical cell with no convection, it is possible to obtain a non-zero-current steady state (in the strictest sense of the term as regards the electrolyte) if all the electroactive species present in the electrolyte have the same stoichiometries in both electrode redox reactions. The average composition of the electrolyte does not change over time, as illustrated in the example in figure 4.36 which shows concentration profiles for a chronopotentiometry with a fast and nernstian redox system in the presence of a supporting electrolyte with no convection [78].

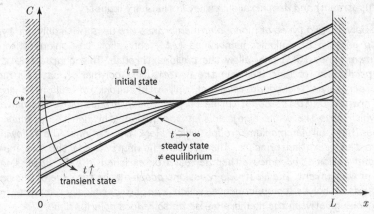

Figure 4.36 - Concentration profile in chronopotentiometry

Here again, in this system without convection, the composition throughout the electrolyte is non-homogeneous, with the diffusion phenomena occupying the entire volume after a transient period.

▶ This case is a simplified version of how a lithium battery works [79]. For example, when a lithium battery has two insertion materials a and b, the electrode phenomena can be written in the following simplified manner:

▶ at the negative electrode (the anode during a discharge):

$$\langle Li \rangle_a \longrightarrow \langle \ \rangle_a + e^- + Li^+_{electrolyte}$$

▶ at the positive electrode (the cathode during a discharge):

$$\langle \ \rangle_b + Li^+_{electrolyte} + e^- \longrightarrow \langle Li \rangle_b$$

Although the two redox couples are not identical, as regards the electrolyte both redox reactions involve a supply or consumption of Li with the same stoichiometry. This applies to a Li-ion battery with two insertion materials or to a Li-polymer battery with Li as the negative electrode and an insertion material at the positive electrode. When a lithium battery is in operation, one must take into account simultaneously the migration and diffusion of the ions in the electrolyte, however, this makes no change in qualitative terms to what has just been illustrated. ◢

---

[78] Appendix A.4.1 adresses in detail the characteristics for such a system at steady state, as well as the establishment of that steady state.

[79] These batteries are described in the illustrated board entitled 'Energy storage: the Li-Metal-Polymer (LMP) batteries'. They are often called 'rocking chair batteries' by the specialists when, during operation, the Li+ ions are de-inserted at one of the interface and simultaneously inserted at the other interface. Lithium is thus tilted from one side to the other during charge or discharge.

ELECTRODIALYSIS

Document written with the kind collaboration of A. SAVALL,
from the Paul Sabatier University in Toulouse, and F. LUTIN, from the company Eurodia

Membrane electrolysis cells have many applications in the food industry (dairy, wine, fruit juice, etc.), water softening, purification or recovering effluents from electroplating and other chemical processes. Possibly the best known processes are desalinating brackish water with a moderate salt content (other processes such as reverse osmosis are used upstream) and demineralising whey in the dairy industry.

In electrodialysis, two types of monopolar membranes are used (with only one type of ion allowed to penetrate). Anionic membranes (AM) only allow the anions through, and cationic membranes (CM) only allow the cations through. This characteristic derives from a specific way of functioning in the constituting polymer. A cationic membrane therefore serves as the conducting material, with sulfonic or sulfonate groups for instance grafted on to the constituting polymer. Such groups block the movement of anions (which have the same sign), and let the cations through the polymer. Anionic membranes have alkylammonium groups which block the cations. The following figure illustrates the operating principle. The solution flowing between the two membranes (anionic and cationic) becomes either depleted or enriched in salt, depending on the direction of the current. The electrode reactions occur at the far ends of the cell, in the anolyte and catholyte compartments which can be chemically distinct from the solutions flowing between the membranes so as to reduce polarisations.

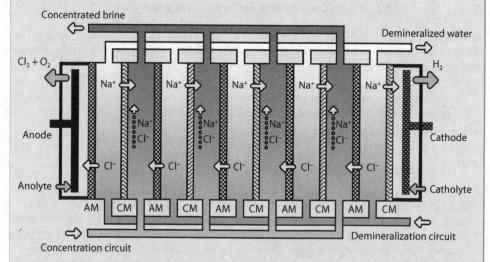

Working diagram of an electrodialysis cell

To attain better results, several modules (pairs of membranes) are connected in parallel as far as the flow of liquids is concerned and in series electrically with the current running in an identical fashion across each compartment. The entire installation has several operating modes: either by single flow through a cell with the possibility of having several cells arranged in series depending on fluid flow or in a loop (recirculation) via a tank.

Industrial electrodialysis installation (photo provided by the company Eurodia, Wissous, France)

The filter-press type configurations are made up of a set of modules which are arranged in alternating anionic and cationic membranes. The thin intermembrane compartments contain spacers to improve the hydrodynamic conditions. High-pressure pumps push the fluids inside the hydraulic circuits so as to overcome any drops in pressure that may occur along the flow line.

Industrial cells consist of one or two hundred modules with membrane surfaces that can reach up to one m^2. These facilities are able to soften brackish water with flow rates spanning from a few hundred up to one thousand m^3 per day, and all with an energy cost of about 1 kWh per m^3. The precise nature of the electrode reactions taking place in the compartments at both ends of the cell plays no direct role in the electrodialysis process. The intermembrane space has a thickness lower than 1 mm, in order to decrease the ohmic drop. However if the solution requires a stronger demineralising effect, then the ohmic drop can be very large because this solution will become poorly conducting.

When the system is in operation, deposits may build up on the membrane surface, namely due to local pH variations which can cause hydroxides to be formed with the metallic cations of the solutions being processed. Such a phenomenon clogs and weakens the membranes which are then difficult to maintain. It is advisable to periodically reverse the polarity so as to eliminate these deposits. Regularly dismantling is still a necessary step to clean the various elements.

If there is a wide distance between the two electrodes in a liquid electrolyte, then it is not possible to overlook the natural convection. In fact, it is this natural convection that homogenises the concentration in the electrolyte's middle zone. For example, if we keep to the same interfacial reactions as those described above and assume that only a short time is needed for the two diffusion layers to reach equal thickness δ, then the changes over time in the concentration profiles illustrated in figure 4.37 show how, yet again, a non-zero-current steady state is achieved.

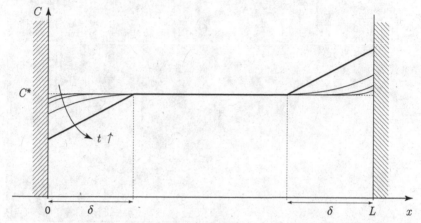

Figure 4.37 - Establishing the steady-state concentration profile in chronopotentiometry involving convection

As in the previous example, the average composition of the electrolyte does not change. The major differences between the two systems lie in the thickness of the diffusion layer, which is found to be lower in the second example: in aqueous media, the thickness of the diffusion layer due to natural convection is estimated at around a few 100 µm. Therefore, the values for the steady-state currents are higher, and less time is needed for the steady state to become settled.

In addition, for open industrial cells involving a circulating electrolyte (the electrolyte is in continuous renewal) forced convection must be taken into account. Under such conditions, a quasi-steady state can be achieved even when the anodic and cathodic reactions do not involve the same redox couple.

In all cases where convection occurs, it is not easy to establish the volume mass balance for each species in simple quantitative terms, unlike in other situations (see section 4.4.2.3 on the HITTORF mass balance).

4.4.2 - CELL WITH TWO SEPARATE COMPARTMENTS

In many industrial processes where soluble species are involved, it is sometimes necessary to keep the homogenisation of the electrolyte to a minimum so as to prevent a species that is produced at one electrode from reacting at the other electrode[80]. Here

---

[80] This phenomenon, which is sometimes called the chemical shuttle, is equally unwelcome in the case of batteries, because it would preclude the possibility of any efficient energy storage.

we will give a few examples to explore electrochemical systems that comprise a device which separates the catholyte from the anolyte[81], though all the while maintaining the electrical contact between them. In particular, we will focus on the role of this separator in terms of mass transport, as well as studying how the nature of the material can be used as a way of influencing the mass balances.

4.4.2.1 - DIFFERENT TYPES OF SEPARATION

Some laboratory devices have compartments separated by a long tube with a small section (capillary). Figure 4.38 shows a simplified view of this type of device. The diffusion coefficients and the mobilities of the various species remain unchanged throughout the electrolyte volume.

To simplify, let us assume that the systems have a unidirectional geometry, although strictly speaking this does not apply to zones where the section areas change. What is peculiar about this type of cell is the form of equations that can be written for the molar fluxes in the cell's various zones. In fact the following equations show how the molar flow rates for each species are preserved:

$$\left(N_{\text{anolyte}} \right)_{x=L} S_{\text{electrodes}} = \left(N_{\text{capillary}} \right)_{x=L} S_{\text{capillary}}$$

$$\left(N_{\text{catholyte}} \right)_{x=L+\ell_{\text{capillary}}} S_{\text{electrodes}} = \left(N_{\text{capillary}} \right)_{x=L+\ell_{\text{capillary}}} S_{\text{capillary}}$$

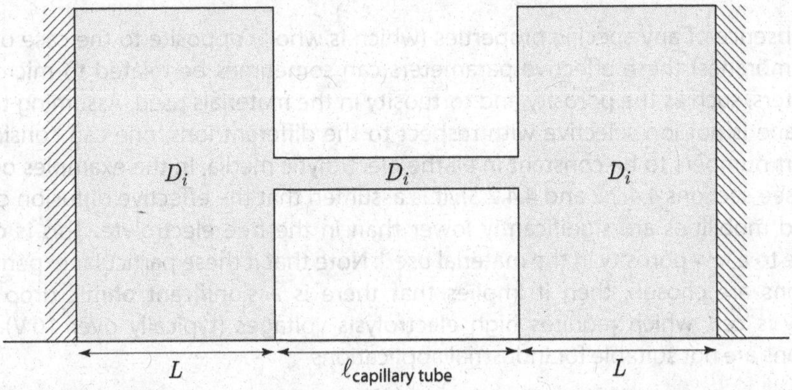

Figure 4.38 – Simplified diagram of a cell with a separating tube

If there is a large enough difference between the section areas, then one can therefore assume that the molar flux densities in the anolyte and the catholyte are low compared to the corresponding quantities within the capillary tube. The impact of this property is addressed in the following sections. On the other hand, using a capillary tube with a narrow section makes the natural convection negligible in this part of the cell.

Therefore, in this instance, the solution chosen is generally based on the use of solid electroactive materials that are insoluble in the electrolyte. This issue will not be dealt with in this work.

[81] The term anolyte refers to the electrolyte on the anodic side. The term catholyte means the electrolyte on the cathodic side.

It is more common to use various types of porous materials (porous frits, polymer membranes) to separate the two compartments in an electrolysis cell[82]. Figure 4.39 gives a simplified view of such a type of cell.

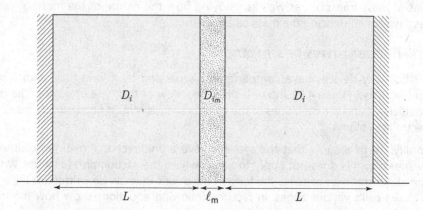

$$\text{Figure 4.39 – Simplified diagram of a two-compartment cell with a membrane}$$

The aim here is not to go into thorough detail about mass transport in porous materials. We will merely describe these phenomena in macroscopic terms using the effective mass transport parameters (diffusion coefficients and mobilities) denoted by the index m and adapted to the material's macroscopic geometry.

In the absence of any specific properties (which is wholly opposite to the case of selective membranes) these effective parameters can sometimes be related to microscopic parameters, such as the porosity and tortuosity in the materials used. Assuming that the membrane is not ion-selective with respect to the different ions, one can consider the transport numbers to be constant in all the electrolytic media. In the examples outlined below (see sections 4.4.2.2 and 4.4.2.3), it is assumed that the effective diffusion coefficients and mobilities are significantly lower than in the free electrolyte. This is due for example to a low porosity in the material used. Note that if these particular experimental conditions are chosen, then it implies that there is a significant ohmic drop in the electrolysis cell, which requires high electrolysis voltages (typically over 10 V). These conditions are not suitable for industrial applications.

In this simplified system, with unidirectional geometry and negligible convection in the porous material, the continuity of the molar flux densities for each species at the two ionic junctions has consequences that are quite similar to those arising when a capillary tube is used. Here again the convection in the central part of the system is overlooked. However, the two systems are not equivalent to each other because in the case where there is separation by a membrane, the equations indicating the time evolution for the concentration at any point (equations deduced from the volume mass balances) are different in the three cell zones. This is not the case when there is separation by a capillary tube.

---

[82] To find an example of an industrial process using membrane electrolysis cells, one can refer to the illustrated board entitled 'Electrodialysis'.

4.4.2.2 - STEADY STATES WITH A NON-ZERO CURRENT

Let us imagine an electrolyte with two monovalent ions, $A^+ X^-$, such that the couple at the anode is identical to that at the cathode, and only A^+ is an electroactive species. A^+ is produced at the anode and consumed at the cathode. Remember that in this case it is possible to attain steady states different from equilibrium states[83]. For example in a chronopotentiometry, after a while the concentration profile ceases to change. Remember also that electroneutrality requires the anion and cation concentrations to be equal at all points throughout the electrolyte. In systems with unidirectional geometry, linear concentration profiles emerge in the zones where there is no convection, and the slopes depend solely on the current and the diffusion coefficient of the electroactive species A^+.

Therefore, when the cell has a membrane, given that the reasoning outlined above applies to each diffusion layer in the electrolyte, then the steady-state concentration profile takes the form of straight segments[84]. If the diffusion coefficient appears to be smaller in the membrane than in the compartments, then significant differences can be seen in the slopes for the various cell zones, as shown in figure 4.40.

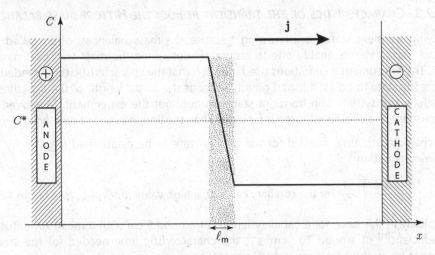

Figure 4.40 - Concentration profile during an electrolysis experiment at steady state for a membrane cell with a diffusion coefficients ratio, D_i / D_{i_m}, equal to 100

At steady state, the electrolyte composition difference between the two compartments on each side of the membrane reaches a steady value. This value depends mainly on the imposed current, the diffusion coefficient of the electroactive ion in the membrane and

---

[83] This example is addressed in detail in appendix A.4.1 for a one-compartment cell.

[84] So as to take into account the natural or forced convection, the electrolyte is divided into 7 zones: the electrolyte inside the membrane (no convection), then each compartment is divided into a central homogeneous zone with two diffusion layers (one is next to the electrode and the other is next to the membrane). Five of these zones are clearly seen in the figure 4.41 in the transient period. The slopes for the steady-state concentration profiles are negligible in the four diffusion layers of the electrolyte outside the membrane. This is due to the difference between the diffusion coefficients chosen. In figure 4.40, these zones cannot be distinguished from the two zones which are homogenised by convection.

the membrane thickness. A limiting current value can also be defined for which the steady-state concentration at the cathode is zero. When dealing with numerical values such that one can disregard the concentration profiles slope in the compartments at each end, the limiting current value is given by the following:

$$I_{lim} = \frac{2D_{+m} \, \mathscr{F} \, S_{electrodes} \, C^*}{\ell_m}$$

Similar reasoning can be applied to a cell with a capillary tube. If the area of the capillary section is much smaller than that of the two compartments on either side, then the concentration profile slope in the electrolyte's intermediate zone is much more pronounced than in the two anolyte and catholyte zones. The limiting current at steady state is then given by the following:

$$I_{lim} = \frac{2D_+ \, \mathscr{F} \, S_{capillary} \, C^*}{\ell_{capillary}}$$

4.4.2.3 - CHARACTERISTICS OF THE TRANSIENT PERIOD: THE HITTORF MASS BALANCE

Studying transient states is interesting because the mass balances, often called the HITTORF mass balance, enable one to accurately determine the mass transport parameters. The experimental conditions used are such that the characteristic time needed for the steady state to be established greatly exceeds the actual length of the experiment. Therefore, the system is in transient states throughout the experiment. Moreover, for this purpose, the current imposed is far larger than the limiting current defined above.

The characteristic time needed for the steady state to be established is given by the following equation[85]:

▸ $\tau = \left(\dfrac{\ell_{capillary}}{2} \right)^2 \dfrac{1}{D_\pm}$ for the capillary cell with a high value for $\ell_{capillary}$ (see figure 4.38).

Typically in this case, for a capillary length of around 5 cm with a mean AX diffusion coefficient[86] of around $10^{-5} \, cm^2 \, s^{-1}$, the characteristic time needed for the steady state to be established is around 10^6 s, that is to say one week.

▸ $\tau = \left(\dfrac{\ell_m}{2} \right)^2 \dfrac{1}{D_{\pm m}}$ for the membrane cell with a small value for $D_{\pm m}$ (see figure 4.39)

Typically in this case, with a membrane thickness of around 2 mm and a mean AX diffusion coefficient of around $10^{-7} \, cm^2 \, s^{-1}$, the characteristic time needed for the steady state to be established is around 10^5 s, that is to say 30 hours.

Figure 4.41 shows the change in the concentration profiles in a membrane cell for the middle zone of the electrolyte (around the membrane) at the beginning of the transient state[84].

---

[85] Since the concentration profiles build up symmetrically on both sides of the capillary or the membrane, then ℓ/2 is to be considered as the characteristic distance for establishing the steady state.

[86] $D_\pm = t_- D_+ + t_+ D_-$, see section 4.2.1.2.

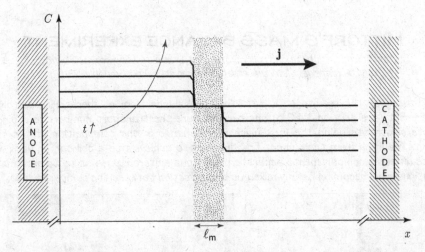

Figure 4.41 - Changes over time of the ions concentration profiles at the beginning of the transient period of a chronopotentiometry with unidirectional geometry (membrane cell)

During the beginning of the transient period, the concentration in the membrane remains unchanged, except in its two diffusion layers, which remain very thin. This means that the major part of the membrane is not a seat for diffusion phenomena. Based on this outcome, it is possible to show that in the transient period, as long as the concentration in the membrane remains constant, the anolyte concentration varies linearly with time, and the corresponding slope is characteristic of the ion transport numbers (see demonstration below: HITTORF's mass balance). The same applies for the catholyte concentration. Therefore, if it is possible to find experimental means to monitor the change in the ion concentrations in the anolyte (and/or the catholyte) over time, or after a given duration, then one can determine the mass transport parameters of both ions.

In general, establishing a HITTORF mass balance comes down to writing the mass balance for an electrolyte volume that is bound on one side by the electrochemical interface and on the other side by a zone in which only migration phenomena play a role in the current flow, as shown in the diagram in figure 4.42 using the example detailed below the figure. In fact, at this surface one simply has the following:

$$j_i = j_{i\,\text{migration}} = t_i\, j$$

In these conditions, the mass balance is given by:

$$dn_i = dn_i^{\text{farad}} - t_i \frac{I\,dt}{z_i\,\mathscr{F}}$$

or, between the initial time and instant t:

$$\Delta n_i = \frac{\nu_i}{\nu_e\,\mathscr{F}} Q - \frac{t_i}{z_i\,\mathscr{F}} Q$$

HITTORF'S MASS BALANCE EXPERIMENT

Document written with the kind collaboration of A. DENOYELLE, Laboratoire d'Electrochimie et Physico-chimie des Matériaux et des Interfaces, at Phelma, Grenoble INP, France

By drawing up the results of a HITTORF mass balance from a laboratory electrolysis experiment (see section 4.4.2.3) one can measure the transport numbers in a single-solute electrolyte using the measurement of change in the composition of the electrolyte resulting from the current flow. In order to achieve this a cell can be used with three different compartments separated by porous sintered glass so as to prevent the solutions from becoming rapidly mixed up by convection between the compartments.

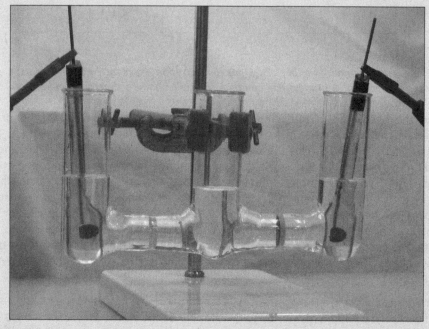

Example of a three-compartment glass cell used for student's practical work at Phelma (Grenoble INP)

To carry the experiment, an electric power supply is needed to impose a constant current so as to simplify determining the amount of charge that has circulated during the experiment. Since porous materials have high resistance to current flow, a high voltage power supply must be used (about a hundred volts).

For example, in the cell pictured above, with a copper sulphate solution with a concentration of 5×10^{-2} mol L^{-1} it takes about two hours of electrolysis with the current set at 20 mA to observe significant concentration changes (about 20%). With an accurate measurement of these variations from complexation or precipitation titration reactions, the transport number of Cu^{2+} ions can be determined: $t_+ = 0.4$ to the nearest few %, close to the values stated in scientific reports. As expected from the HITTORF mass balance, the concentration in the central compartment does not fluctuate a great deal.

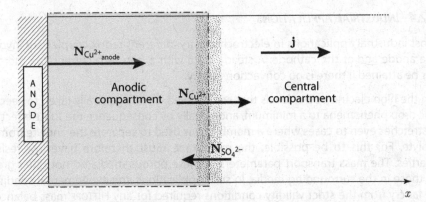

Figure 4.42 - Principle of the particular mass balance called the HITTORF mass balance,
illustrated on the Cu oxidation in a CuSO₄ aqueous solution (see example below)
The electrolyte volume concerned by the mass balance is put in grey and delimited by a dashed and dotted line.

For example, electrolysis can be carried out in a cell with three compartments, separated by two porous frits, filled with an aqueous solution containing copper sulphate $CuSO_4$, and with two copper electrodes in the outer compartments [87]. The amount of substance inside the two porous frits is negligible compared to the large amounts in the volumes contained in each cell compartment. The electrolysis current is adjusted to a moderate constant value so that only the main electrolysis reactions occur, in other words with a faradic yield of 100% at the anode and cathode.

▸ at the anode, oxidation of the copper electrode occurs:
$$Cu \longrightarrow Cu^{2+} + 2\,e^-$$

▸ a copper deposit is produced at the cathode:
$$Cu^{2+} + 2\,e^- \longrightarrow Cu$$

Once the electrolysis has been underway for a few hours, the titration of the three compartments enables one to accurately check that the concentration change in the central compartment is very low (theoretically zero), and also check that the changes in the amount of Cu^{2+} ions are opposite quantities in the two outer compartments. As expected, after such an experiment, the concentration of copper sulphate increases in the anodic compartment while it decreases in the cathodic compartment.

The HITTORF mass balance determines the transport numbers of these two ions from the following experimental data: concentration variations, compartment volumes and the total quantity of charge transferred during electrolysis. In fact, in the anodic compartment you have the following equations:

$$\Delta n_{Cu^{2+}} = \frac{Q}{2\,\mathcal{F}} - \frac{t_+}{2\,\mathcal{F}}Q = t_-\frac{Q}{2\,\mathcal{F}}$$

$$\Delta n_{SO_4^{2-}} = 0 + \frac{t_-}{-2\,\mathcal{F}}Q = t_-\frac{Q}{2\,\mathcal{F}}$$

The same variation is obtained for the amount of each type of ion, since the system remains electrically neutral throughout the whole experiment.

The HITTORF mass balance in the cathodic compartment shows an opposite variation in the amount of substance, in comparison to that in the anodic compartment, since these equations are written in algebraic terms: at the cathode the current, and therefore the charge Q, are both negative.

---

[87] For an illustration of this type of experiment, refer to the illustrated board entitled 'HITTORF's mass balance experiment'.

4.4.2.4 - INDUSTRIAL APPLICATIONS

In most industrial applications in electrochemistry, different redox couples are involved at the anode and at the cathode. A steady state with a non-zero current can therefore never be attained if there is no convection at play.

Given the high electric power levels that are involved, care is generally taken to keep the ohmic drop phenomena to a minimum, and equally by consequence the JOULE effect too. This stretches even to cases where a membrane is used to separate the anolyte from the catholyte. For this to be possible, the membrane must therefore have low resistive properties. The mass transport parameters in these porous media do not differ greatly from those in the surrounding media. In such applications, the experimental conditions are a far cry from the strict validity conditions required for any HITTORF mass balance, as outlined above. In industrial applications that require a separator between the two parts of the cell, the membrane's role in terms of mass transport mainly relates to convection phenomena. The membrane generally prevents or at least slows down the convective mixing process of the anolyte and the catholyte. On the other hand, it is incorrect to assume that there is no diffusion flux through the membrane. HITTORF-type arguments only enable one to make qualitative predictions in terms of materials increase or decrease in the amount of substances in the electrolyte. Nevertheless, they remain interesting from a practical point of view.

To conclude, one should note that the mass balances can also be modified by using permselective or specific membranes in some industrial processes. A good example of such applications includes the use of anionic and cationic membranes to desalinate sea water[88] or other methods of separation or purification.

---

[88] See the illustrated board entitled 'Electrodialysis'.

QUESTIONS ON CHAPTER 4

1 - Complete the following diagram for the interfacial reaction
$$Co + 4\,Cl^- \longrightarrow CoCl_4^{2-} + 2\,e^-$$
by indicating, in qualitative terms, in both phases:
▸ the various molar flux density vectors ($\mathbf{N}_i$)
▸ the various current density vectors ($\mathbf{j}_i$)
▸ the overall current density vector ($\mathbf{j}$)
▸ the vector normal to the surface ($\mathbf{n}$) following the usual sign convention for the current

| METAL (cobalt, Co) | ELECTROLYTE |

2 - When a species is adsorbed at the interface, the interfacial flux and the production rate at steady state are both zero   true   false

3 - In a given electrochemical experiment, where only the current and the potential can vary, a redox couple or reaction can be:
▸ fast or slow, depending on the operating conditions   true   false
▸ reversible or irreversible, depending on the operating conditions   true   false

4 - For an aqueous solution at room temperature containing anions and cations with a concentration of 0.1 mol L^{-1}, what is the order of magnitude and the unit of the electric conductivity?   . . . .

5 - What is the order of magnitude and the unit of the diffusion coefficient of an ion in an aqueous solution at room temperature?   . . . .

6 - The diagram below shows the changes over time in the concentration profile for an electroactive species of a fast redox couple in an experiment:
▸ at steady state   true   false
▸ of voltammetry   true   false
▸ of chronoamperometry   true   false
▸ of chronopotentiometry   true   false

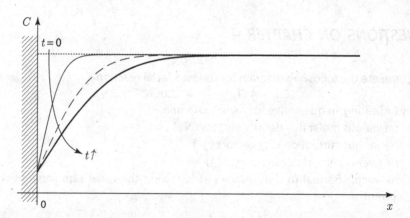

7 - The following diagram shows the concentration profiles for a solution containing
Fe^{3+} (grey) and Fe^{2+} (black) ions. The dotted lines represent the initial instant, and the
solid lines represent instant t. These profiles result from current circulation through
an interface between a platinum electrode and a solution with negligible convection
and migration of the electroactive species, Fe^{3+} and Fe^{2+}.

▸ the case shown corresponds to an oxidation reaction   true   false
▸ what is the quantity δ called?   ....
▸ the diffusion coefficient of Fe^{2+} is larger than that of Fe^{3+} true   false
▸ the two shaded areas are equal   true   false
▸ complete the diagram with a qualitative drawing of the concentration profiles that
one would observe for a larger current, yet with the same value for δ.

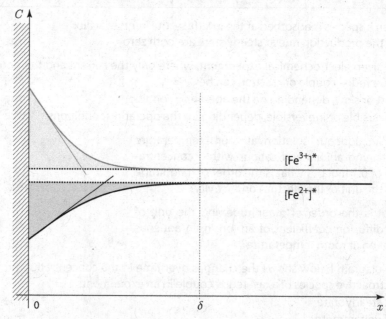

8 - Close to a metal|electrolyte electrochemical interface, one defines the double layer and the diffusion layer. The diffusion layer is usually much thinner than the double layer.   true   false

9 - Look at the following steady-state current-potential curve:

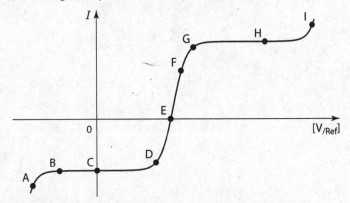

Among the working points indicated above, which ones correspond to the concentration profile shown in the diagram below?

A   B   C   D   E   F   G   H   I

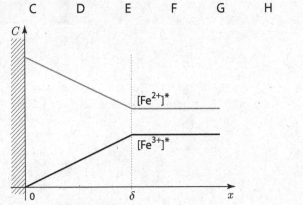

10 - In usual cases, the value of the steady-state limiting anodic current is proportional to the concentration:
  ▶ in the oxidant at the electrode interface   true   false
  ▶ in the oxidant in the bulk solution   true   false
  ▶ in the reductant at the electrode interface   true   false
  ▶ in the reductant in the bulk solution   true   false

APPENDICES

A.1.1 - LIQUID IONIC JUNCTION VOLTAGE WITHOUT CURRENT

When there are ionic junctions involved in an electrochemical chain, in many instances the open-circuit junction voltages (or potential differences, see section 1.5) may be overlooked, although this is not a general rule. In particular, it is rare that this type of approximation can be applied in equilibrium conditions to single-exchange ionic junctions (for example when a solid electrolyte is involved, see section 3.3.4.1), nor indeed in the case of multiple junctions with two electrolytes of a dissimilar nature (for instance two electrolyte solutions in two different solvents, see section 3.3.5). When the liquid junction in question is between two electrolytes in the same solvent, then the ionic junction voltage in thermodynamic equilibrium is zero, because the two solutions are perfectly mixed together in equilibrium (see section 3.3.5). Nevertheless, the result is ultimately immaterial, because the experiments carried out to implement such junctions often make use of devices that are specifically designed to considerably slow down the mixing process of the solutions. This happens with salt bridges, involving a porous material at each contact between one end of the bridge and one of the two different solutions. When approaching the question from an experimental point of view, it is thus very important to focus on the junction voltage's quasi-steady-state value (not at thermodynamic equilibrium) between the two solutions, whose compositions are still not the same. Therefore at open circuit, the species fluxes between the two solutions are very small, although not zero. Significant concentration profiles and potential profiles are only found inside the porous plugs (see below).

Here, let us firstly focus on the conditions required in order to minimise such junction voltages. This section revolves around a series of numerical calculations that are based on what is commonly known as the HENDERSON equation: an equation which makes it possible to evaluate the junction voltage between two solutions with different compositions in the same solvent. Basic elements will then be given about the hypotheses and the reasonings leading to this equation. It is strongly advisable to first read chapter 4 and appendices A.3.4 and A.4.1 in order to gain a better understanding of the second part of this appendix.

THE HENDERSON EQUATION AND ITS IMPACT IN PRACTICAL TERMS

The HENDERSON equation is useful if one is seeking to estimate the junction voltage between two solutions α and β in the same solvent in quasi-steady-state conditions:

$$\varphi_\beta - \varphi_\alpha = \frac{RT}{\mathscr{F}} \frac{\sum_i \frac{\lambda_i}{z_i} \left(C_{i_\beta} - C_{i_\alpha} \right)}{\sum_i \lambda_i \left(C_{i_\beta} - C_{i_\alpha} \right)} \ln \frac{\sum_i \lambda_i C_{i_\alpha}}{\sum_i \lambda_i C_{i_\beta}}$$

261

$$\varphi_\beta - \varphi_\alpha = 0.059 \frac{\sum_i \frac{\lambda_i}{z_i}\left(C_{i\beta} - C_{i\alpha}\right)}{\sum_i \lambda_i \left(C_{i\beta} - C_{i\alpha}\right)} \log \frac{\sum_i \lambda_i C_{i\alpha}}{\sum_i \lambda_i C_{i\beta}} \qquad \text{V at 25°C}$$

Here we will focus on two cases which are important from an experimental point of view:

▸ the first one is a reference electrode with a liquid junction (such as an electrode based on the AgCl/Ag couple or a calomel electrode SCE, see section 1.5.1.2). Here we will study the influence on the junction voltage of the internal solution's composition;

▸ the second one is a salt bridge (implemented for example when measuring the voltage of a DANIELL cell, see section 1.4.1.1). Here we will study the influence on the overall ionic junction voltage of the intermediate electrolyte's composition.

It is assumed in the numerical applications outlined in this section, that the molar conductivities are equal to their values at infinite dilution. These values are all listed in table 4.2 in section 4.2.2.4.

In the first example, the liquid junction voltage will be examined at quasi-steady-state conditions for the following electrochemical chain (which could be part of a reference electrode based on AgCl/Ag):

inner aqueous solution of reference (sol. α) || HCl aqueous solution at $pH = 2$ (sol. β)

To fix the potential reference, one needs to fix the chloride ion concentration in the reference's inner solution (see section 1.5.1.2). However, this concentration is not the only ruling factor to be taken into account designing a high-quality reference electrode, as shown when comparing the four examples outlined below (the ideal reference electrode should have a zero ionic junction voltage). The nature of the solution β being studied (here it is HCl at $pH = 2$) also has an impact on the junction voltage. However, in an experimental situation it is much easier to adapt the composition of solution α.

▸ Solution α: 1 mol L^{-1} KCl aqueous solution

$$\varphi_\beta - \varphi_\alpha = 0.059 \frac{34.98 \times 10^{-2} - 7.35 + 7.63 \times 0.99}{34.98 \times 10^{-2} - 7.35 + 7.63 \times 0.99} \log \frac{7.35 + 7.63}{34.98 \times 10^{-2} + 7.35 \times 10^{-2}}$$
$$= -3.5 \, \text{mV}$$

▸ Solution α: 1 mol L^{-1} NaCl aqueous solution

$$\varphi_\beta - \varphi_\alpha = -20.6 \, \text{mV}$$

▸ Solution α: 1 mol L^{-1} LiCl aqueous solution

$$\varphi_\beta - \varphi_\alpha = -30.9 \, \text{mV}$$

▸ Solution α: 10^{-2} mol L^{-1} KCl aqueous solution

$$\varphi_\beta - \varphi_\alpha = -26.8 \, \text{mV}$$

These results underline the fact that in order to minimise the junction voltage, one needs to choose a highly concentrated electrolyte containing anions and cations whose molar conductivities are very close. Therefore, a KCl electrolyte is a much better choice than LiCl or NaCl when the anion is Cl$^-$.

The second example deals with a DANIELL cell, and shows to what extent the solution that is contained within the salt bridge has an impact on the overall ionic junction voltage. Here the voltage is the algebraic sum of two liquid junction voltages, illustrated by the following electrochemical chain:

$$10^{-1} \text{ mol L}^{-1} \text{ CuSO}_4 \text{ aqueous solution (sol. } \alpha) \,||\, \text{solution in}$$
$$\text{the saline bridge (sol. } \beta) \,||\, 10^{-3} \text{ mol L}^{-1} \text{ ZnSO}_4 \text{ aqueous solution (sol. } \gamma)$$

Due to their chemical nature, solutions α and γ must be slightly acidic in order to prevent hydroxides from forming. Let us fix $pH = 4$ using H_2SO_4, which is completely dissociated into H^+ and SO_4^{2-} at this pH value.

▶ Solution β: solution α (only one liquid junction)

$$\varphi_\gamma - \varphi_\alpha = -11.0 \text{ mV}$$

▶ Solution β: 10^{-3} mol L^{-1} KNO$_3$ aqueous solution

$$\varphi_\gamma - \varphi_\alpha = (-13.0 - 3.6) \text{ mV} = -16.6 \text{ mV}$$

▶ Solution β: 1 mol L^{-1} KNO$_3$ aqueous solution

$$\varphi_\gamma - \varphi_\alpha = (-1.7 + 2.2) \text{ mV} = +0.5 \text{ mV}$$

▶ Solution β: 1 mol L^{-1} HNO$_3$ aqueous solution

$$\varphi_\gamma - \varphi_\alpha = (-50.3 + 120.1) \text{ mV} = +69.8 \text{ mV}$$

As previously highlighted, it is better to select a concentrated electrolyte solution involving ions with very close molar conductivity values. Note that the very fact of inserting this solution (salt bridge) triggers a significant decrease in the overall junction voltage. This is particularly important from an experimental point of view, when this type of electrochemical chain is involved when measuring thermodynamic parameters such as standard potentials, activities, etc.

BASIC ELEMENTS FOR DEMONSTRATING THE HENDERSON EQUATION

Qualitatively speaking, when two solutions made up of different compositions are brought into contact in the same solvent, then quasi-instantaneously a small number of ions are exchanged, as when establishing a thermodynamic equilibrium (see appendix A.3.4). This exchange process leads to (after about 10^{-9} s) a build-up of charge excess on both sides of the junction area, as in the case of a double layer at an electrochemical interface. For instance, when dealing with a single ionic junction directly between the two solutions in a DANIELL cell via a porous plug, a small number of Cu^{2+} ions move from the CuSO$_4$ solution into the ZnSO$_4$ solution and conversely some Zn^{2+} ions are transferred from the ZnSO$_4$ solution into the CuSO$_4$ solution. Here we are talking about extremely small quantities, however the resulting surface charge excess produces a significant and immediate potential difference between the two solutions. The concentration profiles within the porous material then continue to evolve until finally reaching a quasi-steady state with fixed concentrations on each side of the porous plug, which dramatically slows down the mixing of the two solutions (see section 4.4.2). During the time it takes to reach these profiles (a process that may last up to several hours depending on the nature and thickness of the porous material), the fact that electroneutrality prevails in the solutions (although it does not strictly apply) has the effect of forcing the migration and diffusion fluxes of the anions and cations to occur in a combined manner (see appendix A.4.1).

When giving a quantitative description, the first step is to write that the overall current resulting from the ionic fluxes (without convection) is zero at all times (see section 4.2.1.4):

$$0 = -\sum_i D_i\, z_i\, \mathscr{F}\, C_i\, \mathbf{grad}\,(\ln a_i) + \sum_i \lambda_i\, C_i\, \mathbf{E}$$

or:

$$\mathbf{E} = -\,\mathbf{grad}\,\varphi = \sum_i \frac{D_i\, z_i\, \mathscr{F}\, C_i}{\sum_j \lambda_j\, C_j}\, \mathbf{grad}\,(\ln a_i)$$

The NERNST-EINSTEIN equation describes the link between the molar conductivity and the diffusion coefficient in an ideal case. However, the equation can also be stretched to apply to other cases, provided that one remembers that D_i is the true diffusion coefficient and not the apparent one (first FICK's law, see section 4.2.1.4):

$$\lambda_i = D_i\, z_i^{\,2}\, \frac{\mathscr{F}^2}{RT}$$

hence:

$$-\,\mathbf{grad}\,\varphi = \frac{RT}{\mathscr{F}} \sum_i \frac{\lambda_i\, C_i}{z_i \sum_j \lambda_j\, C_j}\, \mathbf{grad}\,(\ln a_i) = \frac{RT}{\mathscr{F}} \sum_i \frac{t_i}{z_i}\, \mathbf{grad}\,(\ln a_i)$$

and:

$$\varphi_\beta - \varphi_\alpha = \frac{RT}{\mathscr{F}} \int_\beta^\alpha \left[\sum_i \frac{t_i}{z_i}\, \mathbf{grad}\,(\ln a_i) \right] dx$$

The common version of the HENDERSON equation is then easily reached using the following approximations: concentrations are used instead of activities, the molar conductivities (or mobilities) are constant and the concentration profiles are considered as being linear.

If referring to usual applications, then the most questionable of all the simplifications listed above is the assumption that the activities are equal to the molar concentrations. Nevertheless, it should be pointed out that only the ratios appear in the final equation, which ends up minimising the impact of such systematic errors.

As for linear concentration profiles, it should be emphasised that the thickness of the zone where these profiles develop does not appear in the calculation result. Therefore it is easy to show that the ionic junction voltage which is reached at quasi-steady state (with a linear profile extending throughout the separator material) is identical to the voltage that is rapidly reached (about 10^{-9} s) before any modifications occur in the bulk concentrations. More complex models even show that the ensuing outcome, known as the HENDERSON equation, can be applied more widely than one would expect when one considers its usual demonstration (the equation can be applied at all times while the quasi-steady state is being reached). It is even possible to use a similar equation when there is a current flowing through the system, though this time the ohmic drop is included.

A.1.2 - POTENTIOSTAT AND GALVANOSTAT

POTENTIOSTAT

A potentiostat is an electronic instrument designed to control the potential difference applied to an electrochemical cell, between the working electrode (WE) where there is a current flow and a reference electrode (Ref) where there is no current flow. A potentiostat needs the use of a 3-electrode device: a working electrode, a reference electrode and a counter-electrode, or auxiliary electrode (CE). This type of instrument must be distinguished from a DC voltage supply, which imposes and controls in an electric circuit the potential difference between two terminals with a current flowing between them, namely the working and counter-electrodes in an electrochemical cell. The potentiostat is a crucial instrument when performing analytical studies in electro-chemistry, because in order to properly understand, analyse and distinguish between the different phenomena at play at each interface, one needs more than simply information on the overall cell voltage.

In order to properly carry out this function, a particular type of software can be used that is driven by a microprocessor. However, here one would need to use electronic interface components (analogue/digital and digital/analogue converters) with extremely short conversion times so as to ensure that the servo-control is fast and accurate, and such converters are particularly expensive.

Alternatively, another possibility would be to use an electronic device with an operational amplifier mounted in servo-control, as illustrated in figure A.1.

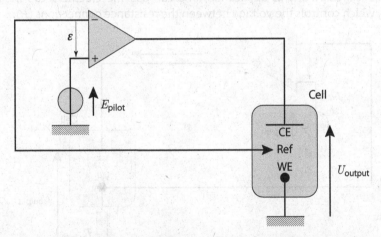

Figure A.1 - Diagram showing the basic principles of an analogue potentiostat

Here is a brief overview of the main theoretical features of an operational amplifier: the voltage gain (G) is infinite between the input and output connections (in practice higher than 10^5), the response time is very short (less than 10^{-6} s), the output impedance is low and the input impedance is infinite (in practice higher than 10^6 Ω, or even 10^{12} to 10^{14} Ω when using field effect transistor technology). Therefore the leakage current is extremely low between the amplifier's – and + input connections, and the pilot generator delivers practically no current.

In this case the voltage can be written as:

$$U_{output} = G\,(E_{applied} - E_{Ref}) \quad \text{or} \quad \varepsilon = E_{applied} - E_{Ref} = U_{output}/G \approx 0$$

In this example above, given that the working electrode is connected to the ground, the following equation can be applied at all times:

$$U_{WE/Ref} = -\,E_{applied}$$

Therefore, this type of device controls the voltage value between the working electrode and the reference electrode that has no current flowing through it. However, the voltage between the counter-electrode and the reference electrode may fluctuate due to experimental factors (for example, bubbles or a passive layer forming, etc.) while leaving the $E_{WE/Ref}$ value unaffected. The imposed voltage $E_{applied}$ can be continuous or it can be modulated by a signal generator (triangular, square or sinusoidal).

It must be pointed out that the U_{output} value is limited by the power supply of the operational amplifier ($+V_{cc}, -V_{cc}$, which are not indicated on the diagram in figure A.1). Therefore, a small laboratory potentiostat matching the simple diagram in figure A.1 has its output power limited by the electronic components being used. More complex devices can drive electrochemical cells that need higher power values. Moreover, some potentiostats are equipped with a device compensating for the ohmic drop that occurs between the reference and the working electrodes.

GALVANOSTAT

A galvanostat is an instrument that controls the current intensity flowing through an electrochemical cell. For this application, one can use the electric setup indicated in figure A.2, which controls the voltage between the resistance connections (R).

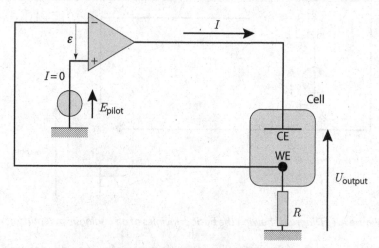

Figure A.2 - Diagram showing an analogue galvanostat

A calculation can be made that is similar to the one applied in the case of the potentiostat, which shows that $E_{applied} - RI \approx 0$. Here, the current flowing through the resistance, which is identical to the current flowing through the cell connected in series, is controlled by the operational amplifier. However, the voltage between the working and the counter-electrodes is not controlled and can fluctuate due to unpredictable experimental factors.

In a setup that uses a galvanostat, it is possible to not use a reference electrode. However, if there is a reference electrode, then it is only used to monitor the voltage between the working electrode and the reference electrode.

A potentiostat can also be used for an experiment that is set up in galvanostat mode. In this case, all that is needed is to insert a resistance in series with the electrochemical cell. The potentiostat is then connected to the assembly by linking the potentiostat's Ref output to the cell's working electrode and linking the WE output (the ground) to the other end of the resistance. The counter-electrode is connected in the usual manner, as shown in figure A.3.

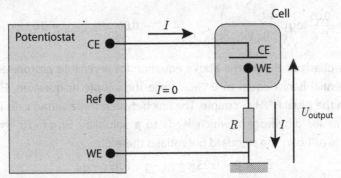

Figure A.3 - Diagram of a galvanostatic setup using a potentiostat

However, for this function to work there is no need to use a high-performance potentiostat. A mere DC current supply can be used, provided that it has adequate capacities (power, rise time, etc.). Some instruments offer the opportunity to use both functions (potentiostat and galvanostat).

A.2.1 - GENERAL SHAPE OF THE CURRENT-POTENTIAL CURVE FOR REDUCING WATER OR PROTONS: THE ROLE OF MASS TRANSPORT KINETICS

When dioxygen and dihydrogen are formed the reactions involved are complex in several ways. However, if they are simplified to a certain degree, then it is possible to define the shape of the current-potential curves corresponding to aqueous solutions with different pH values. The following analysis is confined to situations where the limiting process is the mass transport of electroactive species *via* diffusion. In other words, it only applies when the kinetics of the various steps of the reduction mechanism is much faster than that of mass transport (i.e., mass transport control). To fully grasp the following analysis, the key notions are outlined in the preceding chapters, and a thorough understanding of chapter 4 is particularly essential.

In this section we will first look at dihydrogen production in an acidic aqueous solution, whereby argon is continuously bubbled inside. Here, assuming unidirectional geometry, the steady states are described by the NERNST model: due to convection, the composition remains homogeneous throughout the bulk of the electrolyte, i.e., in the area beyond the thickness layer δ. This composition is identical to the initial composition: the pH is fixed and no dihydrogen is dissolved. Moreover, a supporting electrolyte is added, which has the effect of minimising the migration of electroactive species. It is assumed that the diffusion coefficients are identical, which makes the calculations much easier, yet without limiting in any way the scope of the results that are presented below.

The first difficulty stems from the very nature of the product being formed. To simplify, here we will assume that the dissolved dihydrogen is the only reaction product, even if the solubility limit is often reached in experiments. Since only fast reduction reactions are being considered here, the potential of the inert electrode can be written using the reversibility hypothesis:

$$E = E° + \frac{0.06}{-2} \log \left(\frac{a_{H_{2}dissolved}}{(a_{H^+})^2} \right)_{interface} = E° - 0.06 \, pH_{interface} - 0.03 \log a_{H_{2}interface}$$

It must be emphasized that in the above equation for reversible proton reduction, the standard potential is not equal to $0 \, V_{/SHE}$, since the couple in question, $H^+/H_{2dissolved}$, is different from the usual H^+/H_{2gaz} couple. The link between these values calls into play the HENRY constant for dihydrogen, which leads to a solubility $s_{H_2} = 8 \times 10^{-4} \, mol \, L^{-1}$ for a partial pressure of 1 bar. The standard potential is therefore:

$$E° = 0 - 0.03 \log s_{H_2} = +0.093 \, V_{/SHE}$$

Moreover, again for the sake of simplicity, the values of the activity and concentration for each of the species involved are considered here as being equal.

The second difficulty stems from the autoprotolysis equilibrium of the water or, in other words, from the fact that dihydrogen may form at the interface between the electrode and the electrolyte due to the reduction of H^+ and/or of H_2O (see section 2.1.2.3). In some operating conditions, a priori this leads to non-linear concentration profiles for the H^+ and OH^- ions throughout the diffusion layer, unlike the typical situation found for unidirectional-diffusion steady states.

This phenomenon looks like an anomaly when compared to the conventional case, for example the Fe^{3+}/Fe^{2+} couple, but in electrochemistry the occurrence of chemical equilibrium inside the solution cannot be considered as being exceptional. So, a specific term for volume production must be taken into account in the mass balance for the species involved in this equilibrium. At steady state, the mass balance of the electroactive species can be written as follows (see section 4.1.2):

$$0 = -\operatorname{div} \mathbf{N}_{H^+} + w_{H^+} = -\operatorname{div} \mathbf{N}_{OH^-} + w_{OH^-} = -\operatorname{div} \mathbf{N}_{H_2} + 0$$

The concentration profile for dissolved H_2 is therefore linear (i.e., the molar flux density inside the diffusion layer is constant), though this is not the case a priori for H^+ and OH^-.

However, given that the following water autoprotolysis reaction:

$$H_2O \rightleftharpoons H^+ + OH^-$$

involves H^+ and OH^- with identical stœchiometric numbers, their local production rates per volume unit are equal as written here:

$$w_{H^+} = w_{OH^-}$$

The volume production term can therefore be eliminated by subtracting the first two equations. A linear profile is therefore to be expected throughout the diffusion layer for the quantity $[H^+] - [OH^-]$.

Since the equilibrium constant is very low, it can be shown that H^+ and OH^- do not co-exist in significant quantities. Except for pH values close to 7, the following inequalities always apply:

$$\text{either } [H^+] \gg [OH^-] \qquad \text{or } [H^+] \ll [OH^-]$$

Therefore, depending on the working point in question, the diffusion layer can be divided into two adjoining zones: an alkaline zone and an acidic zone, each showing linear concentration profiles. This can be illustrated by studying the concentration profiles for different working points on the current-potential curve: let us consider the case of an acidic aqueous solution at $pH^* = 4$ in the cathodic domain [1].

▸ When the current is lower than the limiting current related to the simple reduction of H^+, then it suffices to consider that this reduction reaction is the only significant phenomenon at play. The OH^- concentration remains negligible in the diffusion layer. The concentration profile of dissolved dihydrogen is related to the proton concentration profile, taking into account the stœchiometric coefficients.

The limiting current can be expressed by: $I_{\lim} = - D \mathcal{F} S \dfrac{[H^+]^*}{\delta}$

In this zone ($|I| < |I_{\lim}|$), the interfacial concentrations can be expressed as a function of I:

$$[H^+]_{\text{interface}} = [H^+]^* \left(1 - \frac{I}{I_{\lim}} \right) \qquad [H_2]_{\text{interface}} = [H^+]^* \frac{I}{2\,I_{\lim}}$$

Figure A.4 depicts the concentration profiles for two working points that have been selected in this zone: at the half-wave ($I = I_{\lim}/2$; $pH_{\text{interface}} = 4 + \log 2 = 4.3$) and for $pH_{\text{interface}} = 5$, i.e., $I = 0.9\,I_{\lim}$.

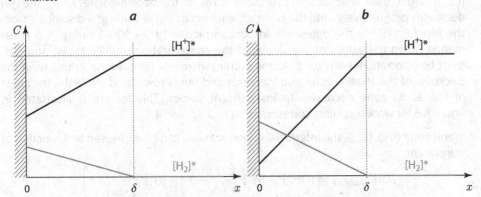

Figure A.4 - Concentration profiles of H^+ (black) and H_2 (grey)
in an aqueous solution, with $pH^* = 4$, for (a) $I = I_{\lim}/2$ and (b) $I = 0.9\,I_{\lim}$
The OH^- concentration is not indicated: it is negligible throughout the solution.

[1] As mentioned in chapter 4, the bulk values are denoted by *, and are equal to the initial values before the current is switched on.

The equation for the current-potential curve for $|I| < |I_{lim}|$, that is plotted in figure A.5, is:

$$E = 0.093 - 0.03 \ pH^* - 0.03 \log\left[\frac{I}{2\,I_{lim}} \left(\frac{I_{lim}}{I_{lim} - I} \right)^2 \right]$$

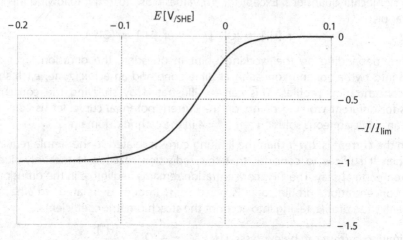

Figure A.5 - *Current-potential curve for reducing an aqueous solution with* $pH^* = 4$ *for working points with* $I/I_{lim} < 0.99$ *(i.e.,* $pH_{interface} < 6$*)*

▸ If H_2O is reduced together with H^+, then it is possible to outstrip the limiting current previously defined. The water reduction leads to dihydrogen and hydroxide ions being produced (linked to water autoprotolysis). Therefore, it leads to a high $pH_{interface}$ value. As previously specified, the diffusion layer is divided into two zones. The first one, next to the electrode, is an alkaline zone where the proton concentration is negligible. The second zone is an intermediate zone, where the proton concentration is once again much larger than that of the hydroxide ions. In this intermediate zone, the pH decreases progressively until the point of reaching its initial value at a distance δ from the interface. Since the difference in concentrations $[H^+] - [OH^-]$ varies in a linear manner, then the values of the slopes of the concentration profiles for H^+ and OH^- must be opposite in each zone. As the current increases (in absolute value), then the thickness of the alkaline zone also increases and tends towards δ, while the thickness of the acidic zone decreases to insignificant values. This feature is illustrated in figure A.6 for working points corresponding to 2 I_{lim} and 4 I_{lim}.

In this zone ($|I| > |I_{lim}|$), the interfacial concentrations can be expressed as a function of the current:

$$[OH^-]_{interface} = [OH^-]^* + [H^+]^* \left(\frac{I}{I_{lim}} - 1 \right) \approx [H^+]^* \left(\frac{I}{I_{lim}} - 1 \right)$$

$$[H^+]_{interface} = \frac{10^{-14}}{[OH^-]_{interface}}$$

$$[H_2]_{interface} = [H^+]^* \frac{I}{2\,I_{lim}}$$

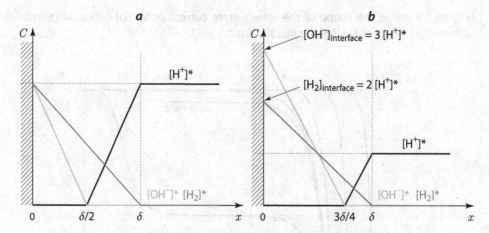

Figure A.6 - *Concentration profiles of H^+ (black), OH^- (clear grey) and H_2 (dark grey) for an aqueous solution with $pH^* = 4$ for (a) $I = 2\,I_{lim}$ and (b) $I = 4\,I_{lim}$ (with a contracted concentration scale)*

The equation for the corresponding part of the current-potential curve ($|I| > |I_{lim}|$), plotted in figure A.7, can be deduced from above and is written as the following:

$$E = 0.093 + 0.09\,pH^* - 0.06 \times 14 - 0.03 \log\left[\frac{I}{2\,I_{lim}}\left(\frac{I_{lim} - I}{I_{lim}}\right)^2\right]$$

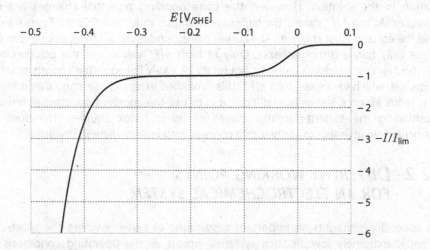

Figure A.7 – *Shape of the current-potential curve for reducing an aqueous solution with $pH^* = 4$*

In this slightly acidic type of solution, the steady-state current-potential curve shows two distinct reaction zones which can be ascribed to a low $pH_{interface}$ value (H^+ is reduced) or to a high $pH_{interface}$ value (H_2O is reduced). Let us recall it is assumed that the interfacial reactions kinetics plays no part here.

Figure A.8 shows the shape of the steady-state current-potential curves obtained for different values of the pH of the initial solution, pH^*.

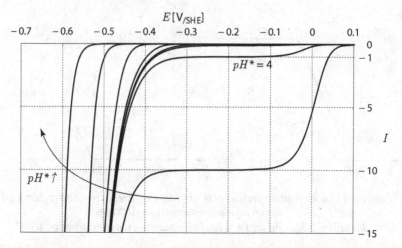

Figure A.8 - Shape of current-potential curves for reducing aqueous solutions with $pH^* = 3, 4, 5, 9, 10, 11, 12$ and 13
The current scale is expressed with an arbitrary unit: -1 corresponds to the limiting current at $pH^* = 4$.

As to be expected from a redox couple, the limiting current changes with the pH^* value in the acidic solutions, since this current is proportional to the proton concentration in the solution. However, the corresponding potential changes are more complex. At low pH^* values, the half-wave potential varies as $- 0.03\ pH^*$ (see figure A5 and the equation just above it). The curves corresponding to pH^* values between 5 and 9 are only barely distinguishable. Only at high pH^* values can the position of the electrochemical window's cathodic boundary be easily linked to the apparent standard potential, which varies as $- 0.06\ pH^*$. This simplified example, assuming that it involves fast redox kinetics, shows how difficult it is to use the apparent standard potential for positioning the current-potential curves of water redox couples. Therefore, it is particularly inadvisable to apply such a concept with intermediate pH^* values.

A.2.2 - DIFFERENT WORKING POINTS FOR AN ELECTROCHEMICAL SYSTEM

This appendix highlights an important case found in systems where the short-circuit current is extremely low. In such systems, almost all the operating conditions with significant currents are in electrolyser mode. For example, imagine a deaerated acidic aqueous electrolyte, with two inert electrodes (platinum and carbon) immersed within. One can assume that proton reduction kinetics is the only characteristic that distinguishes the interfaces. Therefore, the anodic branches of the current-potential curves are viewed as being the same, as long as the water oxidation into dioxygen occurs in the same way. The system's open-circuit voltage equals the difference between two mixed potentials. As presented in figure A.9, platinum is the positive electrode at open circuit.

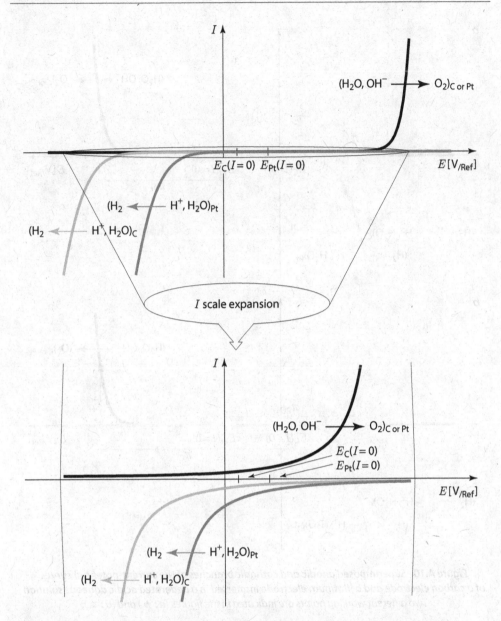

Figure A.9 - Superimposed anodic and cathodic branches of the current-potential curves of a carbon electrode and a platinum electrode immersed in a deaerated acidic aqueous solution

In this system, a slight current variation around 0 causes a sudden voltage variation. This variation reflects the change occurring between an electrolyser mode, where the anode is platinum (the dotted line in figure A.10-a), and another electrolyser mode where the platinum electrode plays the role of cathode (the dashed and dotted line in figure A.10-b).

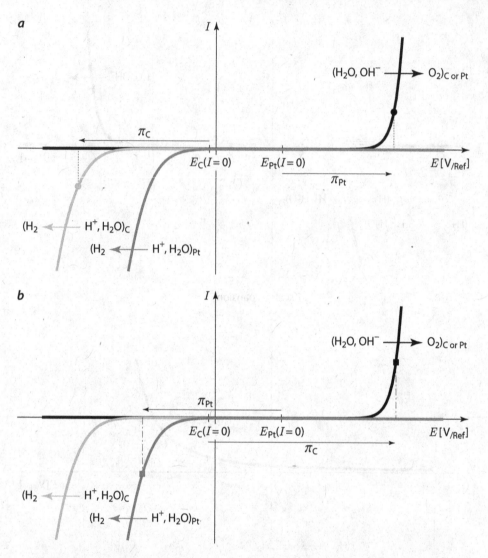

Figure A.10 - Superimposed anodic and cathodic branches of the current-potential curves
of a carbon electrode and a platinum electrode immersed in a deaerated acidic aqueous solution
Two different working points are indicated in the figures (a : ●) and (b : ■).

However, within a certain voltage range the system works as a power source, yet with an insignificant current, as shown in figure A.11. Here, by convention, the working electrode is the electrode that is positive at open circuit, i.e., the platinum electrode. Over and above the particular examples of short circuit and open circuit (points ($U = 0$, I_{sc}) and ($U(I = 0)$, $I = 0$), the two working points depicted in figure A.10 (the dotted line in figure A.10-a and the dashed and dotted line in figure A.10-b) are also indicated in figure A.11.

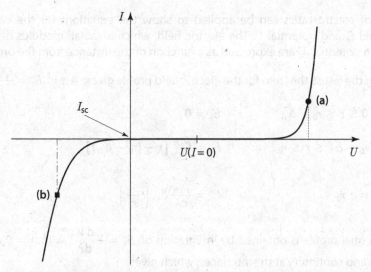

Figure A.11 - I(U) curve showing the working points of a system with a carbon electrode and a platinum electrode immersed in a deaerated acidic aqueous solution

A.3.1 - ELECTRIC POTENTIAL: VOLTA AND GALVANI POTENTIALS

Using a simple example to illustrate the concept of VOLTA and GALVANI potentials, let us consider a charged spherical conductor (radius r_0) in equilibrium and in vacuum. As depicted in figure A.12, only the surface is charged: the volume charge density, ρ_{ch}, is zero. It is assumed that the surface charges, $Q_{surface}$, are uniformly distributed in the form of a thin, spherical crown, with a thickness of δ_{ch}[2]. Inside the crown, the volume charge density is not zero, with the following equation:

$$Q_{surface} = 4\,\pi\,\rho_{ch}\,\delta_{ch}\,r_0^{\,2}$$

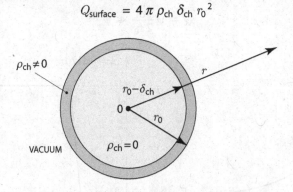

Figure A.12 - Conducting sphere with its charged surface

---

[2] *For electronic conductors, notably semiconductors, this volume is called a 'space charge zone'. It is equivalent to the double layer for an electrochemical interface.*

The laws of electrostatics can be applied to show the equations for the equilibrium electric field $\mathbf{E}$ and potential V. The electric field, which is radial (modulus denoted by E_r), and the potential V are expressed as a function of the distance from the origin:

▸ Applying the GAUSS theorem for the electric field profile gives: $4\pi r^2 E_r = \dfrac{Q_{inner}}{\varepsilon_0}$.

$$0 \leq r \leq r_0 - \delta_{ch} \qquad\qquad E_r = 0$$

$$r_0 - \delta_{ch} \leq r \leq r_0 \qquad\qquad E_r = \frac{\rho_{ch}}{\varepsilon_0}\left[r - (r_0 - \delta_{ch})\right]$$

$$r \geq r_0 \qquad\qquad E_r = \frac{\rho_{ch}}{\varepsilon_0}\delta_{ch}\left(\frac{r_0}{r}\right)^2$$

▸ The potential profile is obtained by integration of $E_r = -\dfrac{dV}{dr}$, with $V = 0$ at infinite distance and continuity at the interfaces, which gives.

$$r \geq r_0 \qquad\qquad V = \frac{\rho_{ch}}{\varepsilon_0}\delta_{ch}\frac{r_0^2}{r} = \frac{Q_{surface}}{4\pi\varepsilon_0 r}$$

$$r_0 - \delta_{ch} \leq r \leq r_0 \qquad\qquad V = -\frac{\rho_{ch}}{2\varepsilon_0}\left\{\left[r - (r_0 - \delta_{ch})\right]^2 - 2r_0\delta_{ch} - \delta_{ch}^2\right\}$$

$$0 \leq r \leq r_0 - \delta_{ch} \qquad\qquad V = \frac{\rho_{ch}}{\varepsilon_0}\delta_{ch}\,r_0\left(1 + \frac{\delta_{ch}}{2r_0}\right)$$

The corresponding profiles are illustrated in figures A.13 and A.14. The two graphs on the right (electric field and potential) are plotted on a dilated scale compared to those on the left. In addition, in figure A.14 the ordinate scales are different for χ and ψ. Consequently, this leads to a break in the slope for r_0, which has no physical meaning since the following equation applies:

$$\chi = \psi\frac{\delta_{ch}}{2r_0} \ll \psi$$

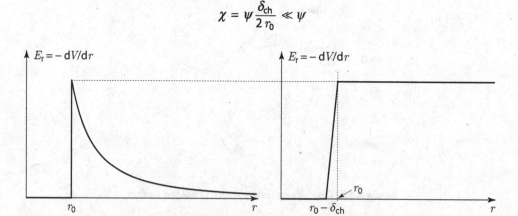

Figure A.13 - Radial component profile of the electric field ($\rho_{ch} > 0$)
In the figure on the right, the r scale is considerably dilated.

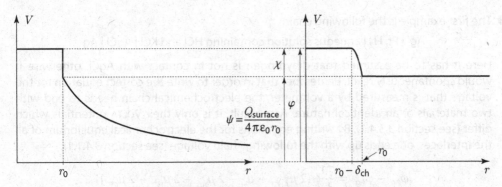

Figure A.14 - Electric potential profile ($\rho_{ch} > 0$)
In the figure on the right, the r scale is considerably dilated.

A.3.2 - MEAN ACTIVITY OF A SOLUTE IN AN ELECTROLYTE

Strictly speaking, it is possible to measure electrochemical potential and activity for any neutral species, though not for an ion: one cannot add a single type of ion to an electrolyte medium because such a modification would not preserve electroneutrality (see section 3.1.2.2). That said, it has to be recognised that the electrochemical potential and the activity of an ion are both conceivable notions. Indeed, they are so widely used in electrochemistry (and especially in this book), that it is easy to forget that, technically speaking, they are non-measurable quantities. This may seem inconsistent with other conventional concepts in electrochemistry, especially with the notion of selective electrodes, which are commonly used as indicators for a specific ion activity. The most famous of these selective electrodes is the pH electrode, although pH is, by definition, impossible to measure from a strictly thermodynamic point of view. The following appendix addresses this apparent contradiction by examining the thermodynamic equilibrium or the quasi-steady state of different electrochemical chains at open circuit.

The first group of examples illustrated deals with electrochemical chains with no ionic junction. In this case the thermodynamic voltage is always written in a straightforward way using an equation that includes the activities of neutral species (in particular the mean activities of electrolytes, see section 3.2.1.1). The second group of examples deals with electrochemical chains with at least one ionic junction. In this case, the equation for the open-circuit voltage usually shows the ion activity as a separate entity. In this section we will show how it is only possible to write the equation in this way when the ionic junction voltage has been disregarded. Let us recall that, in the latter case, the electrochemical interfaces are in thermodynamic equilibrium, whereas the liquid junction are at quasi-steady state (see section 3.3.5 and appendix A.1.1).

ELECTROCHEMICAL CHAINS WITH NO IONIC JUNCTION

It is a hard task to find a realistic example of an electrochemical chain that is both in true thermodynamic equilibrium and with no ionic junction. However, the two examples laid out below come relatively close to this case, provided that an additional assumption is made, which is specified in each case.

▶▶ The first example is the following chain:

Ag' | Pt, H$_2$ | aqueous solution containing HCl and KCl | AgCl | Ag

Here it has to be assumed that dihydrogen is not in contact with AgCl, otherwise it would spontaneously react. Remember that in order to write the correct equation for the voltage that is measured by a voltmeter, the electrochemical chain needs to end with two materials of an identical nature. As a result, it is only their VOLTA potentials which differ (see section 3.3.4.1). By writing equations for the electrochemical equilibrium of all the interfaces, one ends up with the following chain voltage (see section 3.4.1.1):

$$\varphi_{Ag} - \varphi_{Ag'} = \frac{1}{2\,\mathscr{F}}(2\,\mu_{AgCl} + \mu_{H_2} - 2\,\mu_{Ag} - 2\,\mu_{H^+} - 2\,\mu_{Cl^-})$$

$$= U° + \frac{RT}{\mathscr{F}}\ln\sqrt{a_{H_2}} - \frac{RT}{\mathscr{F}}\ln(a_{H^+}a_{Cl^-})$$

$$= U° + \frac{RT}{\mathscr{F}}\ln\sqrt{a_{H_2}} - \frac{2\,RT}{\mathscr{F}}\ln a_{\pm}$$

Therefore the thermodynamic voltage of this cell is mainly a function of the mean activity of HCl, whereas the activities of H$^+$ or Cl$^-$ do not play any part when taken separately. Here, it must be underlined that because the solution contains three types of ions (H$^+$, K$^+$ and Cl$^-$), one therefore needs to include the different H$^+$ and Cl$^-$ concentrations when calculating the mean activity coefficient of HCl.

▶▶ The second example that illustrates a lead-acid battery, presents the following electrochemical chain:

Pb' | PbSO$_4$ | H$_2$SO$_4$ aqueous solution | PbSO$_4$ | PbO$_2$ | Pb

Here, it is assumed that the solvent decomposition reactions (release of dihydrogen and dioxygen) are immeasurably slow. The battery voltage in equilibrium is obtained by writing the electrochemical equilibrium of all the interfaces:

$$\varphi_{Pb} - \varphi_{Pb'} = \frac{1}{2\,\mathscr{F}}\left(\mu_{PbO_2} + \mu_{Pb} + 2\,\mu_{H^+} - 2\,\mu_{HSO_4^-} - 2\,\mu_{H_2O} - 2\,\mu_{PbSO_4}\right)$$

$$= U° - \frac{RT}{\mathscr{F}}\ln a_{H_2O} + \frac{RT}{\mathscr{F}}\ln(a_{H^+}a_{HSO_4^-})$$

$$= U° - \frac{RT}{\mathscr{F}}\ln a_{H_2O} + \frac{2\,RT}{\mathscr{F}}\ln a_{\pm}$$

Therefore the overall battery voltage gives access to the mean activity of H$_2$SO$_4$. Indeed, it can be noted that the data are often expressed in terms of mean activity coefficients. Depending on which author, the results take into consideration either the species H$^+$ and HSO$_4^-$ or H$^+$ and SO$_4^{2-}$. In fact $a_{\pm}$ is the only experimental datum.

ELECTROCHEMICAL CHAINS WITH AN IONIC JUNCTION

▶▶ Let us first consider a chain involving the same electrochemical interfaces as in the first example with no ionic junction, yet this time introducing two aqueous solutions with different compositions which are placed in contact through a porous material:

Ag' | Pt, H$_2$ | HCl solution (sol. α) || KCl solution (sol. β) | AgCl | Ag

The following shows the open-circuit voltage that is obtained by writing the electrochemical equilibria of all the interfaces, with the ionic junction excluded (see section 3.4.1.1):

$$\varphi_{Ag} - \varphi_{Ag'} = \frac{1}{2\mathscr{F}} (2\,\mu_{AgCl} + \mu_{H_2} - 2\,\mu_{Ag} - 2\,\mu_{H^+} - 2\,\mu_{Cl^-}) + \varphi_\beta - \varphi_\alpha$$

$$= U^\circ + \frac{RT}{\mathscr{F}} \ln \sqrt{a_{H_2}} - \frac{RT}{\mathscr{F}} \ln(a_{H^+_\alpha} a_{Cl^-_\beta}) + U_{junction}$$

Using the values in appendix A.1.1, with a solution of $pH = 2$ for α and a 1 mol L^{-1} KCl solution for β, the ionic junction voltage is very small in numerical terms (3.5 mV). Therefore, it would not be incongruous to write [3]:

$$\varphi_{Ag} - \varphi_{Ag'} \approx U^\circ + \frac{RT}{\mathscr{F}} \ln \sqrt{a_{H_2}} - \frac{RT}{\mathscr{F}} \ln(a_{H^+_\alpha} a_{Cl^-_\beta})$$

In this equation, there appears no clear thermodynamic parameter involving ions, the reason being that parameters for the ions belonging to phases α and β are mixed. Yet, this type of equation is commonly found in electrochemistry. In order to gain a full understanding of what is being measured, it is nonetheless worth describing the ionic junction term in precise detail, even though its value is small in numerical terms. By reusing the results from appendix A.1.1, the junction voltage can be written as being an integral over the thickness of the porous material:

$$\varphi_\beta - \varphi_\alpha = \frac{RT}{\mathscr{F}} \int_\beta^\alpha \left(t_{H^+} \, \mathbf{grad}\,(\ln a_{H^+}) + t_{K^+} \, \mathbf{grad}\,(\ln a_{K^+}) - t_{Cl^-} \, \mathbf{grad}\,(\ln a_{Cl^-}) \right) dx$$

Since the sum of the transport numbers is equal to 1 at any point, one can also write:

$$U_{junction} = \frac{RT}{\mathscr{F}} \int_\beta^\alpha \left(t_{H^+} \, \mathbf{grad}\,(\ln(a_{H^+} a_{Cl^-})) + t_{K^+} \, \mathbf{grad}\,(\ln(a_{K^+} a_{Cl^-})) - \mathbf{grad}\,(\ln a_{Cl^-}) \right) dx$$

$$= \frac{RT}{\mathscr{F}} \ln \frac{a_{Cl^-_\beta}}{a_{Cl^-_\alpha}} + \frac{RT}{\mathscr{F}} \int_\beta^\alpha \left(t_{H^+} \, \mathbf{grad}\,(\ln(a_{H^+} a_{Cl^-})) + t_{K^+} \, \mathbf{grad}\,(\ln(a_{K^+} a_{Cl^-})) \right) dx$$

and finally:

$$\varphi_{Ag} - \varphi_{Ag'} = U^\circ + \frac{RT}{\mathscr{F}} \ln \sqrt{a_{H_2}} - \frac{RT}{\mathscr{F}} \ln(a_{H^+_\alpha} a_{Cl^-_\alpha})$$

$$+ \frac{RT}{\mathscr{F}} \int_\beta^\alpha \left(t_{H^+} \, \mathbf{grad}\,(\ln(a_{H^+} a_{Cl^-})) + t_{K^+} \, \mathbf{grad}\,(\ln(a_{K^+} a_{Cl^-})) \right) dx$$

Therefore, one can see that in this more thorough version of the equation, the cell's open-circuit voltage (at quasi-steady state) does not show any isolated ionic activity. Instead, it shows the mean activities of the neutral species HCl and KCl: $\sqrt{a_{H^+} a_{Cl^-}}$ and $\sqrt{a_{K^+} a_{Cl^-}}$ respectively.

▸▸ The case of an ion-selective electrode for a given ion i, produces the same type of equation. If one disregards the ionic junction voltage, due to the fact that a reference

---

[3] *When dealing with a cell with a salt bridge, it would be all the more legitimate to disregard the junction voltage.*

electrode is completing the electrochemical sensor for example, the following type of equation emerges:

$$U \approx \text{Cst} + \frac{v_i RT}{v_e \mathscr{F}} \ln a_i$$

This equation, which is the most common, could indicate that the activity of ion i is a measurable quantity, though this is not strictly exact. The correct expression to give the accurate voltage, with the ionic junction voltages taken into account, always includes the mean activities and their spatial variations.

▸▸ The final example, which also deals with aqueous solutions, gives a result that is commonly found in electrochemistry, when it is not possible to disregard the cell's ionic junction voltage. In this case, there is only one correct way of writing the equation, which involves mean activities as it should be. This applies to a particular type of cell that is sometimes called a ' concentration cell ':

Pt', H_2 | HCl solution (sol. α) | | HCl solution (sol. β) | H_2, Pt

$$\varphi_{Pt} - \varphi_{Pt'} = \frac{RT}{\mathscr{F}} \ln \frac{a_{H^+_\beta}}{a_{H^+_\alpha}} + \varphi_\beta - \varphi_\alpha$$

In this electrolyte with only two ions with identical concentrations, the equation for the junction voltage gets even simpler if one can assume that the ionic transport numbers are identical at all points throughout the system:

$$\varphi_\beta - \varphi_\alpha = \frac{RT}{\mathscr{F}} \int_\beta^\alpha \left(t_{H^+} \, \mathbf{grad} \left(\ln a_{H^+} \right) - t_{Cl^-} \, \mathbf{grad} \left(\ln a_{Cl^-} \right) \right) dx$$

$$= \frac{RT}{\mathscr{F}} \int_\beta^\alpha \left(t_{H^+} \, \mathbf{grad} \left(\ln (a_{H^+} a_{Cl^-}) \right) - \mathbf{grad} \left(\ln a_{Cl^-} \right) \right) dx$$

$$= \frac{RT}{\mathscr{F}} t_{H^+} \, \ln \frac{a_{H^+_\alpha} a_{Cl^-_\alpha}}{a_{H^+_\beta} a_{Cl^-_\beta}} - \frac{RT}{\mathscr{F}} \ln \frac{a_{Cl^-_\alpha}}{a_{Cl^-_\beta}}$$

and finally:

$$\varphi_{Pt} - \varphi_{Pt'} = \frac{RT}{\mathscr{F}} t_{Cl^-} \, \ln \frac{a_{H^+_\beta} a_{Cl^-_\beta}}{a_{H^+_\alpha} a_{Cl^-_\alpha}}$$

This equation for the open-circuit voltage (at quasi-steady state) of a concentration cell is commonly found in electrochemistry, and effectively involves the mean activity of HCl in each compartment. By analogy with the HITTORF mass balance (see section 4.4.2.3), it can also be demonstrated in terms of a thermodynamic energy balance. The energy change that occurs when an extremely low amount of charge flows through the cell is expressed as a function of the transport numbers and of the ionic activities. Therefore, this type of cell is one way of determining transport numbers in an experiment.

A.3.3 - DEBYE-HÜCKEL'S MODEL

Here we will briefly summarise the basic elements that are needed for demonstrating the DEBYE-HÜCKEL laws as a means of estimating the ionic activity coefficients in a solution.

The electrostatic potential φ is the combined result of an individual ion with a charge of z_k that is placed at the origin ($r = 0$) and all the other surrounding ions, including those of the same nature. The solution's average potential is chosen as the reference value ($\varphi_{sol} = 0$). In this case, the following laws and equations are used:

▸ the laws of electrostatics (POISSON's equation in spherical coordinates):

$$\frac{1}{r^2}\frac{\partial}{\partial r}\left(r^2 \frac{\partial \varphi}{\partial r} \right) = -\frac{\rho_{ch}}{\varepsilon} \quad \text{with} \quad \rho_{ch} = \sum_i n_i\, z_i\, |e|$$

where n_i is the concentration of ion i (expressed in the number of ions i per unit volume, i.e., in m^{-3}) and ρ_{ch} is the charge density (expressed in C m^{-3}) at a distance r from the centre;

▸ the BOLTZMANN statistics, indicating how the species concentration is distributed based on their potential energy:

$$n_i = n_i{}^* e^{-\frac{z_i |e| \varphi}{kT}}$$

where $n_i{}^*$ is the mean concentration of ions i;

▸ the electroneutrality equation (based on mean values):

$$\sum_i z_i\, n_i{}^* = 0$$

If one adopts the assumptions underlying this model (see section 3.2.1.3), since $|z_i\, e\, \varphi| \ll kT$, then it is possible to use the TAYLOR expansion around zero of the preceding n_i expression coming from the BOLTZMANN statistics:

$$n_i \approx n_i{}^* \left(1 - \frac{z_i\, |e|\, \varphi}{kT} \right)$$

The equation describing the charge distribution with the potential then becomes:

$$\rho_{ch} = -\sum_i z_i^2\, n_i{}^* \frac{e^2}{kT}\, \varphi$$

Using the following parameter L_D, which has the dimension of a length and is called the DEBYE length (see the physical analogy below):

$$L_D = \sqrt{\frac{\varepsilon\, kT}{\sum_i z_i^2\, n_i{}^*\, e^2}}$$

the POISSON-BOLTZMANN equation can be written in the following form:

$$\frac{1}{r^2}\frac{\partial}{\partial r}\left(r^2 \frac{\partial \varphi}{\partial r} \right) = \frac{\varphi}{L_D^2}$$

By introducing the function $g = r\varphi$, the equation is made much easier to integrate, thus becoming:

$$\frac{\partial^2 g}{\partial r^2} = \frac{g}{L_D^2}$$

with the following boundary conditions:

$$r \to \infty \qquad g \to 0$$

$$r \to 0 \qquad g \to \frac{z_k |e|}{4 \pi \varepsilon}$$

The latter condition relates to a dilute electrolyte solution, whose charges are considered to be point charges. In the area closely surrounding the central ion (r tends towards zero) the only potential that prevails is the one created by the central charge. All the other ions are too far away to have any bearing on the potential.

By integration, one obtains: $\qquad g = A_1\, e^{-\frac{r}{L_D}} + A_2\, e^{+\frac{r}{L_D}}$

with the following equations taking into account the limiting conditions:

$$\begin{cases} A_1 = \dfrac{z_k |e|}{4 \pi \varepsilon} \\[2mm] A_2 = 0 \end{cases}$$

A TAYLOR expansion around zero, i.e., for a short distance to the central ion when compared to the DEBYE length, then gives the following equation:

$$\varphi = \frac{z_k |e|}{4 \pi \varepsilon r} - \frac{z_k |e|}{4 \pi \varepsilon L_D}$$

In this equation, the first term corresponds to the potential that is produced by the point charge of the central ion. The second term corresponds to the potential produced by all the other ions. The parameter L_D, which has the dimension of a length, corresponds to the radius of a sphere whose charge is opposite to that of the central ion (counter-ion). However, it has no direct physical meaning since there is no real charge $-z_k |e|$ placed on this virtual sphere.

By introducing the concept of ionic strength I_s, as defined in section 3.2.1.2, and by using molar notations for the concentrations, the equation for the DEBYE length becomes:

$$L_D = \sqrt{\frac{\varepsilon R T}{2 I_s \mathscr{F}^2}}$$

For example, for an aqueous solution at 25 °C (dielectric permittivity of pure water: $\varepsilon = \varepsilon_r / (36 \times 10^9 \pi)$ with $\varepsilon_r = 78$) of ionic strength 10^{-3} mol L^{-1} ($= 1$ mol m^{-3}), the DEBYE length is about 100 Å, or 10 nm. This means that the previous equation that illustrates the potential surrounding the central ion only applies when dealing with a distance of less than a few Å from this ion, because it stems from a TAYLOR expansion around zero (i.e., for small values of r compared to L_D). However, this suffices to be able to determine the equation for the activity coefficient, as shown in the following reasoning.

From an energy point of view, the deviation from ideality that is caused by the surrounding ions interacting with the central ion, equates to the same value as the energy required to gradually generate a charge of $z_k |e|$ at the point of the central ion, while keeping the place of the other ions:

$$\Delta W_{\text{interactions}} = kT \ln \gamma_k = -\int_0^{z_k|e|} \frac{q}{4\pi\varepsilon L_D} \, dq = -\frac{(z_k e)^2}{8\pi\varepsilon L_D}$$

so:
$$\log \gamma_k = -A z_k^2 \sqrt{I_s}$$

with:
$$A = \frac{\mathscr{F}^3 \sqrt{2}}{8\pi \, \mathscr{N}(\varepsilon RT)^{3/2} \ln 10} \sqrt{10^3} \qquad \text{in } L^{1/2} \, mol^{1/2}$$

For example, using the dielectric permittivity of pure water at 25°C, the value of A is $0.509 \, L^{1/2} \, mol^{1/2}$, which is close to the experimental value determined from the mean activity coefficients in aqueous solutions.

When the ionic strength is higher than $10^{-3} \, mol \, L^{-1}$, then the following extended DEBYE-HÜCKEL equation can be used:

$$\log \gamma_k = -A z_k^2 \sqrt{I_s} \, \frac{L_D}{L_D + a} = -\frac{A z_k^2 \sqrt{I_s}}{1 + a B \sqrt{I_s}}$$

with:
$$B = \sqrt{\frac{2\mathscr{F}^2}{\varepsilon RT}}$$

This equation is based on a calculation similar to the previous one, but in this particular case the volume of the charges is no longer disregarded. The parameter a, which has the dimension of a length, is the minimum distance by which the ions can approach the central ion. The function $g = r\varphi$ is then integrated between a (instead of 0) and the infinite distance. The parameter a is determined experimentally by adjusting the experimental points to fit the specific law in question. For a single solute, a is close to the sum of the two radii of the solvated ions. In reality, its value is slightly lower: there is an intermingling that occurs between the two solvated ions when they are brought into contact. For single-solute electrolytes, the sum of the parameters given for each of the two ions is often used (see the data tables provided in scientific literature: the order of magnitude is a few Å). The value of parameter B for aqueous solutions at 25 °C using the dielectric permittivity of pure water is $3.3 \, L^{1/2} \, mol^{1/2} \, nm^{-1}$. The product aB is then always close to 1. Most often for aqueous solutions at room temperature, a simplified equation is used which is deduced from the previous equation:

$$\log \gamma_k \approx -A z_k^2 \, \frac{\sqrt{I_s}}{1 + \sqrt{I_s}}$$

that is to say, for a solute:
$$\log \gamma_\pm \approx A z_+ z_- \frac{\sqrt{I_s}}{1 + \sqrt{I_s}}$$

Such a law is commonly used for ionic strengths that are lower than $10^{-2} \, mol \, L^{-1}$ (and exceptionally $10^{-1} \, mol \, L^{-1}$). For instance, it leads to a corrective term compared to the limiting law, of about 10% on the logarithm for an ionic strength of $10^{-3} \, mol \, L^{-1}$.

A.3.4 - THERMODYNAMIC EQUILIBRIUM AT A REACTIVE INTERFACE INVOLVING A SINGLE REACTION BETWEEN CHARGED OR NEUTRAL SPECIES

The aim of this section is to compare the different consequences entailed when a state of thermodynamic equilibrium is established between two conducting media. For this purpose we will focus on the case of three different types of reactive interfaces:

▶ the junction between two materials of the same nature, when only neutral molecules M can be exchanged;

▶ the junction between two materials of the same nature, when a single type of charged species can be exchanged: for instance an exchange of cations M^+;

▶ the electrochemical interface between an inert electrode and an electrolyte containing two elements of a redox couple: for example an equimolar mixture of M^{2+} and M^{3+}.

Let us consider two samples of conducting condensed material with identical volumes (two cubes with $\ell = 1$ cm) at room temperature (25°C). In their initial state, they are placed at a distance from each other, and neither their surface nor volume is charged. In the first two cases, both materials are of the same nature and therefore have equal standard chemical potentials. However the initial concentrations of M or M^+ are different:

$$C_{0_\alpha} = 10^{-1} \, \text{mol L}^{-1} \quad \text{and} \quad C_{0_\beta} = 10^{-3} \, \text{mol L}^{-1}$$

The last case involves a metal, with an initial electron concentration of $C_{0_\alpha} = 100 \, \text{mol L}^{-1}$, that is in contact with an equimolar mixture of M^{2+} and M^{3+} in a solution with an initial concentration of $C_{0_\beta} = 10^{-3} \, \text{mol L}^{-1}$.

In the simplified reasoning applied here, the concentrations are used instead of the activities, except in the case of metals where the electron activity is always equal to 1. Moreover, when the equilibrium is established, it is assumed that the surface charges that emerge as a result, are distributed in such a way that a parallel plate capacitor is formed (area is $S = 1$ cm²) [4]. The distance between the planes is $\delta = 1$ nm, with a relative dielectric permittivity $\varepsilon_r = 10$. This capacitor corresponds to the electrochemical double layer.

Therefore, we end up with a first phase α (with a thickness $\ell = 1$ cm perpendicular to the interface) with a surface charge excess σ_{ch}. This interface forms the first plate of the plate capacitor. The second plate, with an excess of opposite charge, is located at $\delta = 1$ nm inside the second phase β (volume thickness $\ell = 1$ cm):

phase α, charge $+ \sigma_{ch}$ | interfacial zone $\delta = 1$ nm | charge $- \sigma_{ch}$, phase β

Moreover, the difference between the surface electric voltages in the two phases is overlooked (see section 3.1.1.2, $\chi_\alpha = \chi_\beta$). Therefore, the VOLTA and GALVANI potential differences between the two phases are considered as being equal ($\varphi_\beta - \varphi_\alpha = \psi_\beta - \psi_\alpha$).

---

[4] This means that the boundary effects are disregarded: the surface charge density is zero on five faces and uniform on the sixth face which is in contact with the other phase. This system can be also represented as part of a system with unidirectional geometry (two conducting volumes with an infinite section and a thickness ℓ).

EXCHANGE OF NEUTRAL SPECIES M

Once both phases have been brought into contact, molecules M are exchanged between them until thermodynamic equilibrium has been reached. The chemical potential of M, expressed by $\mu_M = \mu°_M + RT \ln C$, is higher at the initial state in phase α than in phase β, therefore molecules M move from phase α to phase β until equilibrium is reached.

▸▸ **Variations in concentration and in the amount of substance required to reach equilibrium**

At thermodynamic equilibrium, the equality of chemical potentials leads to the equality of M activities (= concentrations) in both phases:

$$C_{eq_\alpha} = C_{eq_\beta} = \frac{C_{0_\alpha} + C_{0_\beta}}{2} = \frac{0.1 + 0.001}{2} \approx 0.05 \text{ mol L}^{-1}$$

Once the volume of each phase is taken into account, this concentration variation corresponds to the following amount of substance exchanged:

$$\Delta n = n_{eq_\beta} - n_{0_\beta} = -(n_{eq_\alpha} - n_{0_\alpha}) = \frac{C_{0_\alpha} - C_{0_\beta}}{2} \ell^3 = 5 \times 10^{-5} \text{ mol} = 3 \times 10^{19} \text{ molecules}$$

▸▸ **Variations in chemical potential linked to the process of reaching equilibrium**

Between the two phases, the initial difference in chemical potential for M is:

$$\mu_{0_\alpha} - \mu_{0_\beta} = RT \ln \frac{C_{0_\alpha}}{C_{0_\beta}} = +11.4 \text{ kJ mol}^{-1}$$

In each phase, between the equilibrium state and the initial state, the variation in chemical potential for M is:

$$\Delta \mu_\alpha = \mu_{eq_\alpha} - \mu_{0_\alpha} = RT \ln \frac{C_{eq_\alpha}}{C_{0_\alpha}} = -1.7 \text{ kJ mol}^{-1}$$

$$\Delta \mu_\beta = \mu_{eq_\beta} - \mu_{0_\beta} = RT \ln \frac{C_{eq_\beta}}{C_{0_\beta}} = +9.7 \text{ kJ mol}^{-1}$$

EXCHANGE OF CATION M⁺

After contact, M⁺ ions are exchanged between both phases until thermodynamic equilibrium has been reached. Let us recall that initially there is no surface charge, and that the difference in surface voltages is overlooked. In the initial state, the difference in electrochemical potentials is therefore identical to that of chemical potentials, and the electrochemical potential is higher in phase α than in phase β. Therefore, in order to reach the equilibrium state, M⁺ ions move from phase α to phase β.

▸▸ **Variations in concentration and in the amount of substance required to reach equilibrium**

The movement of ions M⁺ from phase α to phase β creates a charge excess in phase β and charge depletion in phase α. In equilibrium, such opposite charge excesses are located on the surface (remember that it is assumed here that they form a plate capacitor).

The related potential difference is given by the following equation:

$$(\psi_{eq\beta} - \psi_{eq\alpha}) \frac{\varepsilon_0\, \varepsilon_r\, S}{\delta} = (n_{eq\beta} - n_{0\beta})\, \mathscr{F} = -(n_{eq\alpha} - n_{0\alpha})\, \mathscr{F}$$

Moreover, with the VOLTA and GALVANI potential differences being taken as equal, the equality of electrochemical potentials yields the following:

$$RT \ln \frac{C_{eq\alpha}}{C_{eq\beta}} = (\psi_{eq\beta} - \psi_{eq\alpha})\, \mathscr{F}$$

This equation can therefore be written using a single unknown $\Delta n = n_{eq\beta} - n_{0\beta}$:

$$RT \ln \frac{C_{0\alpha} - \Delta n/\ell^3}{C_{0\beta} + \Delta n/\ell^3} = \frac{\delta}{\varepsilon_0\, \varepsilon_r\, \ell^2}\, \Delta n\, \mathscr{F}^2 = \frac{\ell \delta\, \mathscr{F}^2}{\varepsilon_0\, \varepsilon_r}\, \frac{\Delta n}{\ell^3}$$

Solving this equation gives:

$$\Delta C = \frac{\Delta n}{\ell^3} = 1.1 \times 10^{-8}\ \text{mol L}^{-1} \ll C_{0\beta}$$

or: $\Delta n = n_{eq\beta} - n_{0\beta} = -(n_{eq\alpha} - n_{0\alpha}) = 1.1 \times 10^{-11}\ \text{mol} = 6.5 \times 10^{12}$ molecules

which corresponds to a surface charge excess of $1.0\ \mu\text{C cm}^{-2}$.

Therefore, what has been demonstrated here is that when a charged species is exchanged, there are only minor variations in concentration and chemical potential in each phase, between the equilibrium and initial states [5]:

$$\begin{cases} C_{eq\alpha} \approx C_{0\alpha} & \text{and} & C_{eq\beta} \approx C_{0\beta} \\ \mu_{eq\alpha} \approx \mu_{0\alpha} & \text{and} & \mu_{eq\beta} \approx \mu_{0\beta} \end{cases}$$

▸▸ Variations in the GALVANI potential difference linked to the process of reaching equilibrium

In equilibrium, the two phases do not have the same GALVANI potential:

$$\varphi_{eq\beta} - \varphi_{eq\alpha} = \frac{\delta}{\varepsilon_0\, \varepsilon_r\, S}\, \Delta n\, \mathscr{F} \approx \frac{RT}{\mathscr{F}} \ln \frac{C_{0\alpha}}{C_{0\beta}} = 118\ \text{mV}$$

with an equal distribution: $\varphi_{eq\beta} = -\varphi_{eq\alpha} = 59\ \text{mV}$.

▸▸ Variations in electrochemical potential linked to the process of reaching equilibrium

Between the two phases, the initial difference in electrochemical potential for M^+ is:

$$\tilde{\mu}_{0\alpha} - \tilde{\mu}_{0\beta} = \mu_{0\alpha} - \mu_{0\beta} = RT \ln \frac{C_{0\alpha}}{C_{0\beta}} = +11.4\ \text{kJ mol}^{-1}$$

---

[5] *This calculation enables one to predict that such an approximation will cease to apply if δ and/or ℓ are strongly decreased: if the product $\ell\,\delta$ decreases, then $\Delta C = \Delta n/\ell^3$ increases and may become significant in comparison to the initial concentrations.*

Between the equilibrium and the initial states, the variation in electrochemical potential for M^+ in each phase is:

$$\Delta\tilde{\mu}_\alpha = \tilde{\mu}_{eq_\alpha} - \tilde{\mu}_{0_\alpha} \approx (\varphi_{eq_\alpha} - \varphi_{0_\alpha})\mathscr{F} = -5.7 \text{ kJ mol}^{-1}$$

$$\Delta\tilde{\mu}_\beta = \tilde{\mu}_{eq_\beta} - \tilde{\mu}_{0_\beta} \approx (\varphi_{eq_\beta} - \varphi_{0_\beta})\mathscr{F} = +5.7 \text{ kJ mol}^{-1}$$

REDOX EQUILIBRIUM AT AN ELECTROCHEMICAL INTERFACE

In order to address this issue, one firstly needs to add an assumption related to the numerical values of the standard potential of various species. It will therefore be assumed here that:

$$\mu^\circ_{M^{2+}} - \mu^\circ_{M^{3+}} - \mu^\circ_e = +100 \text{ kJ mol}^{-1}$$

Although this value has been picked arbitrarily, it nonetheless depicts a realistic order of magnitude for a chemical GIBBS energy of reaction [6].

Once the phases have been brought into contact, the redox reaction can proceed in one direction or in the other, until thermodynamic equilibrium is reached. Given that there is no surface charge in the initial state and that the difference in surface voltages has been overlooked, the GALVANI potential difference is therefore zero and the electrochemical GIBBS energy of reaction is equal to the GIBBS energy of reaction ($\Delta_r\tilde{G} = \Delta_r G + \mathscr{F}(\varphi_\beta - \varphi_\alpha)$ $= \Delta_r G$). Moreover, since the initial mixture of electroactive species is equimolar, the GIBBS energy of reaction is equal to its standard value. Therefore, in the initial state, the following can be written: $(\mu_{M^{2+}} - \mu_{M^{3+}} - \mu_e)_0 > 0$. To conclude, in order for the equilibrium state to be reached, the redox half-reaction proceeds in the direction of oxidation.

▶▶ Variations in concentration and in the amount of substance required to reach equilibrium

The oxidation of M^{2+} ions to M^{3+} leads to an excess of positive charge in phase β and negative charge in phase α (metal). In equilibrium, these charge excesses, which are exact opposites, are located on the surface. Reaching the state of thermodynamic equilibrium leads to the fact that Δn mol of additional electrons appear on the metal surface, together with the transformation of Δn mol of M^{2+} ions into the same amount of M^{3+} ions in the electrolyte.

---

[6] In the case of a M^{3+}/M^{2+} couple this value can be estimated by combining the couple's standard potential value (E°), with the absolute potential of the SHE (estimated as 4.45 $V_{/vacuum}$, see section 1.5.1.1) and the electron work function of platinum (5.32 eV, see section 3.2.2.3). The following combination can therefore be written:

$$M^{3+}_{aq} + \tfrac{1}{2} H_{2,gaz} \rightleftharpoons M^{2+}_{aq} + H^+_{aq} \qquad -|e|E^\circ \text{ eV}$$
$$e^-_{gaz} + H^+_{aq} \rightleftharpoons \tfrac{1}{2} H_{2,gaz} \qquad -4.45 \text{ eV}$$
$$e^-_{Pt} \rightleftharpoons e^-_{gaz} \qquad 5.32 \text{ eV}$$

$$M^{3+}_{aq} + e^-_{Pt} \rightleftharpoons M^{2+}_{aq} \qquad 5.32 - 4.45 - |e|E^\circ$$

Let us examine a few M^{3+}/M^{2+} couples in order to estimate the corresponding order of magnitude. For the Fe^{3+}/Fe^{2+} couple, $E = +0.77 V_{/SHE}$ and the estimated GIBBS energy is 0.10 eV = 10 kJ mol⁻¹. For the couple V^{3+}/V^{2+}, $E^\circ = -0.26 V_{/SHE}$ and the estimated GIBBS energy is 1.13 eV = 107 kJ mol⁻¹. Finally for the couple Cr^{3+}/Cr^{2+}, $E = -0.41 V_{/SHE}$ and the estimated GIBBS energy is 1.28 eV = 122 kJ mol⁻¹. This explains how the numerical value of 100 kJ mol⁻¹ was chosen when estimating the amount of substance variation required to reach equilibrium.

The resulting potential difference is given by the following equation:

$$(\psi_{eq\beta} - \psi_{eq\alpha}) \frac{\varepsilon_0 \, \varepsilon_r \, S}{\delta} = \Delta n \, \mathscr{F} > 0$$

In equilibrium, the electrochemical GIBBS energy of reaction is equal to zero. So, assuming that the VOLTA and GALVANI potential differences are equal, the following is obtained:

$$\mu°_{M^{2+}} - \mu°_{M^{3+}} - \mu°_e + RT \ln \frac{C_{eq\beta}(M^{2+})}{C_{eq\beta}(M^{3+})} = (\varphi_{eq\beta} - \varphi_{eq\alpha}) \mathscr{F}$$

This equation can be written using a single unknown, Δn:

$$\mu°_{M^{2+}} - \mu°_{M^{3+}} - \mu°_e + RT \ln \frac{C_{0\beta} - \Delta n/\ell^3}{C_{0\beta} + \Delta n/\ell^3} = \frac{\delta}{\varepsilon_0 \, \varepsilon_r \, \ell^2} \Delta n \, \mathscr{F}^2 = \frac{\ell \delta \, \mathscr{F}^2}{\varepsilon_0 \, \varepsilon_r} \frac{\Delta n}{\ell^3}$$

Solving this equation gives:

$$\Delta C = \frac{\Delta n}{\ell^3} = 9.5 \times 10^{-8} \text{ mol L}^{-1} \ll C_{0\beta}$$

or: $\Delta n = 9.5 \times 10^{-11} \text{ mol} = 5.7 \times 10^{13} \text{ molecules}$

which corresponds to an excess of surface charge of 9.2 μC cm^{-2}.

Therefore, what has been shown here is that when redox equilibrium is reached, there are only negligible variations in the concentration and chemical potential in each phase, between the equilibrium and initial states [5]:

$$\begin{cases} C_{eq\alpha} \approx C_{0\alpha} & \text{and} \quad C_{eq\beta} \approx C_{0\beta} \\ \mu_{eq\alpha} \approx \mu_{0\alpha} & \text{and} \quad \mu_{eq\beta} \approx \mu_{0\beta} \end{cases}$$

▶▶ Variations in the GALVANI potential difference linked to the process of reaching equilibrium

In equilibrium, the GALVANI potential in the two phases are not the same:

$$\varphi_{eq\beta} - \varphi_{eq\alpha} = \frac{\delta}{\varepsilon_0 \, \varepsilon_r \, S} \Delta n \, \mathscr{F} = 1.15 \text{ V}$$

This difference is divided equally between the two phases: $\varphi_{eq\beta} = -\varphi_{eq\alpha} = 0.58 \text{ V}$.

▶▶ Variations in electrochemical potential linked to the process of reaching equilibrium

Between the equilibrium and initial states, the electrochemical potential variation for each species in its particular phase is:

$$\Delta\tilde{\mu}_{e\alpha} = \tilde{\mu}_{eq,e\alpha} - \tilde{\mu}_{0,e\alpha} \approx - (\varphi_{eq\alpha} - \varphi_{0\alpha}) \mathscr{F} = + 55.7 \text{ kJ mol}^{-1}$$

$$\Delta\tilde{\mu}_{M_\beta^{2+}} = \tilde{\mu}_{eq,M_\beta^{2+}} - \tilde{\mu}_{0,M_\beta^{2+}} \approx 2 (\varphi_{eq\beta} - \varphi_{0\beta}) \mathscr{F} = + 111.4 \text{ kJ mol}^{-1}$$

$$\Delta\tilde{\mu}_{M_\beta^{3+}} = \tilde{\mu}_{eq,M_\beta^{3+}} - \tilde{\mu}_{0,M_\beta^{3+}} \approx 3 (\varphi_{eq\beta} - \varphi_{0\beta}) \mathscr{F} = + 167.0 \text{ kJ mol}^{-1}$$

with: $\Delta \tilde{\mu}_{M_\beta^{3+}} - \Delta \tilde{\mu}_{M_\beta^{2+}} = 55.7 \text{ kJ mol}^{-1} = \Delta \tilde{\mu}_{e_\alpha}$

We can conclude this appendix by stressing that when the equilibrium state is being reached, the process triggers modifications at the level of a reactive interface which are strongly dependent on whether the reactions involve molecules (neutral species) or charged species (ion exchange or redox reaction). This is a key consideration in electrochemical systems, which can be stretched to apply to most cases[7]:

▸ for an interfacial reaction occurring between molecules (e.g., exchange or reaction between molecules) the equilibrium state is reached when a large amount of substance has reacted at the interface, therefore triggering a change in composition on both sides:

$$\begin{cases} C_{eq_\alpha} \neq C_{0_\alpha} & \text{and} & C_{eq_\beta} \neq C_{0_\beta} \\ \mu_{eq_\alpha} \neq \mu_{0_\alpha} & \text{and} & \mu_{eq_\beta} \neq \mu_{0_\beta} \end{cases}$$

▸ for an interfacial reaction involving charged species (e.g., ion exchange or redox reaction), the equilibrium state is reached once an extremely small amount of substance has reacted, therefore triggering no change in the volume composition:

$$\begin{cases} C_{eq_\alpha} \approx C_{0_\alpha} & \text{and} & C_{eq_\beta} \approx C_{0_\beta} \\ \mu_{eq_\alpha} \approx \mu_{0_\alpha} & \text{and} & \mu_{eq_\beta} \approx \mu_{0_\beta} \end{cases}$$

In these cases, the equilibrium state is reached by adjusting the GALVANI potential differences between the two phases in contact with each other.

A.4.1 - HIGHLIGHTING THE ROLE OF THE SUPPORTING ELECTROLYTE IN MASS TRANSPORT AND ITS IMPACT ON AN ELECTROLYSIS CELL

This appendix sets out to illustrate the following concepts, using two simple examples to which one can apply almost all of the reasoning in analytical terms:

▸ steady-state concentration and potential profiles, including estimating the ohmic drop,

▸ distribution of the diffusion and migration mass transport modes, assuming that the NERNST-EINSTEIN equation applies[8],

▸ process of reaching the steady state,

▸ validity limit for the electrolyte electroneutrality.

---

[7] In experimental situations, one needs to call this equation into question mainly in cases where the ratio of phase volumes to the interface area is very low (reflecting a dramatic decrease in the ℓ value in our calculation).

[8] Let us recall here (see section 4.2.1) that it is only possible to split the migration and diffusion contributions in quantitative terms if one first draws up a hypothesis based on their respective properties. The simplest assumption is that the two phenomena have an identical microscopic mechanism, and that the activity and concentration values are equal. This leads to the NERNST-EINSTEIN equation, as used here.

The two examples that have been selected both relate to a case of electrolysis involving two simple and opposite redox reactions[9]. On the one hand a single 1-1 solute is used as electrolyte, and on the other hand the same solute but with a supporting electrolyte. They are dealt with one after the other, but the interest of this exercise also lies in comparing the two situations.

Imagine a parallelepipedic cell (with unidirectional geometry), comprising two planar, parallel, silver electrodes with a surface area of $S = 1$ cm^2, separated by a distance of $L = 2$ mm. They are both successively immersed in one of the two aqueous solutions S_1 or S_2 being studied, and connected to a constant current supply:

solution S_1 :   silver nitrate, AgNO$_3$, with the concentration $C^* = 10^{-3}$ mol L^{-1}

solution S_2 :   silver nitrate, AgNO$_3$, with the concentration $C^* = 10^{-3}$ mol L^{-1}
            + potassium nitrate, KNO$_3$, with the concentration 10^{-1} mol L^{-1} = 100 C^*

The pH is adjusted in both solutions to prevent any silver oxide from precipitating, and the protons and hydroxide ions are assumed to play only a minor role in the mass transport.

At the anode, one can observe the dissolution of the silver sheet: Ag $\longrightarrow$ Ag$^+$ + e$^-$

At the cathode, one can observe silver deposition: Ag$^+$ + e$^-$ $\longrightarrow$ Ag

The origin of the x axis, which is normal to the electrodes, is chosen as the anode plane, as illustrated in figure A.15. In order to build up a comprehensive definition, one needs to choose a unique origin and a unique orientation of the normal axis, and therefore pay particular attention to the signs in the equations.

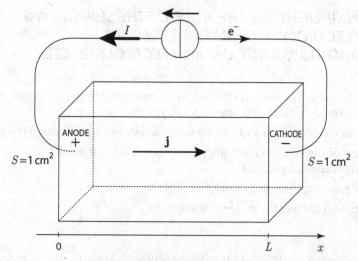

Figure A.15 - Diagram showing the electrolysis cell in question

[9] The advantage of choosing an electrolysis reaction with no overall chemical change is that steady states can be obtained with non zero current. Such steady states are easy to describe in analytical terms. These conditions correspond to an electrolyte whose composition is non-homogeneous, with the diffusion phenomena applying throughout the entire volume, and with a single diffusion layer that measures the same as the inter-electrode distance L.

Assuming that the charge transfer reactions are very fast in comparison to the mass transport phenomena, one can conclude that the charge transfer kinetics plays no part in the voltage value between the electrolysis cell terminals. The voltage is wholly governed by the mass transport phenomena in the electrolyte. The concentration and activity values are considered as equal. It is assumed that both the natural and forced convection phenomena are negligible: only migration and diffusion are taken into account in this analysis[10]. Moreover, the NERNST-EINSTEIN equation is assumed to apply in the system in question. In addition, the characteristic mass transport parameters (diffusion coefficients and mobilities or electrical conductivities) are taken as being independent of the concentration and equal to their values at infinite dilution:

$$\lambda_{Ag^+} = \lambda_+ = 6.19 \text{ mS mol}^{-1} \text{ m}^2 \quad \text{and} \quad D_{Ag^+} = D_+ = 1.65 \times 10^{-9} \text{ m}^2 \text{ s}^{-1}$$

$$\lambda_{NO_3^-} = \lambda_- = 7.14 \text{ mS mol}^{-1} \text{ m}^2 \quad \text{and} \quad D_{NO_3^-} = D_- = 1.90 \times 10^{-9} \text{ m}^2 \text{ s}^{-1}$$

$$\lambda_{K^+} = 7.35 \text{ mS mol}^{-1} \text{ m}^2 \quad \text{and} \quad D_{K^+} = 1.96 \times 10^{-9} \text{ m}^2 \text{ s}^{-1}$$

SOLUTION WITH NO SUPPORTING ELECTROLYTE (SOLUTION S_1)

THE EQUATIONS FOR THE SYSTEM

▸ The electroneutrality of the solution at all points throughout the volume leads to (see the section further on in this appendix dealing with the validity limit of this equation):

$$C_+ = C_- \qquad (\neq C^* \text{ a priori})$$

▸ Because the system is in unidirectional geometry, the molar flux density vectors are all parallel to the x axis and N_i is the algebraic projection on this axis. The volume mass balance in the electrolyte is written in the following way for species i (see section 4.1.2 without homogeneous reaction):

$$\frac{\partial C_i}{\partial t} = -\frac{\partial N_i}{\partial x}$$

and in particular, at steady state:

$$\frac{dN_i}{dx} = 0$$

In other words, the molar flux densities become homogeneous in the overall electrolyte volume at steady state.

▸ The system is also defined in terms of the interfacial mass balance. In this case, NO_3^- does not react at the anode, yet silver is oxidized. Therefore, at this interface the mass balance is written with the anodic current taken as being positive, as usual (see FARADAY's law in section 4.1.4):

$$\begin{cases} (N_+)_{x=0} = \dfrac{I}{\mathscr{F} S} \\ (N_-)_{x=0} = 0 \end{cases}$$

---

[10] If a distance of 2 mm is chosen in an experiment, then a polymer or a gel electrolyte needs to be used in order to eliminate natural convection. In fact, the influence of natural convection cannot be wholly disregarded in a liquid solution. The term 'thin film cell' is sometimes used to describe such a device.

When the FARADAY law is applied at the cathode, the same equations emerge. One must then bear in mind that the conventions usually applied in electrochemistry for a given electrode are not followed when forming the equations for a complete system, since the axis normal to the surface is oriented here at the cathode interface from the electrolyte towards the metal (i.e., outside the system, see figure A.15) [11]. The FARADAY law for the cation should be written as follows:

$$(N_+)_{x=L} = \frac{I}{\mathcal{F} S} < 0$$

with an axis oriented towards the electrolyte. For the same example, yet keeping the same axis orientation for the complete system and specifying $I > 0$ for the current measured in the circuit, we therefore have to write here:

$$(N_+)_{x=L} = \frac{I}{\mathcal{F} S} > 0$$

▸▸ Each different molar flux density can be written as the sum of two terms (diffusion and migration, see section 4.2.1.4):

$$\begin{cases} N_+ = -D_+ \dfrac{\partial C_+}{\partial x} - u_+ \, C_+ \dfrac{\partial \varphi}{\partial x} \\ N_- = -D_- \dfrac{\partial C_-}{\partial x} + u_- \, C_- \dfrac{\partial \varphi}{\partial x} \end{cases}$$

Using the NERNST-EINSTEIN equation: $\dfrac{u_i}{D_i} = \dfrac{|z_i| \, \mathcal{F}}{RT}$

we can write:
$$\begin{cases} N_+ = -D_+ \dfrac{\partial C_+}{\partial x} - D_+ \dfrac{\mathcal{F}}{RT} C_+ \dfrac{\partial \varphi}{\partial x} \\ N_- = -D_- \dfrac{\partial C_-}{\partial x} + D_- \dfrac{\mathcal{F}}{RT} C_- \dfrac{\partial \varphi}{\partial x} \end{cases}$$

STEADY-STATE CONCENTRATION PROFILES

At steady state, the molar flux densities, N_+ and N_-, are constant throughout the whole electrolyte volume and equal to their interfacial values [12].

▸▸ By combining the previous equations so as to eliminate the term showing the electric field, one gets:

$$D_- \, C_- \, N_+ + D_+ \, C_+ \, N_- = - \, D_- \, C_- \, D_+ \frac{\partial C_+}{\partial x} - D_+ \, C_+ \, D_- \frac{\partial C_-}{\partial x}$$

and finally, with $C_+ = C_-$ and $N_- = 0$:

$$N_+ = -2D_+ \frac{\partial C_+}{\partial x} = \frac{I}{\mathcal{F} S}$$

---

[11] See the illustrated board entitled ' Sign convention for current '.

[12] In fact, having two opposite interfacial reactions means that non-zero current steady states can exist in this type of system (see section 4.4.1.2).

that is to say:
$$\frac{\partial C_+}{\partial x} = -\frac{I}{2D_+ \mathscr{F} S} = \text{Cst}$$

Therefore the profiles of C_+ and C_- are linear:

$$C_+ = C_- = (C_-)_{x=L/2} + \frac{I}{2D_+ \mathscr{F} S}\left(\frac{L}{2} - x\right)$$

▸▸ The integration constant (here the concentration at the central plane of the electrolyte, i.e., $x = L/2$) is determined by writing the mass preservation for the Ag^+ or NO_3^- ions. For the nitrate ions, which are neither consumed nor produced at the electrodes, we can write in the volume V:

$$\iiint C_- dV = C^* V$$

The Ag^+ ion preservation leads to an identical equation, because the overall mass balance is also zero: the amount of substance produced at the anode is exactly compensated by the amount consumed at the cathode.

Whatever the current value is, the integration produces the following:

$$(C_-)_{x=L/2} = C^*$$

Therefore, the equation for the concentration profiles (which is identical for both species) is:

$$\frac{C_+}{C^*} = \frac{C_-}{C^*} = 1 + \frac{I}{2D \mathscr{F} S C^*}\left(\frac{L}{2} - x\right) = 1 + \frac{I}{I_{\text{lims}_1}}\left(1 - \frac{2x}{L}\right)$$

by introducing I_{lims_1} with the following definition:

$$I_{\text{lims}_1} = \frac{4D_+ \mathscr{F} S C^*}{L}$$

This quantity corresponds to the limiting current (see section 4.3.3.1, with a thickness of $L/2$ and a diffusion coefficient equal to $2D_+$). Using the numerical data chosen in this example, it is equal to 31.8 μA. Figure A.16 represents the steady-state concentration profile for a current equal to 90% of the limiting current (i.e., 28.6 μA), which gives a slope of -0.9×10^{-3} mol L^{-1} mm^{-1}.

When the current gets very close to the limiting current, the ion concentration in the area next to the cathode becomes very low. Therefore, the solution in this area can no longer be considered as a good conductor. The equations generally used in electrochemistry cannot be applied because the deviations from electroneutrality are too wide. In this case, the equation for expressing electroneutrality cannot be used to make a correct calculation of the concentration profiles in the zone next to the cathode (see section 3.1.1.1). It becomes more complex when one has to describe the system, which will not be addressed here. In short, in order to define the link between the potential and the concentration, one needs to make the same type of calculation, yet this time including the POISSON equation. It is important to keep in mind that the calculation given here no longer applies in quantitative terms once the current value gets close to value of the limiting current.

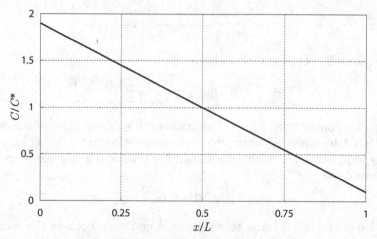

Figure A.16 - Steady-state concentration profile
in the solution with no supporting electrolyte for $I = 0.9\ I_{lim_{s_1}}$

For lower currents, there is also a deviation from electroneutrality when the current flows, though it is very minute in terms of concentration values. This aspect will be examined again after having first dealt with the description of the steady state in the following section.

MOLAR FLUX AND CURRENT DENSITIES AT STEADY STATE

It is possible to depict the diffusion and migration contributions to molar flux densities. Remember that the diffusion flux goes from the most concentrated zone towards the least concentrated (see figure A.16). Keep in mind that the molar flux of the Ag^+ ions is divided into two equal parts between diffusion and migration (see factor $2\ D_+$ above, at steady state, assuming that the NERNST-EINSTEIN equation applies). Finally, since the diffusion coefficient of the NO_3^- ions is higher than that of the Ag^+ ions, the following diagram is obtained:

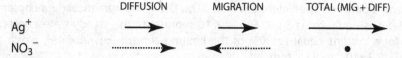

The current densities are in the opposite direction to the molar flux for the nitrate ion because it is an anion:

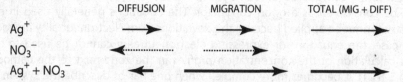

The nitrate ions are electroinactive and do not move macroscopically at steady state. However this is not the case in a transient state, since the steady-state concentration profile has yet to be reached. The interfacial molar flux of nitrate ions is always zero but the transient flux is not constant throughout the whole electrolyte volume, as explained further on.

The overall diffusion current is not zero because of the difference between the anion and cation diffusion coefficients. As shown in an example already described in section 4.2.1.5 the calculation gives:

$$j_{migration} = \frac{I}{2t_+ S} = 1.08 \; j$$

$$j_{diffusion} = 1 - j_{migration} = -0.08 \; j$$

It should be noted that such a distribution between diffusion and migration fluxes remains formal, i.e., based on the assumption that the NERNST-EINSTEIN equation applies.

Given that the molar flux of NO_3^- anion is zero at steady state:

$$\tilde{t}_+ = 1$$

This means that, at steady state, all of the current is carried by the cations, with a half/half distribution of migration and diffusion contributions. This example underlines the difference between the electrochemical transport number and the usual transport number, when the different migration phenomena can be assumed as the only significant phenomena:

$$t_+ = t_{Ag^+} = \frac{\sigma_{Ag^+}}{\sigma} = \frac{\lambda_+}{\lambda_- + \lambda_+} = 0.46$$

In this particular case, whereas the electrochemical transport number has a physical meaning, the migration transport number is only a virtual parameter which is simply the result of a calculation. It does not match the real contributions in the mass transport.

STEADY-STATE POTENTIAL PROFILES

▶▶ Here, by returning to the equations for molar flux densities and combining them together so as to eliminate the term showing the concentration variations, one gets:

$$D_- N_+ - D_+ N_- = -2D_+ D_- \frac{\mathscr{F}}{RT} C_+ \frac{\partial \varphi}{\partial x}$$

Therefore, when using the equations for molar flux densities at steady state, $N_- = 0$ and $N_+ = I/(\mathscr{F}S)$, one obtains:

$$\frac{\partial \varphi}{\partial x} = -\frac{RT}{\mathscr{F}} \frac{I}{2 \mathscr{F} S D_+} \frac{1}{C_+}$$

Given that the concentration is a function of the distance to the electrodes, this equation does not match a linear potential profile. One ends up with:

$$\frac{\partial C_+}{\partial x} = -\frac{I}{2D_+ \mathscr{F} S} \quad \Rightarrow \quad \frac{\partial \varphi}{\partial C_+} = \frac{RT}{\mathscr{F}} \frac{1}{C_+}$$

▶▶ By integrating the previous equation between the abscissa x plane and the middle of the electrolyte $x = L/2$, where the origin of the potential values is taken arbitrarily, one obtains the following:

$$\varphi = \frac{RT}{\mathscr{F}} \ln \frac{C_+}{C^*} = \frac{RT}{\mathscr{F}} \ln \left[1 + \frac{I}{I_{lims_1}} \left(1 - \frac{2 \, x}{L} \right) \right]$$

When the current is low compared to the limiting current, the profile is quasi-linear, as in the case of pure migration phenomena:

$$\varphi \approx \frac{RT}{\mathscr{F}} \frac{I}{I_{lims_1}} \left(1 - \frac{2x}{L}\right)$$

As far as the interfacial voltages are concerned, since the redox couple is considered fast, one can write the following (see section 4.3.2.4):

$$\varphi_{metal} - \varphi_{solution} = Cst + \frac{RT}{\mathscr{F}} \ln [Ag^+]_{interface}$$

Figure A.17 depicts the steady-state potential profile for a current equal to 90% of the limiting current (i.e., 28.6 µA).

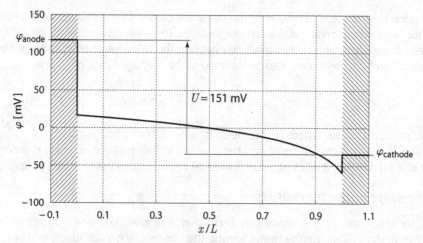

Figure A.17 - Potential profile in the solution with no supporting electrolyte for $I = 0.9\, I_{lims_1}$
The value of the anode potential is fixed arbitrarily.
Yet the value of the cathode potential is deduced from that of the anode.

The electrolysis voltage is therefore equal to:

$$U = \varphi_{anode} - \varphi_{cathode} = \varphi_{x=0} - \varphi_{x=L} + \frac{RT}{\mathscr{F}} \ln \frac{[Ag^+]_{x=0}}{[Ag^+]_{x=L}} = 2 \frac{RT}{\mathscr{F}} \ln \frac{1 + \dfrac{I}{I_{lims_1}}}{1 - \dfrac{I}{I_{lims_1}}}$$

In the example shown in figure A.17 ($I = 0.9\, I_{lims_1}$) the electrolysis voltage is equal to 151 mV (the ohmic drop is 76 mV).

RESISTANCE OF THE ELECTROLYTE AT STEADY STATE

The calculation for the electrolyte's overall resistance is worked out directly using the values of the potential at $x = 0$ and $x = L$:

$$U_{ohmic\ drop} = \varphi_{x=0} - \varphi_{x=L} = R\,I$$

with:
$$R = \frac{RT}{I\,\mathscr{F}} \ln \frac{1 + \dfrac{I}{I_{\text{lims}_1}}}{1 - \dfrac{I}{I_{\text{lims}_1}}}$$

Figure A.18 illustrates the variations in the overall electrolyte resistance as being a function of the current flowing through the cell.

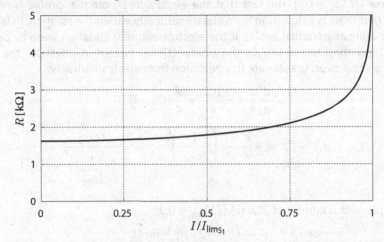

Figure A.18 - Overall resistance of the solution without a supporting electrolyte for different current values

The overall resistance depends on the current value. Therefore, the ohmic drop term is not proportional to the current, as would be the case in a conventional OHM law (for example if there were a supporting electrolyte, as in the case of solution S_2 which is dealt with further on in this section). The overall resistance becomes extremely high when the current approaches the limiting current value, as explained previously.

When I is small compared to I_{lims_1}, a TAYLOR expansion around zero of the general equation can be written, which leads to:

$$R_{I \to 0} \approx \frac{RT}{\mathscr{F}} \frac{2}{I_{\text{lims}_1}} = \frac{L}{S(2\,\lambda_+)\,C^*} = 1.62 \text{ k}\Omega$$

If the electrolyte were to have a homogeneous composition the value of R would be:

$$R_{\text{pure migration}} = \frac{L}{S\,\sigma} = \frac{L}{S(\lambda_- + \lambda_+)\,C^*} = 1.50 \text{ k}\Omega$$

and finally:
$$R_{I \to 0} = \frac{R_{\text{pure migration}}}{2\,t_+}$$

This is to be compared to the equation linking the migration current and the overall current, which is demonstrated above. Indeed, in whatever situation, the OHM law can be written as follows at the local level:

$$j_{\text{migration}} = -\,\sigma\,\frac{\partial \varphi}{\partial x}$$

Due to the difference between the diffusion coefficients, the migration current is slightly higher than the overall current. Therefore the resistance calculated from the potential profile and the overall current (which are the only experimental data available) is slightly higher than the resistance obtained from a calculation using the solution's conductivity.

DEVIATIONS FROM ELECTRONEUTRALITY

In this case ($I/I_{lims_1} = 0.9$), the fact that the electrolyte's potential profile is not linear indicates that there is a deviation from electroneutrality within the solution. In fact, there would be a linear potential profile if the electroneutrality equation were to be strictly applied using the LAPLACE law. In the unidirectional system in question, the POISSON equation gives a mean to estimate this deviation from electroneutrality:

$$\frac{d^2\varphi}{dx^2} = -\frac{\rho_{ch}}{\varepsilon} = \frac{\mathscr{F}}{\varepsilon}(C_- - C_+)$$

$$C_- - C_+ = -\frac{\varepsilon}{\mathscr{F}}\frac{RT}{\mathscr{F}}\left(\frac{I}{I_{lims_1}}\right)^2\frac{1}{L^2}\frac{1}{\left(1 - \frac{x}{L}\frac{I}{I_{lims_1}}\right)^2}$$

For example, with a current of 28.6 µA ($I/I_{lims_1} = 0.9$):

for $x = L$ $C_- - C_+ = -2 \times 10^{-11}$ mol L^{-1}

whereas we have calculated $C_+ = 10^{-4}$ mol L^{-1} at the cathode surface ($x = L$)

for $x = 0$ $C_- - C_+ = -4 \times 10^{-14}$ mol L^{-1}

whereas we have calculated $C_+ = -2 \times 10^{-3}$ mol L^{-1} at the anode surface ($x = 0$).

Therefore, it is proved *a posteriori* that it is legitimate to use electroneutrality to calculate the concentration profiles. However, the LAPLACE law is too approximate in numerical terms to give a correct value for the potential. Therefore, to determine this profile one needs to use the concentration profiles, as already described above.

When the current gets even closer to the limiting current, the Ag^+ and NO_3^- concentrations become very minute in the area next to the cathode. As a result, the deviations from electroneutrality need to be taken into account when simultaneously defining both the concentration and potential profiles (using POISSON's law). Moreover, one also needs to consider the influence of H^+ and OH^- ions in the aqueous solution (concentrations about 10^{-7} mol L^{-1}). In this light, the example previously outlined therefore no longer holds true. However, given that the electrolyte is very resistant, the shape of the curve depicting the resistance related to the current still stands in qualitative terms, showing an abrupt increase when I tends towards the limiting current.

CHARACTERISING THE TRANSIENT PERIOD LEADING TO THE STEADY STATE

If we do not confine ourselves to studying the steady states, then we can no longer assume that the molar flux densities are constant in the electrolyte, nor that the electroinactive ions are immobile. Nevertheless, by using the electroneutrality of the electrolyte, we are then able to go back to the general equations described at the beginning of the appendix (same kind of calculation as in section 4.2.1.2):

$$\begin{cases} \dfrac{\partial C}{\partial t} = -\dfrac{\partial N_+}{\partial x} = -\dfrac{\partial N_-}{\partial x} \\[2mm] D_-N_+ + D_+N_- = -2D_-D_+ \dfrac{\partial C}{\partial x} \end{cases}$$

The same type of calculation as in section 4.2.1.2 gives the following equation:

$$(D_- + D_+)\dfrac{\partial C}{\partial t} = 2D_-D_+ \dfrac{\partial^2 C}{\partial x^2}$$

The progression towards the steady state therefore correlates with the $AgNO_3$ diffusion (i.e., the coupled diffusion of both anions and cations), with a mean diffusion coefficient:

$$D_\pm = \dfrac{2D_-D_+}{D_- + D_+} = 2t_-D_+ = 2t_+D_-$$

The characteristic time, τ, required for establishing the steady state, is defined by:

$$\tau = \dfrac{L^2}{4D_\pm}$$

Here, with the numerical data for the example in question ($D_+ = 1.77 \times 10^{-9}\ m^2\ s^{-1}$), it gives $\tau = 566$ s.

Figure A.19 illustrates how the concentration profiles develop during the transient period while the steady-state profile is being established. It depicts a chronopotentiometry experiment for a constant current equal to 90% of the limiting current, i.e., 28.6 µA.

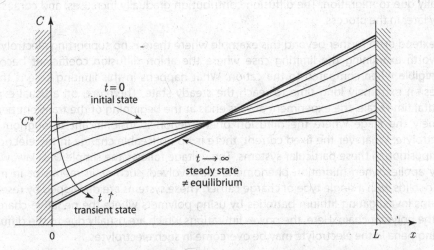

Figure A.19 - Concentration profile change with time towards the steady-state profile during a chronopotentiometry experiment

One must keep in mind that the steady characteristics only depend on parameters of the electroactive ion (here D_+). However, the changes that occur with time during the transient period depend on a parameter, $D_\pm$, which is linked to the properties of both the ions.

The transient period also denotes a change in the distribution between the migration and diffusion processes. At steady state, an equal distribution has been achieved, as already demonstrated above based on the assumption that the NERNST-EINSTEIN equation applies. Figure A.20 illustrates this phenomenon by showing the changing curves for the diffusion and migration molar flux densities ratio for Ag^+, throughout the electrolyte.

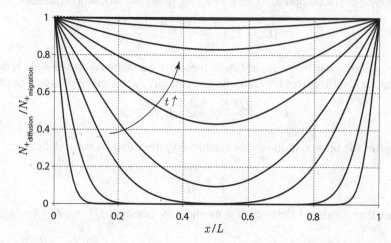

Figure A.20 - Change with time of the diffusion/migration flux densities ratio for Ag^+

The concentration changes in the central part of the electrolyte only appear after a certain time delay. At the beginning of the transient period, the current in this zone is totally due to migration. The diffusion contribution gradually increases, and comes into play later in the process.

To extend even further beyond this example where there is no supporting electrolyte, it is worth examining the limiting case where the anion diffusion coefficient becomes negligible when compared to the cation. What happens in this limiting case is that it takes an extremely long time to reach the steady state. Therefore, on a usual experimental time scale, the experiment always ends at the beginning of the transient period, namely the stage where the diffusion phenomenon is insignificant throughout the electrolyte. Whatever the fixed current, there is no noticeable change in the electrolyte composition in these particular systems. The voltage follows the simple OHM law, which only applies when migration phenomena are involved, such as in metals or in many ionic solids with a single type of charge carrier. These systems are developed by research groups investigating lithium batteries by using polymers where the negative charge is on the polymer chain. Here, the power limitations which are usually due to the diffusion phenomena in the electrolyte may be overcome in such electrolytes.

SOLUTION WITH A SUPPORTING ELECTROLYTE (SOLUTION S_2)

This second part investigates the electrolysis of solution S_2 with a supporting electrolyte. The composition is the following, as indicated at the beginning of this appendix: 10^{-3} mol L^{-1} $AgNO_3$ + 10^{-1} mol L^{-1} KNO_3 aqueous solution.

In the following, the subscripts + and − design Ag^+ and NO_3^- respectively.

THE EQUATIONS FOR THE SYSTEM

▸▸ The equation for expressing electroneutrality throughout the solution volume is:

$$C_+ + C_{K^+} = C_-$$

▸▸ In the case examined here, there is no reaction in the homogeneous phase and the system has unidirectional geometry. The volume mass balance for each species is:

$$\frac{\partial C_i}{\partial t} = -\frac{\partial N_i}{\partial x}$$

Therefore at steady state we have: $\dfrac{dN_i}{dx} = 0$

In other words, at steady state the molar flux densities become constant throughout the electrolyte volume.

▸▸ At the anode, K^+ and NO_3^- do not react, whereas Ag becomes oxidized:

$$\begin{cases} (N_+)_{x=0} = \dfrac{I}{\mathscr{F}\,S} \\[2mm] (N_-)_{x=0} = 0 \\[2mm] (N_{K^+})_{x=0} = 0 \end{cases}$$

The same equations can be obtained by applying the FARADAY law at the cathode (being careful about the signs used; see the previous case).

▸▸ The different molar flux densities can be written as the sum of two terms (diffusion and migration):

$$N_+ = -D_+ \frac{\partial C_+}{\partial x} - u_+ C_+ \frac{\partial \varphi}{\partial x}$$

$$N_- = -D_- \frac{\partial C_-}{\partial x} + u_- C_- \frac{\partial \varphi}{\partial x}$$

$$N_{K^+} = -D_{K^+} \frac{\partial C_{K^+}}{\partial x} - u_{K^+} C_{K^+} \frac{\partial \varphi}{\partial x}$$

or by applying the NERNST-EINSTEIN equation:

$$\begin{cases} N_+ = -D_+ \dfrac{\partial C_+}{\partial x} - D_+ \dfrac{\mathscr{F}}{RT} C_+ \dfrac{\partial \varphi}{\partial x} \\[3mm] N_- = -D_- \dfrac{\partial C_-}{\partial x} + D_- \dfrac{\mathscr{F}}{RT} C_- \dfrac{\partial \varphi}{\partial x} \\[3mm] N_{K^+} = -D_{K^+} \dfrac{\partial C_{K^+}}{\partial x} - D_{K^+} \dfrac{\mathscr{F}}{RT} C_{K^+} \dfrac{\partial \varphi}{\partial x} \end{cases}$$

STEADY-STATE CONCENTRATION PROFILES

At steady state, the molar flux densities N_+, N_- and N_{K^+} are constant throughout the electrolyte volume and are equal to their interfacial values.

▸▸ When the previous equations are combined, the diffusion terms are eliminated to give the following:

$$\frac{N_+}{D_+} + \frac{N_{K^+}}{D_{K^+}} - \frac{N_-}{D_-} = -\frac{\mathscr{F}}{RT}(C_+ + C_{K^+} + C_-)\frac{\partial\varphi}{\partial x}$$

and finally, with $C_+ + C_{K^+} = C_-$ and $N_- = N_{K^+} = 0$,

$$N_+ = -2D_+ \frac{\mathscr{F}}{RT}C_-\frac{\partial\varphi}{\partial x}$$

therefore:
$$N_+ = -D_+\frac{\partial C_+}{\partial x} - D_+\frac{\mathscr{F}}{RT}C_+\frac{\partial\varphi}{\partial x} = -D_+\frac{\partial C_+}{\partial x} + \frac{C_+}{2C_-}N_+$$

Due to the fact that there is a supporting electrolyte, this gives $C_+ \ll C_-$, and therefore:

$$N_+ \approx -D_+\frac{\partial C_+}{\partial x}$$

In other words, with a supporting electrolyte present, the migration of the Ag^+ electroactive ions is negligible when compared to their diffusion.

▸▸ By using the equation coming from the FARADAY law (see above) to express the molar flux densities of Ag^+, it can be shown that the C_+ profile is linear:

$$\frac{\partial C_+}{\partial x} = -\frac{I}{D_+\mathscr{F}S} = \text{Cst}$$

Linear profiles also emerge for the electroinactive ions, again when using the equations involving the molar flux densities. However the slopes are twice as small as those obtained for the electroactive ions, in accordance with the electroneutrality. The equation then becomes:

$$\frac{\partial C_{K^+}}{\partial x} = -\frac{\partial C_-}{\partial x} = \frac{I}{2D_+\mathscr{F}S}$$

▸▸ The integrated mass balance for each type of ion gives the integration constant of the concentration profiles by writing the following:

$$(C_+)_{x=L/2} = C^* \qquad (C_{K^+})_{x=L/2} = 100\ C^* \qquad (C_-)_{x=L/2} = 101\ C^*$$

The new equation for the limiting current in this case can then be introduced (see section 4.3.3.1, with a thickness equal to $L/2$ and a diffusion coefficient equal to D_+):

$$I_{\text{lims}_2} = \frac{2D_+\mathscr{F}S C^*}{L}$$

The limiting current is twice as small as the one obtained with no supporting electrolyte.

The equations for the concentration profiles are the following:

$$\begin{cases} \dfrac{C_+}{C^*} = 1 + \dfrac{I}{I_{\text{lim}s_2}}\left(1 - \dfrac{2\,x}{L}\right) \\[2mm] \dfrac{C_{K^+}}{C^*} = 100 - \dfrac{I}{2\,I_{\text{lim}s_2}}\left(1 - \dfrac{2\,x}{L}\right) \\[2mm] \dfrac{C_-}{C^*} = 101 + \dfrac{I}{2\,I_{\text{lim}s_2}}\left(1 - \dfrac{2\,x}{L}\right) \end{cases}$$

In both solutions S_1 and S_2, the slope of the concentration profile for the electroactive species Ag^+ is proportional to I/I_{lim}. However we must remember that the limiting current is twice as small when there is a supporting electrolyte. Consequently, for a given value of the current I, the fact that there is a supporting electrolyte present causes this slope to increase by a factor of 2, as illustrated in figure A.21. The figure shows the steady-state profiles of Ag^+ for a current equal to 90% of the new limiting current (i.e., 14.3 µA). A slope of -0.9×10^{-3} mol L^{-1} mm^{-1} is obtained for the solution S_2, with a supporting electrolyte, while the slope is twice as small for the solution S_1 with the same current.

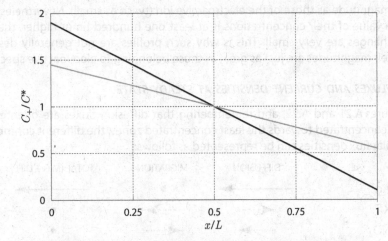

Figure A.21 - Steady-state concentration profiles for Ag$^+$ in the electrolyte for $I = 0.9\ I_{\text{lim}s_2}$
Comparing solution S_1 (without a supporting electrolyte, grey curve)
with S_2 (with a supporting electrolyte, black curve), using the same current.

This difference in behaviour can be explained in qualitative terms. When there is a supporting electrolyte present the current is totally due to diffusion, whereas when there is no supporting electrolyte, half of the steady-state current is carried by migration. One must keep in mind the fact that these distributions between diffusion and migration produce the same total current, i.e., the same amount of substance is transformed per time unit at the electrodes whether they be with or without a supporting electrolyte.

The concentration profiles of electroinactive ions are given in figure A.22 for the same current 14.3 µA.

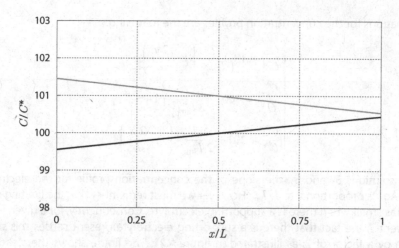

*Figure A.22 - Steady-state concentration profiles
for the supporting electrolyte ions: K^+ (black) and NO_3^- (grey) in the solution S_2 for $I = 0.9\ I_{lim_{S_2}}$*

For these electroinactive ions, the slopes of the concentration profiles share the same order of magnitude as those of the electroactive ion (twice as small). Nevertheless, since the mean value of their concentrations is at least one hundred times higher, then their relative changes are very small. This is why such profiles are not generally described, even if their magnitudes are just as great as those belonging to electroactive species.

MOLAR FLUXES AND CURRENT DENSITIES AT STEADY STATE

Using figures A.21 and A.22, and remembering that diffusion fluxes are oriented from the most concentrated towards the least concentrated zones, the different contributions of the molar flux densities can be represented as follows:

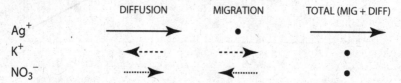

As far as the current densities are concerned, the directions are opposite to the molar flux density in the case of the nitrate ion because it is an anion:

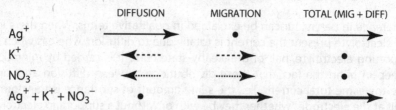

▸▸ Since the molar fluxes of NO_3^- and K^+ are zero at steady state:

$$\tilde{t}_+ = 1$$

This means that the entire current is carried by the Ag^+ ions, as in the case with no supporting electrolyte. In the example studied here almost all of the current is due to diffusion. This value corresponds to the electrochemical transport number. If it were possible to consider the migration phenomena as being the only relevant one, then usual transport numbers would be obtained:

$$t_+ = t_{Ag^+} = \frac{\sigma_{Ag^+}}{\sigma} = \frac{\lambda_+}{101\lambda_- + 100\lambda_{K^+} + \lambda_+} = 0.004$$

and
$$t_{NO_3^-} = 0.49 \qquad t_{K^+} = 0.50$$

STEADY-STATE POTENTIAL PROFILES

▸▸ Previously the following relationship was obtained:

$$\frac{\mathcal{F}}{RT}\frac{\partial\varphi}{\partial x} = -\frac{N_+}{2D_+C_-} = -\frac{I}{2D_+ C \mathcal{F} S}$$

▸▸ Strictly speaking, because the anion concentration is a function of the distance to the electrodes, then the following equation does not relate to a linear potential profile. One has:

$$\frac{\partial\varphi}{\partial C_-} = \frac{RT}{\mathcal{F}}\frac{1}{C_-}$$

▸▸ By integrating the previous equation between the two planes with x and $x = L/2$ abscissae, and with the potential origin being taken arbitrarily at $x = L/2$, one obtains the following:

$$\varphi = \frac{RT}{\mathcal{F}}\ln\left(\frac{C_-}{101C^*}\right) = \frac{RT}{\mathcal{F}}\ln\left[1 + \frac{1}{101}\frac{I}{2\,I_{lims_2}}\left(1 - \frac{2x}{L}\right)\right]$$

Given that there is a supporting electrolyte, even for a current close to the limiting current, then the term $I/(101\,I_{lims_2})$ remains low compared to 1 and the profile is almost linear:

$$\varphi \approx \frac{RT}{2\mathcal{F}}\frac{I}{101\,I_{lims_2}}\left(1 - \frac{2x}{L}\right)$$

This example shows that due to the fact that there is a supporting electrolyte, the potential profile is almost linear even for a current close to the limiting current.

▸▸ Since the redox couple is considered as being fast, the interfacial voltage can be written as follows (see section 4.3.2.4):

$$\varphi_{metal} - \varphi_{solution} = Cst + \frac{RT}{\mathcal{F}}\ln[Ag^+]_{interface}$$

Figure A.23 depicts the steady-state potential profile for a current equal to 90% of the limiting current (i.e., 14.3 µA).

The electrolysis voltage is then equal to:

$$U = \varphi_{anode} - \varphi_{cathode} = \varphi_{x=0} - \varphi_{x=L} + \frac{RT}{\mathcal{F}}\ln\frac{[Ag^+]_{x=0}}{[Ag^+]_{x=L}}$$

$$U \approx \frac{RT}{\mathcal{F}} \ln \frac{[Ag^+]_{x=0}}{[Ag^+]_{x=L}} = \frac{RT}{\mathcal{F}} \ln \frac{1 + \dfrac{I}{I_{lims_2}}}{1 - \dfrac{I}{I_{lims_2}}}$$

In the example shown in figure A.23 ($I = 0.9\, I_{lims_2}$) the electrolysis voltage is equal to 76 mV (the ohmic drop value is 0.2 mV).

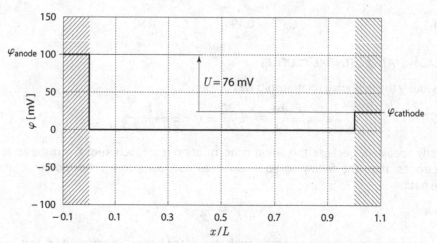

Figure A.23 - *Potential profile within the electrolyte for $I = 0.9\, I_{lims_2}$*
The anode potential value is arbitrarily fixed.
Yet the cathode potential is deduced from that of the anode.

ELECTROLYTE RESISTANCE AT STEADY STATE

The overall electrolyte resistance is calculated from the potential values at $x = 0$ and $x = L$: $U_{ohmic\ drop} = \varphi_{x=0} - \varphi_{x=L} = R\,I$.

$$R = \frac{\varphi_{x=0} - \varphi_{x=L}}{I} = \frac{RT}{I\,\mathcal{F}} \ln \left(\frac{1 + \dfrac{1}{101} \dfrac{I}{2\, I_{lims_2}}}{1 - \dfrac{1}{101} \dfrac{I}{2\, I_{lims_2}}} \right)$$

A TAYLOR expansion around zero of the previous equation gives, even for a current close to I_{lims_2}:

$$R \approx \frac{RT}{\mathcal{F}} \frac{1}{101\, I_{lims_2}} = \frac{L}{S(2\,\lambda_+)\,101\,C^*} = 16.0\ \Omega$$

If the composition of the electrolyte were homogeneous, then the resistance would be:

$$R_{pure\ migration} = \frac{L}{S\,\sigma} = \frac{L}{S(101\,\lambda_- + 100\,\lambda_{K^+} + \lambda_+)\,C^*} = 13.7\ \Omega$$

and finally,
$$R \approx \frac{R_{pure\ migration}}{2\,t_+} \frac{C_+^*}{C_-^*}$$

The resistance remains almost constant when the current varies when there is a supporting electrolyte. Moreover it is much lower than in the previous case, i.e., with no supporting electrolyte. However, the voltage between the system's terminals also tends towards infinity when the current approaches the limiting current: since the Ag^+ concentration tends towards zero, the interfacial voltage at the cathode tends towards infinity.

CHARACTERISING THE TRANSIENT PERIOD LEADING TO THE STEADY STATE

If we do not confine ourselves here to studying the steady states, then the molar flux densities can no longer be considered as constant in the electrolyte, and the electro-inactive ions can no longer be considered as immobile. Nevertheless, by using the expression for electroneutrality, one can reuse the general equations previously written:

$$\frac{\partial C_+}{\partial t} = -\frac{\partial N_+}{\partial x} \approx D_+ \frac{\partial^2 C}{\partial x^2}$$

The progression towards the steady state therefore relates to the Ag^+ diffusion, with a characteristic time, τ, defined by:

$$\tau = \frac{L^2}{4 D_+}$$

In the particular case examined here, by putting $D_+ = 1.65 \times 10^{-9} \, m^2 \, s^{-1}$, one ends up with $\tau = 606 \, s$.

At this point, the fact that there is a supporting electrolyte or not has no significant impact, provided that D_+ and D_- share the same order of magnitude.

A.4.2 - CONCENTRATION PROFILES AT AN INTERFACE

Concentration profiles are very useful for understanding electrochemical phenomena. Yet it should be kept in mind that they make up only one part of those phenomena, since only diffusion is visualized. For example, in the case described in appendix A.4.1, the profile of the NO_3^- anion does not appear flat, although this ion is not consumed at the electrodes. Indeed, even if diffusion has a tendency to carry this species from anode to cathode, the migration compensates for this phenomenon in exact terms. Finally, at steady state, there is no overall mass transport for these ions. Another difficulty arises from the different interpretations of these concentration profiles, in terms of mass consumption or production in transient or steady states.

To illustrate this difference, let us consider the example of a system with Fe^{2+} and Fe^{3+} ions, a supporting electrolyte, two inert electrodes, in unidirectional geometry (see figure A.15 in appendix A.4.1). As in the example chosen in appendix A.4.1, both redox reactions are exact opposites (with the same reaction in both forward and reverse directions):

▸ at the anode ($x = 0$), the reaction is:

$$Fe^{2+} \longrightarrow Fe^{3+} + e^-$$

▸ at the cathode ($x = L$), the reaction is:

$$Fe^{3+} + e^- \longrightarrow Fe^{2+}$$

The migration fluxes of the electroactive ions can be disregarded, and the supporting electrolyte's ions are not examined here either. The mass transport parameters of the two electroactive ions involved are considered as being constant and equal to[13]:

$$D_{Fe^{2+}} = 0.72 \times 10^{-9} \ m^2 \ s^{-1} \qquad D_{Fe^{3+}} = 0.40 \times 10^{-9} \ m^2 \ s^{-1}$$

Here, a chronopotentiometry experiment is studied in the successive conditions below:

▸ semi-infinite unidirectional diffusion (with no convection nor migration of the electro-active species, and with the two electrodes set far apart from each other),

▸ steady-state unidirectional diffusion (with no convection nor migration of the electro-active species, and with the two electrodes set close to each other),

▸ diffusion-convection as defined using the NERNST model (with no migration of the electroactive species).

Finally, to complete the study, a detailed description is given to specify the interfacial conditions involved in chronoamperometry, in the case where there are unidirectional diffusion conditions with a fast redox couple (a reversible reaction) and where there are different diffusion coefficients for the electroactive species.

CHRONOPOTENTIOMETRY WITH SEMI-INFINITE UNIDIRECTIONAL DIFFUSION

The first case has already been described in qualitative terms in section 4.3.1.2. In this section, we will complete figure 4.15 by returning to the analysis given in section 4.3.1.3, figure 4.19. At a given time t, one is able to calculate the concentration profiles as they appear in figure A.24 on the anodic side, and in figure A.25 on the cathodic side. The FARADAY law produces an equation for the interfacial concentrations at the anode:

$$-\frac{I}{\mathcal{F} S} = -D_{Fe^{2+}} \left(\frac{\partial [Fe^{2+}]}{\partial x} \right)_{x=0} = -D_{Fe^{2+}} \frac{[Fe^{2+}]^* - [Fe^{2+}]_{x=0}}{2\sqrt{\dfrac{D_{Fe^{2+}} t}{\pi}}}$$

$$[Fe^{2+}]_{x=0} = [Fe^{2+}]^* - \frac{2I}{\mathcal{F} S} \sqrt{\frac{t}{D_{Fe^{2+}} \pi}}$$

and similarly:
$$[Fe^{3+}]_{x=0} = [Fe^{3+}]^* + \frac{2I}{\mathcal{F} S} \sqrt{\frac{t}{D_{Fe^{3+}} \pi}}$$

As described in section 4.3.1.2, the two grey surfaces are identical in absolute value: they represent the amount of substance produced or consumed between the initial time and time t. Equally, the same result can be reached by using the equations previously established. Indeed, one can even determine the order of magnitude of these two areas by building a calculation based on the algebraic area of triangles that is bounded by the interfacial slopes at time t.

---

[13] For Fe^{2+}, the value is taken from table 4.2 in section 4.2.2.4, using the NERNST-EINSTEIN law, and the value for Fe^{3+} is underestimated (since its real value is $0.60 \times 10^{-9} \ m^2 \ s^{-1}$, deduced from the values in table 4.2), so as to make it easier to visualize the impact made by the difference between these two values.

These are given below:

$$A_{Fe^{2+}} = \frac{1}{2}\left([Fe^{2+}]_{x=0} - [Fe^{2+}]^*\right) 2\sqrt{\frac{D_{Fe^{2+}}t}{\pi}} = -\frac{2I}{\mathscr{F}S}\frac{t}{\pi}$$

$$A_{Fe^{3+}} = \frac{1}{2}\left([Fe^{3+}]_{x=0} - [Fe^{3+}]^*\right) 2\sqrt{\frac{D_{Fe^{3+}}t}{\pi}} = +\frac{2I}{\mathscr{F}S}\frac{t}{\pi} = -A_{Fe^{2+}}$$

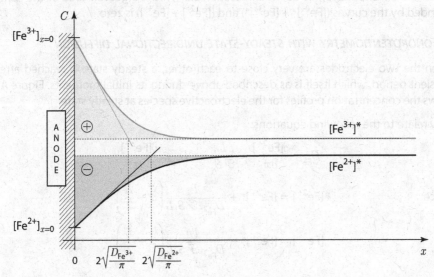

Figure A.24 - Concentration profiles for Fe^{3+} (grey) and Fe^{2+} (black) on the anodic side

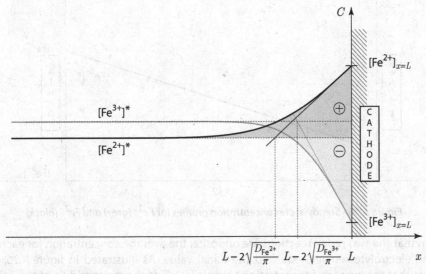

Figure A.25 - Concentration profiles for Fe^{3+} (grey) and Fe^{2+} (black) on the cathodic side

It is also possible to visualize the mass preservation (equal amounts of species produced and consumed in this particular example) by plotting the curve that shows the profile of the sum $[Fe^{2+}] + [Fe^{3+}]$. At the interface this quantity is greater than the sum $[Fe^{2+}]^* + [Fe^{3+}]^*$.

It then decreases to lower than $[Fe^{2+}]* + [Fe^{3+}]*$, before reaching a minimum, whereupon it increases again, tending towards $[Fe^{2+}]* + [Fe^{3+}]*$. If the two diffusion coefficients were equal, then the sum $[Fe^{2+}] + [Fe^{3+}]$ would remain constant and equal to $[Fe^{2+}]* + [Fe^{3+}]*$ throughout the whole electrolyte, reflecting a local application of the mass balance. However, when the two diffusion coefficients are different, then mass preservation can no longer apply at all points. Nonetheless, an overall preservation applies throughout the whole diffusion layer, as defined in section 2.2.1.1: the algebraic area of the surface bounded by the curves $([Fe^{2+}]* + [Fe^{3+}]*)$ and $([Fe^{2+}] + [Fe^{3+}])$ is zero.

CHRONOPOTENTIOMETRY WITH STEADY-STATE UNIDIRECTIONAL DIFFUSION

When the two electrodes are very close to each other, a steady state is reached after a transient period, which itself is as described above during its initial moments. Figure A.26 shows the concentration profiles for the electroactive species at steady state.

They relate to the following equations:

$$D_{Fe^{2+}} \frac{\partial [Fe^{2+}]}{\partial x} = \frac{I}{\mathscr{F} S} = -D_{Fe^{3+}} \frac{\partial [Fe^{3+}]}{\partial x}$$

hence:

$$[Fe^{2+}] = [Fe^{2+}]* + \frac{I}{D_{Fe^{2+}} \mathscr{F} S L} \left(x - \frac{L}{2} \right)$$

$$[Fe^{3+}] = [Fe^{3+}]* - \frac{I}{D_{Fe^{3+}} \mathscr{F} S L} \left(x - \frac{L}{2} \right)$$

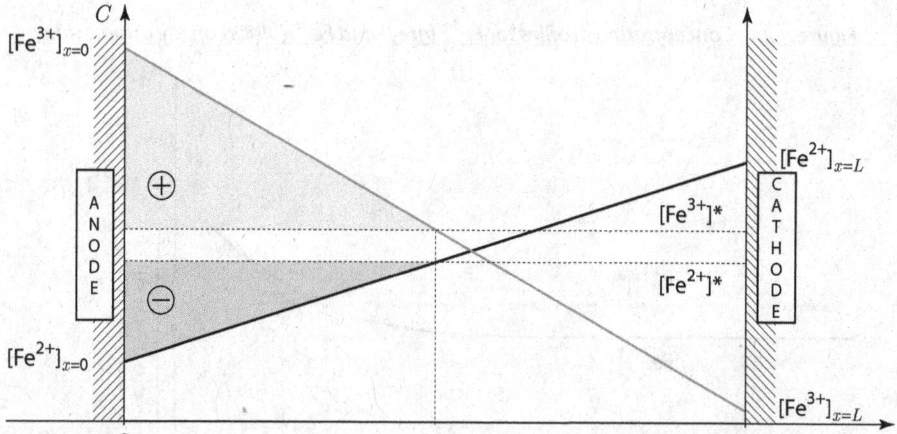

Figure A.26 - Steady-state concentration profiles for Fe^{3+} (grey) and Fe^{2+} (black)

Given that the two redox reactions are opposite, the average concentration for each ion in the electrolyte remains equal to the initial value. As illustrated in figure A.26, this means that the average concentration value is found at the exact middle of the cell, at $x = L/2$. The grey surfaces have different areas because they no longer represent the amounts of substance consumed between two specific moments in time. This disparity observed at steady state is explained by the fact that Fe^{2+} and Fe^{3+} ions do not take the same time to reach steady state, because their diffusion coefficients are different.

CHRONOPOTENTIOMETRY WITH DIFFUSION-CONVECTION ACCORDING TO THE NERNST MODEL

The concept of forced convection is introduced here in simple terms. The steady state that emerges is defined by using the NERNST model (see section 4.3.1.4), and the corresponding concentration profiles are shown in figure A.27.

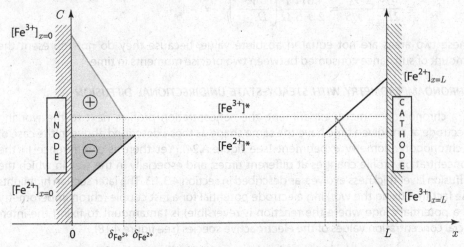

Figure A.27 - Steady-state concentration profiles for Fe^{3+} (grey) and Fe^{2+} (black) for a diffusion-convection mode according to the NERNST model

The concentration profiles on the anodic side are described by the following equations:

$$[Fe^{2+}] = [Fe^{2+}]^* + \frac{I}{D_{Fe^{2+}} \mathscr{F} S}\left(\frac{x}{\delta_{Fe^{2+}}} - 1\right)$$

$$[Fe^{3+}] = [Fe^{3+}]^* - \frac{I}{D_{Fe^{3+}} \mathscr{F} S}\left(\frac{x}{\delta_{Fe^{3+}}} - 1\right)$$

To give an even more precise description, the LEVICH law can be applied to express the thickness of the NERNST layer in experiments using a rotating disk electrode:

$$\delta_i = 1.611\, D_i^{1/3}\, v^{1/6}\, \Omega^{-1/2}$$

with : δ_i thickness of the NERNST layer   [m]
D_i diffusion coefficient   [$m^2\, s^{-1}$]
v kinematic viscosity of the electrolyte [14]   [$m^2\, s^{-1}$]
Ω rotating speed of the electrode [15]   [$rad\, s^{-1}$]

Therefore, as shown in figure A.27, the NERNST layer for Fe^{2+} ions is thicker than that for Fe^{3+} ions.

---

[14] *Be careful not to mix up kinematic viscosity v (in $m^2\, s^{-1}$) with dynamic viscosity η (in Pa s), as used for example in the case of the STOKES law (see section 4.2.2.4). The link between these two parameters is expressed as: $v = \eta / \rho$, where ρ is the density of the medium.*

[15] *1 rpm = $(2\pi/60)\, rad\, s^{-1}$*

The following gives the algebraic areas of the triangles bounded by the steady state and the initial profiles:

$$A_{Fe^{2+}} = \frac{1}{2}\left([Fe^{2+}]_{x=0} - [Fe^{2+}]^*\right)\delta_{Fe^{2+}} = -\frac{I(\delta_{Fe^{2+}})^2}{2D_{Fe^{2+}}\,\mathscr{F}\,S} = -\frac{1.611^2 I}{2\,\mathscr{F}\,S\,\Omega}\left(\frac{v}{D_{Fe^{2+}}}\right)^{1/3}$$

$$A_{Fe^{3+}} = \frac{I(\delta_{Fe^{3+}})^2}{2D_{Fe^{3+}}\,\mathscr{F}\,S} = \frac{1.611^2 I}{2\,\mathscr{F}\,S\,\Omega}\left(\frac{v}{D_{Fe^{3+}}}\right)^{1/3} \neq -A_{Fe^{2+}}$$

These two areas are not equal in absolute value, because they do not represent the amount of substance consumed between two precise moments in time.

CHRONOAMPEROMETRY WITH STEADY-STATE UNIDIRECTIONAL DIFFUSION

In a chronoamperometry experiment, the concentration profiles next to the working electrode at any given time have the same shape as those described above in the case of a chronopotentiometry experiment (see figure A.24). Yet there is a difference in the concentration profile changes at different times, and especially in the way in which the diffusion layer thickness evolves, as described in section 4.3.1.3. The later section highlights the fact that fixing the working electrode potential for a fast couple (chronamperometry in a potential range where the reaction is reversible) is tantamount to fixing the interfacial concentration values of the electroactive species (see figure 4.18).

It is easy to demonstrate this result when the two diffusion coefficients are equal. In fact, by combining the equations for Fe^{3+} and Fe^{2+} (mass balance, initial and boundary conditions), the following system can be obtained:

$$\frac{\partial}{\partial t}\left([Fe^{3+}] + [Fe^{2+}]\right) = D\frac{\partial^2}{\partial x^2}\left([Fe^{3+}] + [Fe^{2+}]\right)$$

with $t \leq 0$ $[Fe^{3+}] + [Fe^{2+}] = [Fe^{3+}]^* + [Fe^{2+}]^*$

        $t > 0,\, x = 0$ $D\dfrac{\partial}{\partial x}\left([Fe^{3+}] + [Fe^{2+}]\right) = 0$

        $t > 0,\, x = L$ $D\dfrac{\partial}{\partial x}\left([Fe^{3+}] + [Fe^{2+}]\right) = 0$

The unique solution of this system is:

$$\forall t,\, \forall x \qquad\qquad [Fe^{3+}] + [Fe^{2+}] = [Fe^{3+}]^* + [Fe^{2+}]^*$$

This equation applies in particular at the working electrode surface. By combining it with the reversibility hypothesis for the interfacial reaction:

$$\frac{[Fe^{3+}]_{x=0}}{[Fe^{2+}]_{x=0}} = e^{\xi}$$

one can obtain equations that give the interface concentrations, showing that they are not time-dependent:

$$[Fe^{2+}]_{x=0} = \frac{[Fe^{3+}]^* + [Fe^{2+}]^*}{1 + e^{\xi}}$$

$$[Fe^{3+}]_{x=0} = \left([Fe^{3+}]^* + [Fe^{2+}]^*\right)\frac{e^{\xi}}{1 + e^{\xi}}$$

The fact that the interfacial concentrations for a fast couple are determined by the electrode potential still applies when the diffusion coefficients are different, however the demonstration then becomes much more complicated[16]. In this case the equations are:

$$[Fe^{3+}] + [Fe^{2+}] \neq [Fe^{3+}]^* + [Fe^{2+}]^*$$

and
$$[Fe^{2+}]_{x=0} = \frac{[Fe^{3+}]^* + \sqrt{\dfrac{D_{Fe^{2+}}}{D_{Fe^{3+}}}}[Fe^{2+}]^*}{\sqrt{\dfrac{D_{Fe^{2+}}}{D_{Fe^{3+}}}} + e^{\xi}}$$

$$[Fe^{3+}]_{x=0} = e^{\xi}[Fe^{2+}]_{x=0}$$

Therefore, in a chronamperometry experiment, the interface concentrations are fixed, though this rule does not apply for a slow redox couple.

---

[16] The results given here can be demonstrated using the LAPLACE transformation.

SUMMARY TABLES

1 - BASIC NOTIONS

2 - SIMPLIFIED DESCRIPTION OF THE ELECTROCHEMICAL SYSTEMS

3 - THERMODYNAMIC FEATURES

4 - CURRENT FLOW: A NON-EQUILIBRIUM PROCESS

1 - BASIC NOTIONS

WHAT NEEDS TO BE KNOWN	RESPONSE ELEMENTS

Introduction

Definition of *electrochemistry*	**1.1.1** Exchange of electric and chemical energies
Principal types of electrochemical applications	**1.1.3** Electrosynthesis - Surface treatments - Energy storage and conversion Analysis and measurements - Environment - Corrosion - Bioelectrochemistry

Oxidation-reduction

Definition of: *oxidant, reductant, oxidation, reduction*	**1.2.1** **Oxidant** (or oxidized form, written Ox): form able to **fix electrons** **Reductant** (or reduced form, written Red): form able to **give electrons** $$\text{Red} \xrightleftharpoons[\text{reduction}]{\text{oxidation}} \text{Ox} + n\,e^-$$
Calculating an *oxidation number*	**1.2.2** The usual o.n. of **H** is equal to $+$**I**. The usual o.n. of **O** is equal to $-$**II**. The usual o.n. of **X** (**halogen**) is equal to $-$**I**. The usual o.n. of **M** (**alkaline**) is equal to $+$**I**.
Writing a balanced *redox half-reaction*	**1.2.3** Preserving the amounts of elements and charge The number of exchanged electrons results from the difference of o.n. in Ox and Red

Current

Different types of conductors and interfaces	**1.3.2** Electronic, ionic, mixed conductors Electronic, ionic junctions, electrochemical and mixed interfaces
Definition of: *anode, cathode*	**1.3.3** Electrochemical system: heterogeneous system including at least two electrodes and one electrolyte: at **anode, oxidation** at **cathode, reduction**

Electrochemical chain

Sign convention for the current	**1.4.1** At anode: $I > 0$ at cathode: $I < 0$ (this comes down to having the normal vector to the interface oriented from the metal to the electrolyte)
Use of the property of conservative current	**1.4.1** Conservative current flux resulting from the electroneutrality of the conducting media: $|I| = \langle j_{an} \rangle S_{an} = -\langle j_{cat} \rangle S_{cat}$ j: current density (A m^{-2})
Representing an electrochemical system in *electrolyser* mode (or for a *recharging battery*)	**1.4.2** 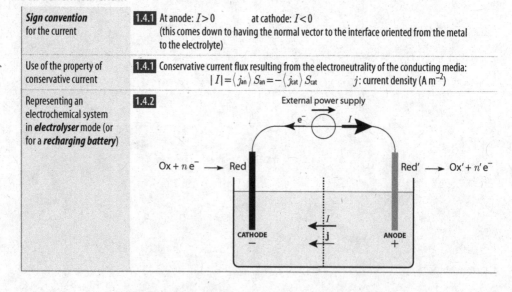

WHAT NEEDS TO BE KNOWN	RESPONSE ELEMENTS

Representing an electrochemical system in **power source** mode (or for a **discharging battery**)

1.4.3

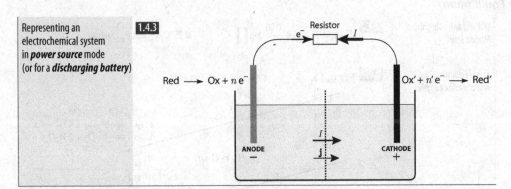

Potential - Voltage - Polarisation

Defining the **standard hydrogen electrode**	1.5.1 SHE is a virtual half-cell (which cannot be achieved by experiment) used as potential reference. It corresponds to the H^+/H_2 couple in its standard state
Knowing the main experimental **reference** systems	1.5.1 ▸ HE, NHE, ▸ Silver chloride electrode (AgCl/Ag couple) or calomel electrode (Hg_2Cl_2/Hg couple): the potential relative to SHE depends on the chloride ion activity in the electrolyte, ▸ SCE: calomel electrode with a KCl-saturated solution
Definition of **polarisation** and **overpotential**	1.5.2 $\pi_+ = E_{+/Ref} - E_{+/Ref}(I=0)\ldots$ Polarisation (π) is the general term. Overpotential (η) is used when the system at zero current is in equilibrium

Experiments

Types of **control**	1.6.2 Potentiometry: I is imposed (possibly $I=0$) and U or E is measured. Amperometry: U or E is imposed and I is measured.
Denomination of electrodes	1.6.2 ▸ Working electrode (WE) ▸ Counter-electrode (or auxiliary electrode, CE) ▸ Reference electrode (Ref)
Defining principles of a **potentiostat**	1.6.2 Laboratory device using 3 electrodes and enabling the control of the WE potential *vs* Ref (no current flows through the reference electrode)
Definition of a **steady**, **quasi-steady** state	1.6.4 Quantities (I, U, concentrations…) independent of time ▸ strictly speaking, for a true steady state ▸ numerically, on the observation time scale, for a quasi-steady state

2 - SIMPLIFIED DESCRIPTION OF THE ELECTROCHEMICAL SYSTEMS

| WHAT NEEDS TO BE KNOWN | RESPONSE ELEMENTS |

Equilibrium

Writing and using the **NERNST law**	**2.1.2** $E_{/Ref} = E°_{/Ref} + \dfrac{0.06}{v_e} \log \prod_i a_i^{v_i}$ at 25 °C, v_i and v_e algebraic
E/pH diagram of **water redox couples**	**2.1.2** 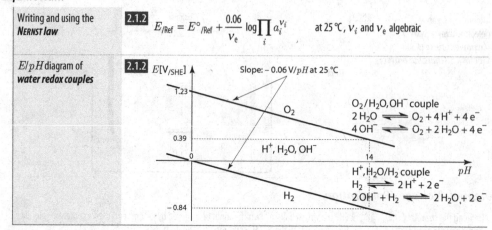

$E[V_{/SHE}]$ Slope: $- 0.06\ V/pH$ at 25 °C

1.23

O_2

$O_2/H_2O, OH^-$ couple
$2\,H_2O \rightleftharpoons O_2 + 4\,H^+ + 4\,e^-$
$4\,OH^- \rightleftharpoons O_2 + 2\,H_2O + 4\,e^-$

0.39

H^+, H_2O, OH^-

$H^+, H_2O/H_2$ couple pH
$H_2 \rightleftharpoons 2\,H^+ + 2\,e^-$
$2\,OH^- + H_2 \rightleftharpoons 2\,H_2O + 2\,e^-$

H_2

-0.84

Current

Main phenomena occurring when a current flows	**2.2.1** Transport within the volume: migration, diffusion (diffusion layer), convection At the interfaces: redox or chemical reaction, exchange, accumulation
Writing and using the **FARADAY law**	**2.2.2** $\Delta n_i^{farad} = \dfrac{v_i}{v_e \mathcal{F}} Q^{farad} = \dfrac{v_i}{v_e \mathcal{F}} j^{farad} S\,\Delta t$ $\Delta n_i, v_i, v_e, Q$ and j algebraic, $1\,\mathcal{F} = \mathcal{N}\,|e| \approx 96\,500\ C\,mol^{-1}$
Defining and using the **faradic yield**	**2.2.2** Faradic yield: proportion of current used for a given half-reaction Additivity of faradic currents if several simultaneous reactions occur at the same interface
Describing the **potential profile** in an electrochemical system	**2.2.3** $U(I \neq 0) = E_+ - E_- = U(I=0) + \pi_+ - \pi_- + \Sigma\,U_{ohmic\ drop}$ **2.4.3** **2.4.4** ▸ system in electrolyser mode: $U(I \neq 0) \geq U(I=0) + U_{ohmic\ drop} \geq U(I=0)$ ▸ system in power source mode : $U_{supplied} \leq U(I=0) - |U_{ohmic\ drop}| \leq U(I=0)$
Conduction by pure migration: **conductivity, transport number**	**2.2.4** $\mathbf{j} = \sigma\mathbf{E} = \dfrac{1}{\rho}\mathbf{E}$ or $U = RI$, if σ is homogeneous in the volume $\sigma = \sum_i \sigma_i = \sum_i \lambda_i C_i$ $t_i = \dfrac{I_i}{I} = \dfrac{j_i}{j} = \dfrac{\sigma_i}{\sigma}$ λ_i molar ionic conductivity $\approx 10^{-2}\,S\,m^2\,mol^{-1} = 10^2\,S\,cm^2\,mol^{-1}$ in aqueous solution σ_i ionic conductivity $\approx 1\,S\,m^{-1}$ for a concentration of about 0.1 mol L^{-1} t_i transport number of species i: $0 < t_i < 1$
Definition of a **supporting electrolyte**	**2.2.4** Non-electroactive ions in much higher concentrations than those of electroactive ions The migration current of electroactive species is then negligible compared to the overall migration current. Moreover, the ohmic drop is most often insignificant in this situation

WHAT NEEDS TO BE KNOWN		RESPONSE ELEMENTS

Current-potential curves

Knowing the general shapes of (E, I) *curves*	2.3.1	Curves normally increase, with $\pi I > 0$ or else: $\pi_{an} > 0$ $(I > 0)$ and $\pi_{cat} < 0$ $(I < 0)$
		For a redox couple, there is only one branch if either the Red or Ox is absent in the system
Influence of the *mass transport kinetics*	2.3.2	Existence of a limiting current, generally proportional to the consumed species concentration within the bulk electrolyte
Influence of the *redox kinetics*	2.3.3	The absolute value of the overvoltage increases when the redox reaction kinetics gets slower
Definition of the *electrochemical window*	2.3.6	Potential range of non-electroactivity of a half-cell when only the solvent and the supporting electrolyte are in contact with a given electrode; it is also called the redox stability window of this half-cell

Predicting reactions

Predicting the *spontaneous* evolution of an electrode *in open circuit*	2.4.1	Basing logic on the current-potential curves (including the kinetic and thermodynamic aspects) and not only on a thermodynamic potential scale
		A zero current in a non-equilibrium state is characterized by a mixed potential: at least two different half-reactions occur at the interface (oxidation and reduction)
Predicting the reactions during *forced current flow*	2.4.3	When there are several possible reactions at an interface, the main half-reaction is that presenting the lowest polarisation (in absolute value) for the same current. The main overall reaction is therefore that which requires the lowest imposed voltage
Predicting the reactions during *spontaneous current flow*	2.4.4	When there are several possible reactions at an interface, the main half-reaction is the one which presents the lowest polarisation (in absolute value) for the same current. The main overall reaction is therefore that which delivers the highest voltage, i.e., the greatest energy amount to an external circuit.

3 - THERMODYNAMIC FEATURES

WHAT NEEDS TO BE KNOWN   **RESPONSE ELEMENTS**

Potential

VOLTA and **GALVANI** potentials	**3.1.1** φ: internal or GALVANI potential, not measurable ψ: external or VOLTA potential, measurable
Writing the **electrochemical potential** and the **activities**	**3.1.2** $\widetilde{\mu}_i = \mu_i + z_i \mathscr{F} \varphi$ with $\mu_i = \mu^\circ{}_i + RT \ln a_i$ For a solution : $a_{solvent} \approx 1$ $a_i = \gamma_i \dfrac{C_i}{C^\circ}$
Conventions in thermodynamic tables	**3.1.2** For pure elements : the chemical potentials are zero To define $\mu^\circ{}_i$ of ion i in solution : $\forall T$ $\mu^\circ{}_{H^+} = 0\,\text{J mol}^{-1}$

Monophasic system

Use of the **mass action law**	**3.2** $K_{eq}(T) = \prod_i a_i^{\nu_i}$ with $\Delta_r G^\circ + RT \ln K_{eq} = 0$ K_{eq}, a_i and ν_i are dimensionless numbers (ν_i algebraic)
Use of the **mean activity** of a solute	**3.2.1** For a solute $A_{p_+} B_{p_-}$: $a = a_\pm^p = a_-^{p_-} a_+^{p_+}$ wih $p = p_+ + p_-$ (for the simple case AB, $a_\pm = \sqrt{a_+ . a_-}$ is measurable whereas a_+ and a_- are not individually measurable)
Defining and calculating the **ionic strength** of an electrolyte	**3.2.1** $I_s = \dfrac{1}{2} \sum_i C_i z_i^2$
Use of the **DEBYE-HÜCKEL limiting law** (including validity limits)	**3.2.1** $\log \gamma_\pm = A z_+ z_- \sqrt{I_s}$ with $A \approx 0.5\ \text{L}^{1/2}\ \text{mol}^{-1/2}$ (aqueous solutions at 25 °C) valid for $I_s < 10^{-3}$ to $10^{-2}\ \text{mol L}^{-1}$

Interface

Describing the electrochemical **double layer**	**3.3.1** The electrochemical double layer is the thin layer (about 10 Å thick) where electroneutrality does not apply due to charge accumulation on both sides of an electrochemical interface (comparable to a capacitor, this zone is also called the space charge zone). Separating this zone into two layers: HELMOLTZ's layer and diffuse layer
Describing the equilibrium at a **reactive interface**	**3.3.2** $\Delta_r \widetilde{G} = \sum_i \nu_i \widetilde{\mu}_i = 0$ taking care to indicate which phase each species i belongs to
Describing the equilibrium at an **ionic junction** where only one charged species is exchanged	**3.3.4** The equilibrium corresponding to $\widetilde{\mu}_{i_\alpha} = \widetilde{\mu}_{i_\beta}$, is set by the exchange of a very low quantity of ions (there is an insignificant concentration difference between the initial and the equilibrium states) This results in a variation in the junction voltage
Describing the equilibrium at a reactive **electrochemical interface**	**3.3.4** There is an interfacial voltage in equilibrium: $\varphi_{metal} - \varphi_{electrolyte} = \dfrac{1}{\mathscr{F}} \left(\mu_{e_{metal}} + \dfrac{1}{\nu_e} \sum_i \nu_i \mu_i \right) = \text{Cst} + \dfrac{RT}{\nu_e \mathscr{F}} \ln \prod_i a_i^{\nu_i}$

WHAT NEEDS TO BE KNOWN | RESPONSE ELEMENTS

Electrochemical systems

WHAT NEEDS TO BE KNOWN	RESPONSE ELEMENTS
Indicating the relationship between *emf* (in equilibrium) and *thermodynamic data*	**3.4.1** $\Delta_r G = \nu_e \mathcal{F} U = \nu_e \mathcal{F}(\varphi_{WE} - \varphi_{CE})$ when ν_e represents the algebraic stoichiometric number of electrons involved in the half-reaction at the interface of the working electrode.
Making qualitative use of an E/pH *diagram*	**3.4.2** For $pH < pH_i$ $Ox_1 + Red_2 \rightleftharpoons Ox_2 + Red_1$ For $pH > pH_i$ $Ox_1 + Red_2 \rightleftharpoons Ox_2 + Red_1$ For $pH = pH_i$, equilibrium $Ox_1 + Red_2 \rightleftharpoons Ox_2 + Red_1$

4 - CURRENT FLOW: A NON-EQUILIBRIUM PROCESS

WHAT NEEDS TO BE KNOWN   **RESPONSE ELEMENTS**

Mass balance

Relationships between *molar flux densities and current densities*	**4.1.1** $\mathbf{j} = \sum_i z_i \mathscr{F} \mathbf{N}_i$
Writing the *local mass balance in volume*, using the concept of a reaction rate	**4.1.2** $\dfrac{\partial C_i}{\partial t} = -\,\mathrm{div}\,\mathbf{N}_i + w_i$ in unidirectional geometry: $\dfrac{\partial C_i}{\partial t} = -\dfrac{\partial N_{x,i}}{\partial x} + w_i$ $w_i = \displaystyle\sum_{\text{reactions}} \nu_{i_r}\, v_r$ with ν_{i_r} the algebraic stoichiometric number of the species i involved in the reaction r, with the rate v_r
Writing the *interfacial mass balance*	**4.1.3** $(N_i)_{\text{interface}_\beta} - (N_i)_{\text{interface}_\alpha} = -\dfrac{\partial \Gamma_i}{\partial t} + w_{S_i}$ normal oriented from α to β
Writing the mass balance at an electrochemical interface: *capacitive and faradic currents*	**4.1.3** For a species i, when it is mobile in the electrolyte (with the normal oriented from the metal to the electrolyte): $(N_i)_{\text{interface}}$ $\quad = \quad - \quad \dfrac{\partial \Gamma_i}{\partial t} \quad + \quad w_{S_i}$ $\qquad\qquad\qquad\qquad$ capacitive term $\qquad$ faradic term $\qquad\qquad\qquad$ (supporting electrolyte) (electroactive species)
Using the *FARADAY law*	**4.1.4** $j^{\text{farad}} = \mathscr{F}\,\dfrac{\nu_e}{\nu_i}(N_i^{\text{farad}})_{\text{interface}}$ for a species i mobile in the electrolyte

Mass transport

Writing the *3 components of molar flux densities* or of current densities	**4.2.1** $\mathbf{j}_i = \underbrace{-\,D_i\, z_i\, \mathscr{F}\,\mathbf{grad}\,C_i}_{\mathbf{j}_{i\,\text{diffusion}}} + \underbrace{\lambda_i\, C_i\, \mathbf{E}}_{\mathbf{j}_{i\,\text{migration}}} + \underbrace{C_i\, z_i\, \mathscr{F}\,\boldsymbol{\omega}_{\text{medium}}}_{\mathbf{j}_{i\,\text{convection}}}$ $\mathbf{N}_i = \underbrace{-\,D_i\,\mathbf{grad}\,C_i}_{\mathbf{N}_{i\,\text{diffusion}}} + \underbrace{\dfrac{z_i}{|z_i|}\, u_i\, C_i\, \mathbf{E}}_{\mathbf{N}_{i\,\text{migration}}} + \underbrace{C_i\,\boldsymbol{\omega}_{\text{medium}}}_{\mathbf{N}_{i\,\text{convection}}}$ At a given point, the overall current density of convection is zero (electroneutrality) Moreover, if the diffusion coefficients of charge carriers have close values, then the overall current density of diffusion is insignificant and therefore the overall current is close to the migration current.
The *NERNST-EINSTEIN* equation	**4.2.1** $D_i = \tilde{u}_i\, \mathrm{R}T = \dfrac{u_i}{|z_i|\mathscr{F}}\mathrm{R}T$ or $\lambda_i = D_i\, z_i^2\, \dfrac{\mathscr{F}^2}{\mathrm{R}T}$ This equation results from the link made between migration and diffusion phenomena (assuming there are identical mechanisms at the microscopic level).
KOHLRAUSCH's law	**4.2.2** $\Lambda = \Lambda^0 - \mathrm{Cst}\,\sqrt{C}$

WHAT NEEDS TO BE KNOWN	RESPONSE ELEMENTS

Interface

Drawing the shape of the **concentration profiles** and defining the **diffusion layer**	**4.3.1** 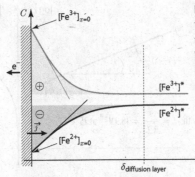 $Fe^{2+} \longrightarrow e^- + Fe^{3+}$ The current is proportional to the slope of the concentration profile at the interface, when a supporting electrolyte is present (first Fick's law and FARADAY's law) $$\left(N_{Fe^{2+}}\right)_{x=0} = -D_{Fe^{2+}}\left(\frac{\partial [Fe^{2+}]}{\partial x}\right)_{x=0}$$ $$= -\frac{I}{\mathcal{F}S}$$
	In the particular case of a limiting current, the interfacial concentration of the consumed species is negligible.
Writing the **rate laws** in the simplest case of the E redox mechanism	**4.3.2** $$v_{oxidation} = k^\circ \exp\left[+\alpha\frac{n\mathcal{F}}{RT}(E-E^\circ)\right][Red]_{x=0} = k^\circ e^{\alpha\xi}[Red]_{x=0}$$ $$v_{reduction} = k^\circ \exp\left[-(1-\alpha)\frac{n\mathcal{F}}{RT}(E-E^\circ)\right][Ox]_{x=0} = k^\circ e^{-(1-\alpha)\xi}[Ox]_{x=0}$$
Definition of the concepts of **fast/slow** redox couples	**4.3.2** Fast redox couple : $\dfrac{k^\circ}{m} \gg 1$ slow redox couple : $\dfrac{k^\circ}{m} \ll 1$ k° is the standard redox reaction rate constant. It is intrinsic to the couple. m characterizes the mass transport rate and therefore depends on the particular experimental conditions in each case. For an experiment in steady state with an RDE in aqueous solution ($m = D/\delta$), a redox couple following an E mechanism is: fast if $k^\circ > 10^{-2}$ cm s^{-1} and slow if $k^\circ < 10^{-4}$ cm s^{-1}
Definition of the concepts of **reversible/irreversible** reactions	**4.3.2** ▶ A step is called **reversible** if, and only if, the overall reaction rate is very low in comparison to the forward and backward reaction rates: $v \ll v_\rightarrow$ and $v \ll v_\leftarrow$ Consequently, the forward and backward reaction rates are almost equal: $v_\rightarrow \approx v_\leftarrow$ ▶ A step is called **irreversible** in the forward direction (for example) if, and only if, the backward reaction rate is negligible compared to the forward reaction rate: $v_\rightarrow \gg v_\leftarrow$ Consequently the overall rate is almost equal to the forward reaction rate: $v \approx v_\rightarrow$
Analytical expression of the steady-state current-potential curve of a **fast couple**	**4.3.3** $$E = E^\circ + \frac{RT}{n\mathcal{F}}\ln\frac{[Ox]_{x=0}}{[Red]_{x=0}} \text{ (nernstian system), thus } E = E_{1/2} + \frac{RT}{n\mathcal{F}}\ln\frac{I - I_{lim_{cat}}}{I_{lim_{an}} - I}$$ with: $E_{1/2} = E^\circ + \dfrac{RT}{n\mathcal{F}}\ln\dfrac{m_{Red}}{m_{Ox}}$ and $I_{lim} = -\dfrac{v_e}{v_i}\mathcal{F}S\,m_i\,C_i{}^*$ (i consumed species)

WHAT NEEDS TO BE KNOWN   **RESPONSE ELEMENTS**

The *TAFEL* plot of the steady-state current-potential curve of a **slow couple**	4.3.3 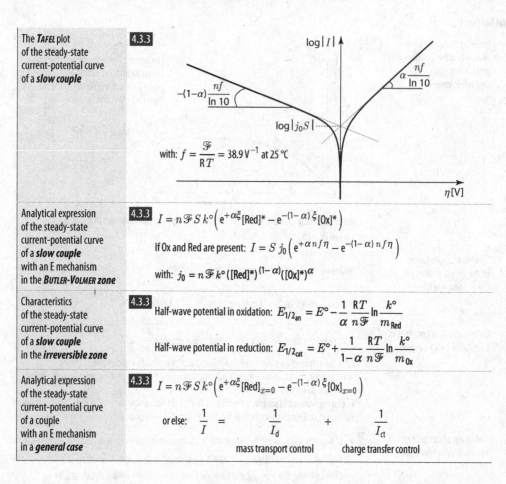 with: $f = \dfrac{\mathscr{F}}{RT} = 38.9\,\text{V}^{-1}$ at 25 °C
Analytical expression of the steady-state current-potential curve of a **slow couple** with an E mechanism in the *BUTLER-VOLMER* zone	4.3.3 $I = n\,\mathscr{F}\,S\,k°\left(e^{+\alpha\xi}[\text{Red}]^* - e^{-(1-\alpha)\,\xi}[\text{Ox}]^*\right)$ If Ox and Red are present: $I = S\,j_0\left(e^{+\alpha\,nf\eta} - e^{-(1-\alpha)\,nf\eta}\right)$ with: $j_0 = n\,\mathscr{F}\,k°\,([\text{Red}]^*)^{(1-\alpha)}([\text{Ox}]^*)^{\alpha}$
Characteristics of the steady-state current-potential curve of a **slow couple** in the **irreversible zone**	4.3.3 Half-wave potential in oxidation: $E_{1/2_{an}} = E° - \dfrac{1}{\alpha}\dfrac{RT}{n\,\mathscr{F}}\ln\dfrac{k°}{m_{\text{Red}}}$ Half-wave potential in reduction: $E_{1/2_{cat}} = E° + \dfrac{1}{1-\alpha}\dfrac{RT}{n\,\mathscr{F}}\ln\dfrac{k°}{m_{\text{Ox}}}$
Analytical expression of the steady-state current-potential curve of a couple with an E mechanism in a **general case**	4.3.3 $I = n\,\mathscr{F}\,S\,k°\left(e^{+\alpha\xi}[\text{Red}]_{x=0} - e^{-(1-\alpha)\,\xi}[\text{Ox}]_{x=0}\right)$ or else: $\dfrac{1}{I} = \underbrace{\dfrac{1}{I_d}}_{\text{mass transport control}} + \underbrace{\dfrac{1}{I_{ct}}}_{\text{charge transfer control}}$

ANSWERS

1 - BASIC NOTIONS

1.2.1 1 - An anion is always negatively charged   ✓ **true**   false

1.2.1 2 - An oxidant is always a cation   true   ✓ **false**

1.2.1 3 - In the following half-reaction:
1.2.2
$$Co + 4\,Cl^- \longrightarrow CoCl_4{}^{2-} + 2\,e^-$$
indicate:
- the redox couple involved   **$CoCl_4{}^{2-}$/Co**
- the oxidized species of the couple   **$CoCl_4{}^{2-}$**
- the (algebraic) charge number of the oxidant   **−2**
- the (algebraic) stoichiometric number
  of the reducing agent   **−1**
- the element undergoing oxidation   **Co**
- the oxidation number of the oxidized element   **+II**
- the direction of the reaction   ✓ **oxidation**   reduction

1.2.1 4 - An anion can be reduced at the cathode   ✓ **true**   false
1.3.3

1.2.2 5 - What is the usual oxidation number of oxygen in a compound?   **−II**

Among the following compounds, circle where oxygen features:
- at its usual oxidation number

   H_2O   **FeO** H_2O_2 **OH^-**   **$ClO_4{}^-$** F_2O **CO_2**   **CO**

- at a higher oxidation number

   H_2O FeO   **H_2O_2** OH^- $ClO_4{}^-$ **F_2O** CO_2 CO

1.2.2 6 - What is the oxidation number of oxygen in O_3?

     −III   −I   **0**   +I   +III

1.2.3 7 - Write the redox half-reaction of the SiO_2/Si couple in an acidic medium

$$Si + 2\,H_2O \rightleftharpoons SiO_2 + 4\,H^+ + 4\,e^-$$

1.3.2 8 - An electrolyte is:
- an ionically conducting medium   ✓ **true**   false
- a vessel used for performing electrolysis   true   ✓ **false**
- a compound that dissolves in a solvent giving rise to ions   ✓ **true**   false
- a man performing electrolysis   true   ✓ **false**
- an electrocuted person   true   ✓ **false**

1.3.2 9 - Molten salts are media with mainly electronic conduction — true ✓ **false**

1.3.2 10 - Semiconductors are media with an electronic conduction type — ✓ **true** false

1.3.2 11 - An electrolyte can exist in a .... state:
- ▸ solid — ✓ **true** false
- ▸ liquid — ✓ **true** false
- ▸ gas — true ✓ **false**

1.3.3 12 - The usual order of magnitude for the thickness of a metal | aqueous electrolytic solution interfacial zone is a few micrometres — true ✓ **false**

1.3.3 13 - In electrochemistry, the cathode:
1.4.4
1.5.1
- ▸ is always the negative electrode of the system — true ✓ **false**
- ▸ always has a negative potential *vs* SHE — true ✓ **false**
- ▸ is always a reduction site — ✓ **true** false

1.4.1 14 - An electrolysis process is carried out between an electrode with a surface of 1 m² where the current density is equal to 1 mA cm⁻² and an electrode whose active surface is a 10 cm × 10 cm square. The absolute value of the current density at this second electrode is

$$10^{-3}\,A\,m^{-2} \qquad 1\,A\,m^{-2} \qquad \boxed{10^3\,A\,m^{-2}}$$

1.4.3 15 - Considering the electrode reactions given below, complete the following diagram, by specifying:
- ▸ the positions of the anode and the cathode
- ▸ the direction of the current (or of the current density)
- ▸ the type of the external circuit component
(indicate your answers by replacing the question marks on the diagram)

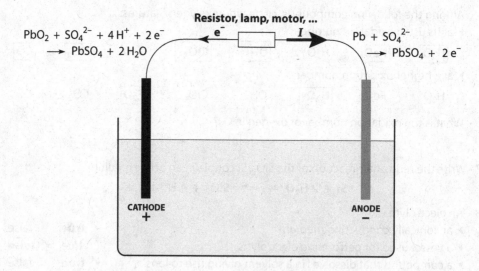

1.4.1
1.6.2
16 - In a 3-electrode setup, these electrodes are called:
- ▸ **working electrode**
- ▸ **counter-electrode** or **auxiliary electrode**
- ▸ **reference electrode**

What is the name of the electronic device generally
used in the lab in this case? **potentiostat**

1.5.1
17 - In electrochemistry, an electrode playing a specific role is the SHE.
- ▸ What is this specific role? **a system setting the origin of the potentials**
- ▸ What do the initials stand for? **Standard Hydrogen Electrode**
- ▸ What is the redox couple involved? H^+/H_2
- ▸ A potential difference exists between SHE and NHE ✓ **true** false

1.5.1
18 - Cite two types of reference electrodes of experimental use, and specify the redox
couple involved.
- ▸ **SCE : Hg_2Cl_2/Hg**
- ▸ **silver chloride electrode: $AgCl/Ag$**

1.5.1
19 - A silver wire coated with silver chloride is dipped into an
aqueous solution containing copper nitrate. This electrode
can be used as reference electrode for measuring potentials
that can be spotted in the potential scale true ✓ **false**

1.5.2
20 - When a system, not equilibrium at open circuit,
is crossed by a current, then one must exclusively use the term ✓ **polarisation** overpotential

1.6.5
21 - Complete the diagram by showing the appropriate shape of the curves that would
indicate the variations of the voltage and the current as a function of time, in a
simple chronoamperometry experiment.

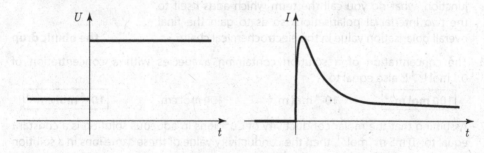

2 - SIMPLIFIED DESCRIPTION OF ELECTROCHEMICAL SYSTEMS

2.2.1 1 - The three mass transport processes are:
- **diffusion**
- **migration**
- **convection**

2.2.1 2 - One studies an interface between cobalt (metal) and an Na^+Cl^- solution in acetonitrile (an organic solvent) where the following reaction occurs:
$$Co + 4\,Cl^- \longrightarrow CoCl_4^{2-} + 2\,e^-$$

▶ The interface is reactive	✓ **true**	false
▶ By convention the current sign is positive	✓ **true**	false

2.2.2 3 - FARADAY's law expresses, for a redox reaction, the amount of substance transformed as a function of the amount of electric charge which crosses the interface in question. The coefficient of proportionality at the numerator involves:

▶ the temperature	true	✓ **false**
▶ the FARADAY constant	true	✓ **false**
▶ the number of electrons	true	✓ **false**
▶ the stoichiometric number of the species in question	✓ **true**	false

2.2.2 4 - In an industrial aluminium production plant, the main cathodic reaction involves the Al(III)/Al couple with a faradic yield of 90%. The amount of aluminium produced per hour in an electrolysis cell working with a current of 300 000 A is:

10^3 mol   **3.4×10^3 mol** 3.6×10^3 mol 10^4 mol 3.4×10^6 mol

2.2.3 5 - The overall polarisation of an electrochemical chain can be split into different terms. In a system with no ionic junction, what do you call the term which adds itself to the two interfacial polarisations so as to gain the final overall polarisation value in the electrochemical chain? **the ohmic drop**

2.2.4 6 - The concentration of a solution containing a species with a concentration of $0.1\ mol\,L^{-1}$ is also equal to

$100\ mol\ m^{-3}$ $10^{-4}\ mol\ m^{-3}$ $100\ mol\ cm^{-3}$ **$10^{-4}\ mol\ cm^{-3}$**

2.2.4 7 - Assuming that the molar conductivity of Cu^{2+} ions in aqueous solution is a constant equal to $10\ mS\ m^2\ mol^{-1}$, then the conductivity value of these same ions in a solution with a concentration of $0.1\ mol\,L^{-1}$ is:

$1\ S\ cm^{-1}$ **$10^{-2}\ S\ cm^{-1}$**   **$1\ S\ m^{-1}$** $10\ S\ m^{-1}$

2.2.4 8 - Adding a supporting electrolyte to an electrochemical system causes, for the electroactive ions, the decrease in:

▶ their transport numbers	✓ **true**	false
▶ their ionic conductivities	true	✓ **false**

2.3.1 9 - In the following diagram:
  ▸ hatch the half-plane corresponding to an anodic operating mode
  ▸ indicate the half-reactions occurring in each half-plane with the usual writing conventions, taking the example of the Fe^{3+}/Fe^{2+} couple

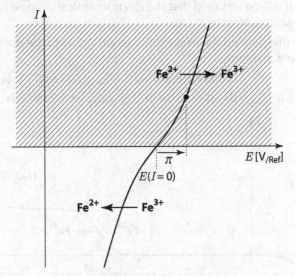

  ▸ what does the black arrow represent for the working point which is identified by a black dot?

  The anodic polarisation (or, in the example here, overpotential) of the electrode considered

2.3.1 10 - Except in very specific cases, one can predict the signs for each of the two interfacial polarisations in a given system. Therefore, in most cases, one can say that:
  ▸ the polarisation of the positive electrode is positive   true ✓ **false**
  ▸ the polarisation of the anode is positive   ✓ **true**   false

2.3.3 11 - On the following diagram, draw the shape of the steady-state current-potential curves of three systems with the same open-circuit potential and the same diffusion limiting currents: a fast system (**a**), a slow system (**b**) and a very slow system (**c**).

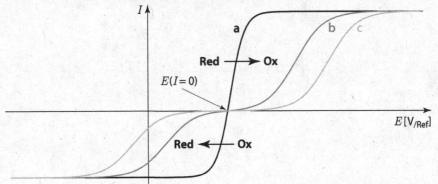

2.3 12 - On the following diagram, plot the steady-state current-potential curve of a system containing an inert working electrode dipped in a deaerated acidic aqueous solution ($pH = 0$), with no Fe^{3+} ions and an amount of Fe^{2+} ions befitting the existence of a limiting current. The reference electrode is a saturated calomel electrode. It will be assumed that the electrochemical window is determined by the fast half-reactions of water.

Indicate in the diagram the relevant numerical values of the potentials as well as the half-reactions involved.

$$E(SCE) = +0.24 \text{ V}_{/SHE} \quad E^{\circ}_{Fe3+/Fe2+} = +0.77 \text{ V}_{/SHE}$$

$$E^{\circ}_{H+/H_2} = 0 \text{ V}_{/SHE} \text{ (at } pH = 0) \quad E^{\circ}_{O_2/H_2O} = +1.23 \text{ V}_{/SHE} \text{ (at } pH = 0)$$

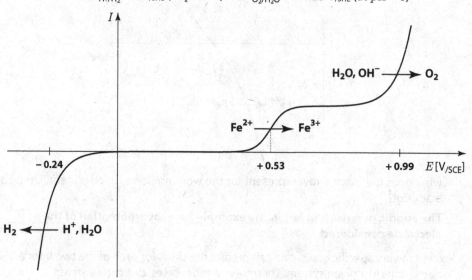

2.4.2 13 - Using arrows, complete the following diagram (which represents the steady-state current-potential curves of the two electrodes in a given electrochemical cell) to indicate the following:

▸ the system's open-circuit voltage, $U(I = 0)$
▸ the polarisation π_- of the negative electrode at the working point indicated by a black dot
▸ the polarisation π_+ of the positive electrode at the corresponding working point of this electrode (indicate this second dot in the diagram)
▸ the corresponding working voltage, $U(I \neq 0)$

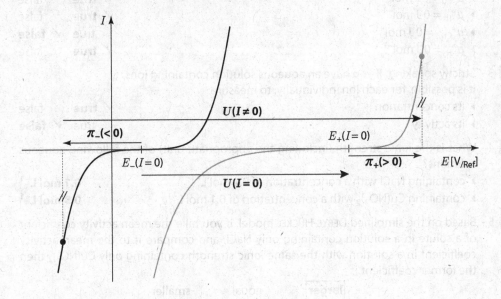

▸ the operating mode represented in the diagram corresponds to electrolysis   ✓ **true**   false

3 - *THERMODYNAMIC FEATURES*

3.1   1 - In an electrochemical system:
- ▶ at thermodynamic equilibrium, the current is always zero   ✓ **true**   false
- ▶ if the current is zero, then the system is always
  in thermodynamic equilibrium   true   ✓ **false**

3.1.2   2 - For the following species in their standard state:
- ▶ $\mu^{\circ}_{Cu} = 0$ J mol^{-1} ✓ **true**   false
- ▶ $\mu^{\circ}_{H^+} = 0$ J mol^{-1} ✓ **true**   false
- ▶ $\mu^{\circ}_{Cu^{2+}} = 0$ J mol^{-1} true   ✓ **false**
- ▶ $\mu^{\circ}_{H_2} = 0$ J mol^{-1} ✓ **true**   false

3.2.1   3 - Strictly speaking, if you have an aqueous solution containing ions, it is possible, for each ion individually, to measure:
- ▶ its concentration   ✓ **true**   false
- ▶ its activity   true   ✓ **false**

3.2.1   4 - What is the ionic strength (including the appropriate unit) of the following aqueous solutions?
- ▶ containing NaCl with a concentration of 0.1 mol L^{-1} **0.1 mol L^{-1}**
- ▶ containing Cu(NO$_3$)$_2$ with a concentration of 0.1 mol L^{-1} **0.3 mol L^{-1}**

3.2.1   5 - Based on the simplified DEBYE-HÜCKEL model, if you take the mean activity coefficient of a solute in a solution containing only NaCl, and compare it to the mean activity coefficient in a solution with the same ionic strength containing only Cu(NO$_3$)$_2$ then the former coefficient is

larger   equal   smaller

3.2.2   6 - For a metal, it is possible to measure:
- ▶ the electrochemical potential of free electrons   ✓ **true**   false
- ▶ the chemical potential of free electrons   true   ✓ **false**
- ▶ the GALVANI potential   true   ✓ **false**
- ▶ the VOLTA potential   ✓ **true**   false

3.3.3
3.3.4   7 - On both sides of a single-exchange junction (i.e., with interfacial reaction involving only one species) between two media in which the species studied share the same standard chemical potential, identical concentrations of exchangeable species are always seen in thermodynamic equilibrium, when the latter is:
- ▶ an ion   true   ✓ **false**
- ▶ a neutral species   ✓ **true**   false

3.3.4
3.4.1   8 - The thermodynamic equilibrium of an interface involving the Cu^{2+}/Cu couple can be illustrated in the following equation: $\varphi_{metal} - \varphi_{electrolyte} = Cst + \dfrac{RT}{2\mathscr{F}} \ln \dfrac{a_{Cu^{2+}}}{a_{Cu}}$

whereby the constant is the standard potential
of the couple in question, relative to SHE   true   ✓ **false**

3.4.1 9 - Fill in the missing numbers (with the appropriate sign) in the equations below, which characterize the thermodynamic equilibrium of the following electrochemical chain: $\quad$ Cu' | Pt, H_2 | aqueous solution containing H_2SO_4 and $CuSO_4$ | Cu

Cu is chosen as the working electrode and Cu' as the counter-electrode ($U = \varphi_{Cu} - \varphi_{Cu'}$). The sign for the GIBBS energy of reaction corresponds to the overall reaction, written as: $\quad\quad Cu^{2+} + H_2 \rightleftharpoons Cu + 2\,H^+$

$$\Delta_r G = \boxed{-2}\,\mathscr{F}U = \boxed{+1}\,\mu_{Cu} \boxed{+2}\,\mu_{H^+} \boxed{-1}\,\mu_{Cu^{2+}} \boxed{-1}\,\mu_{H_2}$$

3.4.1 10 - Complete the simplified POURBAIX diagram for iron below. It has been plotted for an overall iron element concentration equal to C_0, ($[Fe^{3+}] + [Fe^{2+}] = C_0$) in aqueous solution. You must indicate the following:

▸ the areas of either thermodynamic stability or predominance for the following species: Fe, $Fe(OH)_3$, Fe^{2+}, Fe^{3+}

▸ the point symbolizing the potential (vs SHE) of a piece of iron immersed in a solution of $pH = pH_1$, containing Fe^{2+} ions with the concentration C_0

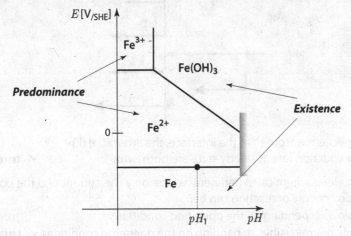

In addition:

▸ where would you locate the point symbolizing the potential (vs SHE) of a piece of iron that is immersed in a solution of $pH = pH_1$ containing Fe^{2+} ions with a concentration of $C_0/100$, in relation to the previous point?

$\quad$ on the right $\quad\quad$ in the same place $\quad\quad$ above $\quad\quad$ **below**

▸ what reaction occurs in the previous system if a 0 V_{SHE} potential is imposed at this metal interface?

$$Fe \longrightarrow Fe^{2+} + 2\,e^-$$

▸ a piece of iron is stable in a solution containing Fe^{3+} ions with the concentration C_0 $\quad\quad\quad\quad\quad$ true $\quad$ ✓ **false**

3.4.2 11 - If a reference electrode Ag, AgCl | KCl 1 mol L^{-1} | has been stored with its tip immersed in distilled water, then calibration would show that its potential after storage

$\quad\quad$ **has increased** $\quad\quad\quad$ has not changed $\quad\quad\quad$ has diminished

4 - CURRENT FLOW: A NON-EQUILIBRIUM PROCESS

4.1.1 1 - Complete the following diagram for the interfacial reaction
$$Co + 4\,Cl^- \longrightarrow CoCl_4^{2-} + 2\,e^-$$
by indicating, in qualitative terms, in both phases:
- ▸ the various molar flux density vectors ($\mathbf{N}_i$)
- ▸ the various current density vectors ($\mathbf{j}_i$)
- ▸ the overall current density vector ($\mathbf{j}$)
- ▸ the vector normal to the surface ($\mathbf{n}$) following the usual sign convention for the current

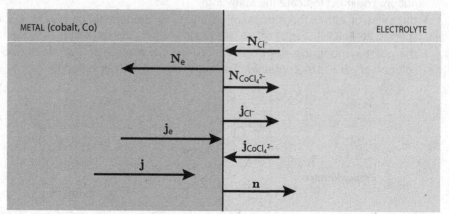

4.1.3 2 - When a species is adsorbed at the interface, the interfacial flux and the production rate at steady state are both zero   ✓ **true**   false

4.2.2 3 - In a given electrochemical experiment, where only the current and the potential can vary, a redox couple or reaction can be:
- ▸ fast or slow, depending on the operating conditions   true ✓ **false**
- ▸ reversible or irreversible, depending on the operating conditions ✓ **true**   false

4.2.2 4 - For an aqueous solution at room temperature containing anions and cations with a concentration of 0.1 mol L^{-1}, what is the order of magnitude and the unit of the electric conductivity? $1\,S\,m^{-1} = 10^{-2}\,S\,cm^{-1}$

4.2.2 5 - What is the order of magnitude and the unit of the diffusion coefficient of an ion in an aqueous solution at room temperature? $10^{-9}\,m^2\,s^{-1} = 10^{-5}\,cm^2\,s^{-1}$

4.3.1 6 - The diagram below shows the changes over time in the concentration profile for an electroactive species of a fast redox couple in an experiment:
- ▸ at steady sate   true ✓ **false**
- ▸ of voltammetry   true ✓ **false**
- ▸ of chronoamperometry   ✓ **true**   false
- ▸ of chronopotentiometry   true ✓ **false**

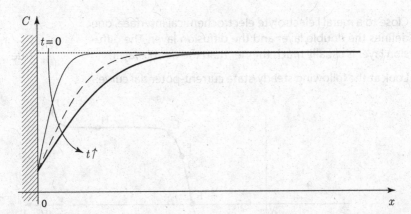

4.3.1 7 - The following diagram shows the concentration profiles for a solution containing Fe^{3+} (grey) and Fe^{2+} (black) ions. The dotted lines represent the initial instant, and the solid lines represent instant t. These profiles result from current circulation through an interface between a platinum electrode and a solution with negligible convection and migration of the electroactive species, Fe^{3+} and Fe^{2+}.

▸ the case shown corresponds to an oxidation reaction   ✓ **true**   false

▸ what is the quantity δ called?   **diffusion layer thickness**

▸ the diffusion coefficient of Fe^{2+} is larger than that of Fe^{3+} ✓ **true**   false

▸ the two shaded areas are equal   ✓ **true**   false

▸ complete the diagram with a qualitative drawing of the concentration profiles that one would observe for a larger current, yet with the same value for δ.

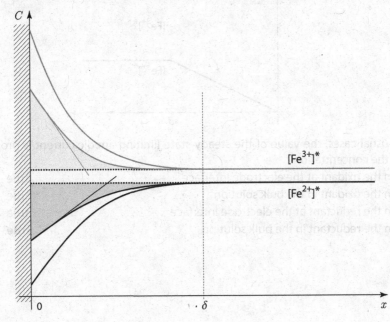

4.3.1 8 - Close to a metal | electrolyte electrochemical interface, one
defines the double layer and the diffusion layer. The diffu-
sion layer is usually much thinner than the double layer.   true ✓ **false**

4.3.3 9 - Look at the following steady-state current-potential curve:

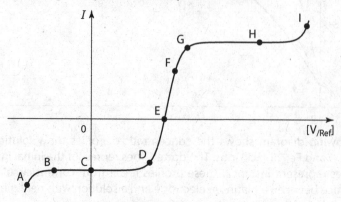

Among the working points indicated above, which ones correspond to the
concentration profile shown in the diagram below?

| A | | B | | C | | D | | E | | F | | G | | H | | I |

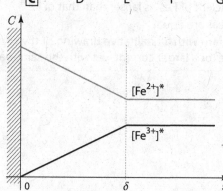

4.3.3 10 - In usual cases, the value of the steady-state limiting anodic current is proportional
to the concentration:

▸ in the oxidant at the electrode interface   true ✓ **false**
▸ in the oxidant in the bulk solution   true ✓ **false**
▸ in the reductant at the electrode interface   true ✓ **false**
▸ in the reductant in the bulk solution   ✓ **true**   false

BIBLIOGRAPHY

IN FRENCH

▸ *Atlas d'équilibres électrochimiques.* M. POURBAIX. Gauthier-Villars (Paris) 1963

One of the key reference books in the field of electrochemical thermodynamics, covering a wide number of inorganic systems in aqueous solution: standard chemical potentials, redox potentials and *E/pH* diagrams. It also includes data on the kinetic parameters involved for reducing protons in aqueous medium. Particularly useful for understanding corrosion in aqueous medium.

▸ *Electrochimie. Principes, méthodes et applications.* A.J. BARD, L.R. FAULKNER. Traduction : Masson (Paris) 1983

Refer to the comments shown for the English edition.

▸ *Usuel de chimie générale et minérale.* M. BERNARD, F. BUSNOT. Dunod, Bordas (Paris) 1984

A collection of numerical data covering a relatively large number of quantities used in physical chemistry and thermodynamics, mainly for inorganic species: for example acidity constants (pK_a), including those found in non-aqueous solvents, solubility constants and complexation constants. Regarding electrochemistry, you can find the redox potentials for numerous couples, the molar conductivities for the main ions in aqueous solution, the activity coefficients for electrolytes, as well as a small number of kinetic features (exchange current density, and transfer coefficient, etc.).

▸ *L'oxydoréduction.* J. SARRAZIN, M. VERDAGUER. Ellipses (Paris) 1991

By describing a set of simple experiments in accessible terms, this work brings together the fundamental concepts linked to redox reactions, mainly in an aqueous phase, focusing on various parameters (*pH*, complexation and precipitation phenomena, etc.). Centring on the fields of thermodynamics and kinetics, it lays out the key concepts of electrochemistry. This book is geared towards students at beginner level, but also to more experienced scientists, such as teachers and lecturers.

▸ *Electrochimie analytique et réactions en solution* (tomes 1 et 2). B. TREMILLON. Masson (Paris) 1993

A didactical work written for researchers, as well as engineers working in chemistry and physical chemistry, that spans a wide spectrum of research fields. Numerous exercises and theoretical problems to solve.

Volume 1: Case analyses involving organic, aqueous solutions, with dissolved salts.

Volume 2: Theoretical aspects of electrochemistry, applications, and methods used in electrochemical analysis. It includes additional information on insertion materials, solid electrolytes, semiconducting electrodes, photoelectrochemistry and the thermoelectric effect.

▸ *Electrochimie des solides.* C. DÉPORTES, M. DUCLOT, P. FABRY, J. FOULETIER, A. HAMMOU, M. KLEITZ, E. SIEBERT, J.L. SOUQUET. Presses Universitaires de Grenoble (Grenoble) 1994

Aimed at master's degree or PhD students as well as researchers and specialist engineers, this work focuses on electrochemical systems using electrolytes in solid phases (ionic crystals, ceramics, different types of glass and polymers). The fundamental concepts of electrochemistry are laid out (the thermodynamics of point defects and amorphous phases, transport mechanisms, mixed conduction, and gas electrode reactions) alongside the specific research methods used. Several applications are also described.

▶ *Cinétique électrochimique.* C. MONTELLA, J.P. DIARD, B. LE GORREC. Hermann (Paris) 1996

Lays out the fundamental concepts of electrochemistry, but with a particular focus on the theoretical aspects involved in the kinetics of electrode reactions. Covers various methods, yet with strong emphasis placed on impedance spectroscopy as well as voltamperometric methods. The main systems examined include redox reactions, electrosorption, insertion and the VOLMER-HEYROVSKY mechanism. For master's degree level, engineering students and researchers. In 2000 and 2005 the same authors and publishing house produced two books compiling exercises on electrochemical kinetics (steady state and insertion method).

▶ *Chimie physique expérimentale.* B. FOSSET, C. LEFROU, A. MASSON, C. MINGOTAUD. Hermann (Paris) 2000 (2nd édition en 2006)

Targeted mainly at students preparing for competitive teaching french exams it combines both descriptions and commentaries for 110 physical chemistry experiments, with at least 40 of these linked to the field of electrochemistry. Moreover, the first part provides descriptions for different techniques, in particular related to electrochemistry.

▶ *Electrochimie physique et analytique.* H.H. GIRAULT. Presses polytechniques et universitaires romandes (Lausanne) 2001 (2nd édition en 2008)

Electrochemistry as a subject is approached using the fundamental concepts of physical chemistry and physics, and the theoretical aspects are dealt with from a strictly mathematical point of view. Focusing on aqueous media, this book describes the particular phenomena involved at the electrolyte and interfaces. Various research methods are listed (amperometry, impedancemetry, and voltamperometry). Specifically targeted at students in higher education, and even stretching to researchers.

▶ *L'indispensable en électrochimie.* V. BERTAGNA, M. CHEMLA. Bréal (Rosny-sous-Bois) 2001

The fundamental concepts of electrochemistry are collected together in 24 technical tables, divided by theme, featuring most of the key formulas (100 pages), with examples given for aqueous solutions. Specifically aimed at students on foundation courses, and first-year university students (including French 'IUT' students carrying out a two-year technology BA), this book can equally be used as a data form or glossary.

▶ *Electrochimie. Des concepts aux applications.* F. MIOMANDRE, S. SADKI, P. AUDEBERT, R. MEALLET-RENAULT. Dunod (Paris) 2005

An overview of the different concepts involved in electrochemistry, spanning from basic theory (thermo-dynamics, transport and electrode kinetics, etc.) to the main applications (batteries corrosion, electrosynthesis and sensors). In addition, it covers the research methods used in electrochemistry, as well as giving an insight into organic electrochemistry. For master's degree level and engineering schools. A dozen practical experiments are described in brief, with a few problems to solve (corrections provided on the website).

▶ *De l'oxydoréduction à l'électrochimie.* Y. VERCHIER, F. LEMAÎTRE. Ellipses (Paris) 2006

For first-year university students. This book bases itself initially on redox reactions as an introduction, covering the amounts and *E/pH* diagrams, before moving on to present the basics of electrochemistry (the thermo-dynamics and kinetics of electrode reactions), adapting its approach to the target reader. It includes a large number of basic exercises.

▶ *Capteurs électrochimiques. Fonctionnement, utilisation, conception. Cours et exercices corrigés.* P. FABRY, C. GONDRAN. Ellipses (Paris) 2008

Focused on the electrochemical applications used to design sensors for ionic, molecular or gas species. Targeted at BA level and master's degree students, as well as students at engineering school, it is also adapted to self-training. This book provides a recap of the basic concepts, before then applying them to different types of transduction (amperometric, conductimetric and potentiometric sensors). There is also a selection of problems to solve with the corrections provided.

▸ Articles quoted from the monthly chemical review '*Actualité Chimique*"

   nov-déc 1989 : 143-147, J.F. FAUVARQUE.

   janv-fév 1992 : 1-148, numéro spécial *L'électrochimie*

   janvier 2003 : 31-41, D. DEVILLIERS, E. MAHÉ.

   février-mars 2009 : 9-119, numéro coordonné par F. BEDIOUI.

▸ Articles quoted from the French engineering document source site '*Techniques de l'Ingénieur*'

IN ENGLISH

▸ *Instrumental methods in electrochemistry*. Southampton Electrochemistry Group.
   Ellis Horwood Limited (Chichester) 1985

   For master's degree and PhD level, this work describes the main methods used in the field of electrochemistry (steady-state and non-steady-state) and applies them to various concepts, including kinetic models for electronic transfer, double layer and electrocatalysis. There are particular chapters focused on electro-crystallization, optical and spectroscopic methods as well as designing an electrochemical experiment, covering the suitable instruments required.

▸ *Industrial electrochemistry*. D. PLETCHER, F.C. WALSH.
   Chapman and Hall (London) 2nd edition 1990

   For master's degree level and engineers. Basic concepts, electrochemical engineering, with an exhaustive study of industrial applications, including surface treatment, corrosion and sensors.

▸ *Electrochemical systems*. J.S. NEWMAN.
   Prentice Hall Inc. (Englewood Cliffs, New Jersey) 2nd edition 1991

   This work presents electrochemistry from a macroscopic viewpoint, and is divided into 4 parts: the thermo-dynamics of electrochemical cells, electrochemical kinetics, transport processes, and finally current distribution and mass transfer in electrochemical systems (including porous electrodes and semiconducting electrodes). Problems to solve are presented at the end of each chapter, without the answers.

▸ *Principles of electrochemistry*. J. KORYTA, J. DVORAK & L. KAVAN.
   John Wiley & Sons (Chichester) 2nd edition 1993

   For master's degree level in physical chemistry, biology and materials science. Divided into 6 chapters: equi-libria in electrolyte solutions, transport, charge transfer, double layer, kinetics (including a concise outline of the different methods used in electrochemical analysis), membrane electrochemistry and bioelectrochemistry.

▸ *Fundamentals of electrochemical science*. K.B. OLDHAM & J.C. MYLAND.
   Academic Press Inc. (San Diego) 1994

   For master's degree and PhD level. This book offers a highly mathematical approach, and is divided into 11 chapters, comprising essential reminders on the key basics in physics and chemistry, as well as in the thermodynamics of ionic solutions. It touches on the full list of notions and methods related to modern electrochemistry. Problems to solve are presented at the end of each chapter, without the answers.

▸ *Electrochemistry, principles, methods and applications*. C.M.A. BRETT, A.M. OLIVEIRA BRETT.
   Oxford University Press (Oxford) 1994

   Divided into three parts, spanning from BA level to master's degree level. Chapters 2 to 6: thermodynamics and electrochemical kinetics. Chapters 7 to 12: experimental strategy and methods used in electrochemical analysis. Chapters 13 to 17: applications (sensors and industrial electrochemistry).

▶ *Principles and applications of electrochemistry.* D.R. CROW.
Blakie Academic & Professional, Chapman & Hall (London) 1994

Master's degree level. Focusing on the aspect of physical chemistry related to solutions, equilibria, ion transport, double layer, NERNST, kinetics, applications, industrial applications and sensors. Problems to solve are presented at the end of each chapter, with the corrections provided.

▶ *Electrochemical process engineering. A guide to the design of electrolytic plant.*
F. GOODRIDGE, K. SCOTT. Plenum Press (New York) 1995

A work targeted at chemistry and engineering students at master's degree and PhD level. Covering concepts related to electrochemical engineering, kinetics, two-phase flow electrocatalysis and reactor modeling.

▶ *Electrochemical methods. Fundamentals and applications.* A.J. BARD, L.R. FAULKNER.
John Wiley & Sons (New York) 2nd edition 2001

For master's degree and PhD level. Presenting the scientific concepts of electrochemistry, methods used in electrochemical analysis, instruments, techniques used in physics for analysing interfaces and photoelectrochemistry. Problems to solve are presented at the end of each chapter, without the answers.

▶ *Modern electrochemistry.* J.O'M. BOCKRIS *et al.*
Plenum Press (New York) 2nd edition

An introduction to modern electrochemistry: each notion is based on a unique theory aimed at opening out on to a deeper exploration of the subject. Mathematical elements are provided in the appendix. For master's degree and PhD level.

Vol. 1 : Ionics. J.O'M. BOCKRIS, A.K.N. REDDY, 1998.
A historical presentation followed by a description of the key notions of physical chemistry which are essential for understanding the properties of electrolytes (ion-solvent and ion-ion interactions, ion transport and ionic liquids).

Vol. 2A : Fundamentals of electrodics. J.O'M. BOCKRIS, A.K.N. REDDY, M. CAMBOA-ALDECO, 2000
Description of electrified interfaces, electrochemical kinetics, transient analysis methods and quantum mechanics applied to electrochemical kinetics.

Vol. 2B : Electrodics in chemistry, engineering, biology and environmental science. J.O'M. BOCKRIS, A.K.N. REDDY, 2000
Fundamental aspects of photoelectrochemistry, electrochemical energy conversion and storage, bioelectrochemistry and electrochemistry related to the environment.

▶ *Fundamentals of electrochemistry.* V.S. BAGOTSKY.
John Wiley & Sons, Wiley-Interscience (Hoboken, New Jersey) 2nd edition 2006

The first three parts approach the classic aspects of electrochemistry on a BA/master's degree level. The fourth part touches on cutting-edge developments in the field of modern research (electrochemistry of solids, conducting polymers, physical methods for analysis, electrocatalysis, photoelectrochemistry, bioelectrochemistry, electrokinetics, interfaces between immisicible liquids, numerical simulations and nanoelectrochemistry, etc.)

▶ *Electrochemistry.* C.H. HAMANN, A. HAMNETT, W. VIELSTICH.
John Wiley & Sons, Wiley Vch Verlag GmbH (Weinheim) 2nd edition 2007

A reasonably exhaustive study of the physical chemistry involved in electrolytes and electrochemistry (kinetics and methods of analysis), and the applications, placing particular emphasis on the methods used in physics (spectroscopy and optics) for examining electrochemical interfaces.

Symbol	Meaning	Usual unit	Comment
a_i	activity of ion i	dimensionless	see section 3.1.2
$a_\pm$	mean activity of a solute	dimensionless	see section 3.2.1
A	parameter for the DEBYE-HÜCKEL law	$L^{1/2}\,mol^{-1/2}$	$\approx 0.5\ L^{1/2}\,mol^{-1/2}$ at 25 °C see section 3.2.1
C_i (or C) or $[i]$	concentration of species i	$mol\,L^{-1}$	[1]
$C°$	reference concentration (standard state)	$mol\,L^{-1}$	$= 1\ mol\,L^{-1}$ see section 3.1.2
C^*	bulk concentration (in the zone far from the interface)	$mol\,L^{-1}$	
$C_{x=0}$	interfacial concentration ($x=0$)	$mol\,L^{-1}$	see section 4.1.3
$C°_P$	standard molar heat capacity	$J\,K^{-1}\,mol^{-1}$	
D_i (or D)	diffusion coefficient of species i	$cm^2\,s^{-1}$	see section 2.2.1 [2]
D_i^0	diffusion coefficient of species i at infinite dilution	$cm^2\,s^{-1}$	see section 4.2.1
e	electron charge	C	$= -1.6\times10^{-19}\ C$
E	electrode potential (half-cell)	$V_{/Ref}$	relative to the chosen reference see section 1.5.1
$E°$	standard potential of redox couple	$V_{/Ref}$	
E	electric field	$V\,m^{-1}$ (modulus)	
f	$f = \dfrac{\mathscr{F}}{RT}$	V^{-1}	$38.9\ V^{-1}$ at 25 °C
$\mathscr{F}$	FARADAY constant	$C\,mol^{-1}$	$\approx 96\,500\ C\,mol^{-1}$
G	GIBBS energy	J	
$\widetilde{G}$	GIBBS energy in an electrochemical system	J	

[1] Unit conversion for a concentration: $C = y\ mol\,L^{-1} = y\times10^3\ mol\,m^{-3} = y\times10^{-3}\ mol\,cm^{-3}$.

[2] Unit conversion for a diffusion coefficient: $D = y\ cm^2\,s^{-1} = y\times10^{-4}\ m^2\,s^{-1}$.

H	enthalpy	J	
I	current (intensity)	A	sign convention: see section 1.4.1
I_{lim}	limiting current (diffusion)	A	see section 2.3.2
I_s	ionic strength of a medium	mol L^{-1}	see section 3.1.2
$\mathbf{j}\,(j)$	current density (modulus)	A m^{-2} (modulus)	see section 1.3.1 [3]
k	BOLTZMANN constant	J K^{-1}	$= 1.38 \times 10^{-23}$ J K^{-1}
k°	standard rate constant of a redox reaction (E mechanism)	cm s^{-1}	[4] see section 4.3.2
K_{eq}	equilibrium constant	dimensionless	
ℓ, L	distance between two points, width	cm	
L_D	DEBYE length	Å	see section 3.3.1 [5]
L_H	thickness of the HELMOLTZ layer	Å	see section 3.3.1
m_i	mass transport rate constant of species i	cm s^{-1}	see section 4.3.2
$m_i\,(m^\circ)$	molality of species i (standard state)	mol kg^{-1}	see section 3.1.2
M_i (or M)	molar mass of species i	g mol^{-1}	
n	number of electrons exchanged in a redox half-reaction	dimensionless	positive number see section 1.2.1
n_i	amount of substance of species i	mol	
$\mathbf{n}$	vector normal to the surface	dimensionless	
$\mathcal{N}$	AVOGADRO constant	mol^{-1}	$= 6.02 \times 10^{23}$ mol^{-1}
$\mathbf{N}_i\,(N_i)$	molar flux density of species i (modulus)	mol m^{-2} s^{-1} (modulus)	see section 4.1.1 [6]
o.n.	oxidation number (or degree)	dimensionless	
$P\,(P^\circ)$	pressure (standard state)	bar	$P^\circ = 1$ bar [7]
P_i	partial pressure of species i	bar	
pH	potential of Hydrogen	dimensionless	$pH = -\log a_{H^+}$

---

[3] Unit conversion for a current density: $j = y$ A m$^{-2} = y \times 10^{-4}$ A cm^{-2}.

[4] Unit conversion for a redox rate constant: $k = y$ cm s$^{-1} = y \times 10^{-2}$ m s^{-1}.

[5] Unit conversion for a distance: $\ell = y$ Å $= y \times 10^{-10}$ m $= y \times 10^{-1}$ nm $= y \times 10^{-4}$ µm $= y \times 10^{-8}$ cm.

[6] Unit conversion for a molar flux density: $N = y$ mol m^{-2} s$^{-1} = y \times 10^{-4}$ mol cm^{-2} s^{-1}.

[7] Unit conversion for a pressure: $P = y$ bar $= y \times 10^5$ Pa $= (y/1.013)$ atm.

pK_i	relative to the equilibrium constant K_i	dimensionless	$pK_i = -\log K_i$
q, Q	electric charge	C	
r_{farad}	faradic yield	dimensionless	see section 2.2.2
R	ideal gas constant	$J\,K^{-1}\,mol^{-1}$	$= 8.314\ J\,K^{-1}\,mol^{-1}$
R	resistance	Ω	
$S\,(S^\circ)$	molar entropy (standard state)	$J\,K^{-1}\,mol^{-1}$	
S	S surface (or section) area	cm^2	[8]
t	time	s	
T	temperature	°C	[9]
t_i	transport number of species i	dimensionless	see section 2.2.4
$\tilde{t}_i$	electrochemical transport number of species i	dimensionless	see section 4.1.1
u_i	electric mobility of species i	$m^2\,s^{-1}\,V^{-1}$	see section 4.2.1
$\tilde{u}_i$	electrochemical mobility of species i	$mol\,m\,s^{-1}\,N^{-1}$	or $mol\,s\,kg^{-1}$ $mol\,m^2\,J^{-1}\,s^{-1}$ see section 4.2.1
U	potential difference (voltage)	V	
V	electric potential (general sense)	V	
v_r	volume reaction rate per unit volume	$mol\,m^{-3}\,s^{-1}$	see section 4.1.2
v_{S_r}	surface reaction rate per unit area	$mol\,m^{-2}\,s^{-1}$	see section 4.1.3
w_i	local volume production rate of species i	$mol\,m^{-3}\,s^{-1}$	see section 4.1.2
w_{S_i}	local surface production rate of species i	$mol\,m^{-2}\,s^{-1}$	see section 4.1.3
x	spatial coordinate	m	
x_i	molar fraction of species i	dimensionless	
y	insertion rate	dimensionless	see section 2.1.2
z_i	charge number of species i	dimensionless	algebraic number see section 1.2.1

[8]   *Unit conversion for a surface area:* $S = y\ cm^2 = y \times 10^{-4}\ m^2$.

[9]   *Unit conversion for a temperature:* $T = y\ °C = (y + 273)\ K$.

GREEK SYMBOLS

Symbol	Meaning	Usual unit	Comment
α	symmetry factor	dimensionless	see section 4.3.2
δ	thickness (e.g., diffusion layer)	μm	
$\Delta_r H\,(\Delta_r H°)$	enthalpy of reaction (standard enthalpy of reaction)	$J\,mol^{-1}$	
$\Delta_r G\,(\Delta_r G°)$	GIBBS energy of reaction (standard GIBBS energy of reaction)	$J\,mol^{-1}$	
γ_i	activity coefficient of species i	dimensionless	see section 3.1.2
Γ_i	surface concentration of species i	$mol\,m^{-2}$	see section 4.1.3
ε	dielectric permittivity of a medium	$J^{-1}\,C^2\,m^{-1}$	
ε_0	vacuum dielectric permittivity	$J^{-1}\,C^2\,m^{-1}$	$= 8.855 \times 10^{-12}$ $J^{-1}\,C^2\,m^{-1}$
η	overpotential (or overvoltage) of a cell or an electrode	V	see section 1.5.2
λ_i	molar conductivity of species i	$S\,cm^2\,mol^{-1}$	see section 2.2.4 [10]
λ_i^0	molar conductivity of species i at infinite dilution	$S\,cm^2\,mol^{-1}$	see table 4.2, section 4.2.2
Λ	molar conductivity of a solute	$S\,cm^2\,mol^{-1}$	see section 2.2.4
$\mu_i\,(\mu_i°)$	chemical potential of species i (standard state)	$J\,mol^{-1}$	see section 3.1.2
$\widetilde{\mu}_i$	electrochemical potential of species i	$J\,mol^{-1}$	see section 3.1.2
ν_i	stoichiometric number of species i in a reaction	dimensionless	algebraic value see section 1.2.1
ξ	dimensionless potential $\xi = \dfrac{n\,\mathscr{F}}{R\,T}(E - E°) = nf(E - E°)$	dimensionless	see section 4.3.2
π	polarisation of a cell or an electrode	V	see section 1.5.2
ρ	electric resistivity of a medium	$\Omega\,cm$	see section 2.2.4
ρ_{ch}	charge density	$C\,m^{-3}$	

---

[10] *Unit conversion for a molar conductivity:* $\lambda = y\,S\,cm^2\,mol^{-1} = y \times 10^{-4}\,S\,m^2\,mol^{-1}$.

σ	electric conductivity of a medium	$S\,cm^{-1}$	see section 2.2.4 [11]
σ_{ch}	surface charge density	$C\,m^{-2}$	
τ	time constant	s	
φ	GALVANI potential or internal potential	V	see section 3.1.1
χ	surface electric voltage	V	see section 3.1.1
ψ	VOLTA potential or external potential	V	see section 3.1.1
$\omega_i\,(\omega_i)$	local velocity of species i (modulus)	$m\,s^{-1}$ (modulus)	see section 4.1.1

[11] *Unit conversion for an electric conductivity:* $\sigma = y\,S\,cm^{-1} = y \times 10^2\,S\,m^{-1}$.

INDEX

THE AUTHORS

Christine LEFROU
is a graduate of ENS (Ecole Normale Supérieure), the elite French institution of higher education and research, and currently a university lecturer at the PHELMA engineering school (Physics, Applied Physics, Electronics and Materials Science), part of the Grenoble Institute of Technology (INP). She teaches electrochemistry on core education courses, as well as on a wide array of continuing education courses. Her research work to date has mainly focused on applying the concept of modeling material transport to the field of electrochemistry (batteries and electroanalysis).

Pierre FABRY
is a university-trained physicist, who was formerly a professor at Grenoble University (Université Joseph Fourier). He has taught electrochemistry and the structure of materials at university level, (undergraduate and master's degrees) as well as at engineering schools, and on adult training courses. His research work has focused specifically on the subject of electrochemical solids for high-temperature energy storage systems and electrochemical sensors for biomedical and environmental applications.

Jean-Claude POIGNET
was formerly a Professor of electrochemistry at the Grenoble Institute of Technology (INP). After completing a thesis on the structure and transport properties of molten salts, he then focused his research career on studying low-temperature ionic liquids, before turning his attention towards electrochemistry of molten salts between 450 and 1000°C: electrode and electrolyte materials for thermal batteries, Li or Na solutions dissolved in molten LiCl or NaCl, the cathodic separation of lanthanides and actinides and the electrosynthesis of Na, Al, Nb and Pu.

Printed in the United States
By Bookmasters